Manipulation of
the Avian Genome

Edited by
Robert J. Etches, Ph.D., D.Sc.
Ann M. Verrinder Gibbins, Ph.D.

University of Guelph
Guelph, Ontario, Canada

CRC Press
Taylor & Francis Group
Boca Raton   London   New York

CRC Press is an imprint of the
Taylor & Francis Group, an **informa** business

CRC Press
Taylor & Francis Group
6000 Broken Sound Parkway NW, Suite 300
Boca Raton, FL 33487-2742

First issued in paperback 2019

ISBN-13: 978-0-8493-4216-5 (hbk)
ISBN-13: 978-0-367-40258-7 (pbk)

Library of Congress Card Number 92-25659

Library of Congress Cataloging-in-Publication Data

Manipulation of the avian genome / edited by Robert J. Etches, Ann M.
  Verrinder Gibbins.
     p. cm.
  Includes bibliographical references and index.
  ISBN 0-8493-4216-3
  1. Chickens--Breeding--Congresses. 2. Chickens--Genetics-
-Congresses. 3. Chickens--Genetic engineering--Congresses.
  4. Birds--Genetics--Congresses. 5. Birds--Genetic engineering-
-Congresses. 6. Genomes--Congresses. I. Etches, Robert J.
  II. Gibbins, Ann M. Verrinder (Ann Margaret Verrinder)
  SF492.M36   1992
  636.5'0821--dc20   92-25659
                                                                CIP

Visit the Taylor & Francis Web site at
http://www.taylorandfrancis.com

and the CRC Press Web site at
http://www.crcpress.com

Preface

For several years, avian biologists have been viewing with envy the exciting developments that have occurred in the production of transgenic mammals, fishes, insects, and plants, with the important benefits for basic and applied biology that these unique organisms afford. Because of certain characteristics of the avian ovum, zygote, and embryo, the genome of these structures is difficult to access, and successful strategies for the production of transgenic birds are only now emerging. Interception of the normal development of the embryo requires a detailed knowledge of the components involved — from the gametes through to the multicomponent embryo, and from the reproductive tract of the hen to the incubated egg. In addition, conjectures on the production of transgenic birds are incomplete without detailed knowledge of metabolic and structural processes that might be modified productively. Since many of these modifications could ultimately be of value to the poultry industry, the requirements and concerns of the industry should be borne in mind by applied biologists so that the sophisticated and expensive research that is required can be pursued effectively. Transgenic strategies for the improvement of commercial stocks are still viewed by the poultry industry as long-term challenges for the future, but more immediate gain is anticipated from nontransgenic molecular methods that are being developed to allow selection of superior birds by poultry breeders.

All of these related issues were discussed at a colloquium, ''Manipulation of the Avian Genome'', held under the auspices of the Keystone Symposia Organization at the Granlibakken Conference Center, Lake Tahoe, California, in March 1991. The unprecedented gathering of basic biologists and members of the poultry industry, applied biologists from medicine and agriculture, and scientists from biotechnology companies and multinational pharmaceutical organizations, from 11 different countries and in a milieu that encouraged a lively interchange of ideas, resulted in heightened awareness of the challenges being faced by all of us in coming to terms with the biology of the bird and its practical implications. Underlying everything, though, was the excitement caused by appreciation of the progress that is being made in all aspects of the manipulation of the avian genome. This volume contains a compilation of papers written by the principal speakers at the meeting, and it is our belief that it conveys the promise of this burgeoning area of biology.

We wish to extend our sincere appreciation to the sponsors of the Lake Tahoe meeting, whose names are listed on an adjoining page, and to the long-suffering personnel at the Keystone Symposia Organization, who ministered patiently to our foibles. The Granlibakken Conference Center provided us with splendid accommodation, and California was kind enough to give us plenty of snow! Our research group provided us, as always, with essential and patient support, for which we sometimes forget to thank them. The principal speakers have conveyed the spirit of the meeting admirably in their papers, and it is a pleasure to us to list on an adjoining page the names of all of the participants, who provided such a creative and happy atmosphere.

Rob Etches and Ann Gibbins

The Editors

Robert Etches, Ph.D., D.Sc., is a Professor in the Department of Animal and Poultry Science and a Professor in the Department of Zoology at the University of Guelph, Guelph, Ontario, Canada. Dr. Etches received his B.Sc. (Agr.) from the University of British Columbia in 1970 and his M.Sc. from McGill University, Montreal, Canada in 1972. He was awarded a Commonwealth Scholarship to study at the University of Reading, England and obtained his Ph.D. and D.Sc. from this institution in 1975 and 1985, respectively. He was appointed as an Assistant Professor at the University of Guelph in 1975 and became an Associate Professor in 1980 and Professor in the Departments of Animal and Poultry Science and Zoology in 1985 and 1988, respectively. During a period of one year (1981 to 1982), Dr. Etches was a Visiting Scientist at the Station de Recherches Avicoles of the Institut National de la Recherche Agronomique at Tours, France. In 1988, he was awarded a Visiting Professorship at the University of Nagoya in Japan and, in 1991, was a Visiting Professor at the University of California, Davis.

Dr. Etches is a member of the World's Poultry Science Association, the Poultry Science Association, the American Physiological Society, the Canadian Society of Animal Science, and Sigma Xi. He has received the Poultry Science Research Award in 1980 and was the recipient of the Sigma Xi *Excellence in Research* Award in 1990. He has published more than 100 refereed papers describing his research and is a frequent speaker at universities and research institutes. Dr. Etches' work has been funded by the Natural Sciences and Engineering Research Council of Canada, the Ontario Ministry of Agriculture and Food, the Ontario Egg Board and Agriculture Canada, and the University of Guelph. His current research interests are in early development of the chick embryo and manipulation of the genome of chickens.

Ann Gibbins, Ph.D., is an Associate Professor in the Department of Animal and Poultry Science, Ontario Agricultural College, and an Adjunct Professor in the Department of Biomedical Sciences, Ontario Veterinary College, at the University of Guelph in Guelph, Ontario, Canada. Dr. Gibbins received her B.Sc. (First Class Honours) degree in Biochemistry from the University of Birmingham, England in 1960. She obtained her M.Sc. and Ph.D. degrees in Microbiology (1971) and Genetics (1980), respectively, from the University of Guelph. After doing post doctoral work in the Department of Biomedical Sciences in the Ontario Veterinary College, University of Guelph, she was appointed as an Assistant Professor in this department in 1982. In 1986, she assumed her present position as Associate Professor in the Department of Animal and Poultry Science. From 1982 to 1992, Dr. Gibbins was the recipient of a Natural Sciences and Engineering Research Council of Canada (NSERC) University Research Fellowship.

Dr. Gibbins is a member of the Poultry Science Association, World's Poultry Science Association, Canadian Society of Animal Science, Canadian Society for Cell Biology, American Association for the Advancement of Science, and the honorary society Sigma Xi. She is a member of the International Advisory Board for the publications of the Asian Network for Biotechnology in Animal Production and Health. Dr. Gibbins' research interests are in the application of molecular genetics to animal science and veterinary medicine. Currently, the majority of her effort is expended in a collaborative program for the development of transgenic birds. She has been the recipient of research grants from NSERC, Agriculture Canada, Ontario Ministry of Colleges and Universities, University of Guelph, Canadian Association of Animal Breeders, Ontario Milk Marketing Board, Canadian Veterinary Research Trust Fund, Ontario Egg Producers' Marketing Board, NATO, and the British Council. She has published more than 40 refereed research papers and has given numerous guest lectures.

Contributors

Murray R. Bakst, Ph.D.
LPSI, Germplasm and Gamete
  Physiology Laboratory
Agricultural Research Service
U.S. Department of Agriculture
Beltsville, Maryland

J. James Bitgood, Ph.D.
Department of Poultry Science
University of Wisconsin at Madison
Madison, Wisconsin

Stephen E. Bloom, Ph.D.
Department of Avian and Aquatic
  Animal Medicine
Cornell University
Ithaca, New York

James A. Bradac, Ph.D.
ABL - Basic Research Center
NCI/Frederick Cancer Research
  Center
Frederick, Maryland

Cynthia L. Brazolot
Department of Animal and Poultry
  Science
University of Guelph
Guelph, Ontario, Canada

**Ruth A. Burtch-Wright,
  Hons. B. Sc.**
Cancer Research Laboratories
Queen's University
Kingston, Ontario, Canada

Avigdor Cahaner, Ph.D.
Department of Genetics
The Hebrew University
Rehovot, Israel

Mary Ellen Clark
Department of Animal and Poultry
  Science
University of Guelph
Guelph, Ontario, Canada

Lyman B. Crittenden, Ph.D.
Department of Microbiology
Michigan State University
East Lansing, Michigan

Ariel Darvasi, M.S.
Department of Genetics
The Life Sciences Institute
Hadassah Hebrew University
Givet Ram Campus
Jerusalem, Israel

Roger G. Deeley, Ph.D.
Stauffer Research Professor
Cancer Research Laboratories
Queen's University
Kingston, Ontario, Canada

Mary E. Delany, Ph.D.
Department of Avian and Aquatic
  Animal Medicine
Cornell University
Ithaca, New York

**Francoise Dieterlen-Lièvre,
  Dr. es Sc.**
Institut d'Embryologie Cellulaire et
  Moleculaire
CNRS
Nogent Sur Marne, France

E. Ann Dunnington, Ph.D.
Virginia Polytechnic Institute and
  State University
Blacksburg, Virginia

Alan Emsley, Ph.D.
Department of Layer Genetics ISA
Ithaca, New York

Robert J. Etches, Ph.D., D. Sc.
Department of Animal and Poultry
  Science
University of Guelph
Guelph, Ontario, Canada

Hefzibah Eyal-Giladi, Ph.D.
Department of Cell and Animal
  Biology
Hebrew University
Jerusalem, Israel

Abraham Fainsod, Ph.D.
Department of Cellular
  Biochemistry
The Hebrew University - Hadassah
  Medical School
Jerusalem, Israel

Mark J. Federspiel, Ph.D.
ABL - Basic Research Center
NCI/Frederick Cancer Research
  Center
Frederick, Maryland

Ayala Frumkin
Department of Cellular
  Biochemistry
The Hebrew University - Hadassah
  Medical School
Jerusalem, Israel

Orit Gal, M.S.
Department of Genetics
The Hebrew University
Rehovot, Israel

**Jan S. Gavora, D.Sc.,
  C. Sc. Ing.**
Department of Agriculture
Centre for Food and Animal
  Research
Ottawa, Ontario, Canada

Hazel Gilhooley, B.Sc.
Department of Cellular and
  Molecular Biology
AFRC Institute of Animal
  Physiology and Genetics Research
Edinburgh Research Station
Roslin, Midlothian,
United Kingdom

Christopher Goddard, Ph.D.
Department of Cellular and
  Molecular Biology
AFRC Institute of Animal
  Physiology and Genetics Research
Edinburgh Research Station
Roslin, Midlothian,
United Kingdom

Caroline E. Grant, Ph.D.
Cancer Research Laboratories
Queen's University
Kingston, Ontario, Canada

Alexander Gray, Ph.D.
Department of Cellular and
  Molecular Biology
AFRC Institute of Animal
  Physiology and Genetics Research
Edinburgh Research Station
Roslin, Midlothian,
United Kingdom

Clare Gribbin, Ph.D.
Department of Reproduction and
  Development
AFRC Institute of Animal
  Physiology and Genetics Research
Edinburgh Research Station
Roslin, Midlothian,
United Kingdom

Yosef Gruenbaum, Ph.D.
Department of Genetics
The Life Sciences Institute
Hadassah Hebrew University
Givet Ram Campus
Jerusalem, Israel

Alon Haberfeld, Ph.D.
Department of Animal and Poultry
  Science
Virginia Polytechnic Institute and
  State University
Blacksburg, Virginia

Jossi Hillel, Ph.D.
Department of Genetics
The Hebrew University
Rehovot, Israel

Pamela Hoodless, Ph.D.
Laboratory of Molecular and Cell
  Biology
The Rockefeller University
New York, New York

Stephen Hughes, Ph.D.
LBI - Basic Research Program
NCI/Frederick Cancer Research
  Institute
Frederick, Maryland

Hasan Khatib
Department of Genetics
The Life Sciences Institute
Hadassah Hebrew University
Givet Ram Campus
Jerusalem, Israel

Ryota Kunita, Master Agric.
Department of Applied Biological
  Chemistry
Faculty of Agriculture
Tohoku University
Sendai, Japan

Susan J. Lamont, Ph.D.
Department of Animal Science
Iowa State University
Ames, Iowa

Uri Lavi, Ph.D.
Agricultural Research Organization
Bet Dagan, Israel

Nicole Le Douarin, Dr. es Sc.
Institut d'Embryologie Cellulaire et
  Moleculaire
College de France/CNRS
Nogent Sur Marne, France

Guodong Liu
Department of Animal and Poultry
  Science
University of Guelph
Guelph, Ontario, Canada

Yael Margalit
Department of Cellular
  Biochemistry
The Hebrew University - Hadassah
  Medical School
Jerusalem, Israel

H. L. Marks, Ph.D.
Southern Regional Poultry Genetics
  Laborators
U.S.D.A.
Athens, Georgia

Christine Mather, M.Sc.
Department of Reproduction and
  Development
AFRC Institute of Animal
  Physiology and Genetics Research
Edinburgh Research Station
Roslin, Midlothian
United Kingdom

Shigeki Mizuno, Dr. Agric.
Department of Applied Biological
  Chemistry
Faculty of Agriculture
Tohoku University
Sendai, Japan

David Morrice, B.Sc.
Department of Reproduction and
  Developement
AFRC Institute of Animal
  Physiology and Genetics Research
Edinburgh Research Station
Roslin, Midlothian
United Kingdom

Donna E. Muscarella, Ph.D.
Department of Avian and Aquatic
  Animal Medicine
Cornell University
Ithaca, New York

Katsuhiko Nishimori, Dr. Agric.
Department of Applied Biological
  Chemistry
Faculty of Agriculture
Tohoku University
Sendai, Japan

Osamu Nomura, Master Agric.
Department of Applied Biological
  Chemistry
Faculty of Agriculture
Tohoku University
Sendai, Japan

Kohei Ohtomo
Department of Applied Biological
  Chemistry
Faculty of Agriculture
Tohoku University
Sendai, Japan

Iris E. O'Neill
Department of Cellular and
  Molecular Biology
AFRC Institute of Animal
  Physiology and Genetics Research
Edinburgh Research Station
Roslin, Midlothian
United Kingdom

Hiroyuki Ono, Dr. Agric.
Bio-Organic Chemistry Laboratory
Institute for Biomedical Research
SUNTORY Limited
Osaka, Japan

William S. Payne, B.S.
Department of Microbiology
Michigan State University
East Lansing, Michigan

Margaret Perry, B.Sc.
Department of Reproduction and
  Development
AFRC Institute of Animal
  Physiology and Genetics Research
Edinburgh Research Station
Roslin, Midlothian
United Kingdom

James N. Petitte, Ph.D.
Department of Poultry Science
North Carolina State University
Raleigh, North Carolina

Christos J. Petropoulous, Ph.D.
GENENTECH
South San Francisco,
California

Yoram Plotsky, Ph.D.
Department of Genetics
The Hebrew University
Jerusalem, Israel

Zehava Rangini, Ph.D.
Cardiovascular Research Center
Massachusetts General Hospital
Charlestown, Massachusetts

Elie S. Revel, B.Sc.
Department of Genetics
The Life Sciences Institute
Hadassah Hebrew University
Givet Ram Campus
Jerusalem, Israel

Aimee K. Ryan, M.Sc.
Cancer Research Laboratories
Queen's University
Kingston, Ontario, Canada

Hisato Saitoh, Dr. Agric.
Department of Cell Biology and
    Anatomy
Johns Hopkins University School of
    Medicine
Baltimore, Maryland

Yasushi Saitoh, Dr. Agric.
Department of Molecular Biology
    and Biochemistry
Geneva University
Geneva, Switzerland

Helen Sang, Ph.D.
Department of Reproduction and
    Development
AFRC Institute of Animal
    Physiology and Genetics Research
Edinburgh Research Station
Roslin, Midlothian,
United Kingdom

Donald W. Salter, Ph.D.
Department of Microbiology
Michigan State University
East Lansing, Michigan

Timothy J. Schrader, Ph.D.
Mutagenesis Section
Environmental Health Centre
Health and Welfare Canada
Ottawa, Ontario, Canada

Robert N. Shoffner, Ph.D.
Department of Animal Science
University of Minnesota
St. Paul, Minnesota

Paul B. Siegel, Ph.D.
Virginia Polytechnic Institute and
    State University
Blacksburg, Virginia

**Ann M. Verrinder Gibbins,
Ph.D.**
Department of Animal and Poultry
    Science
Ontario Agricultural College
    and
Department of Biomedical Sciences
Ontario Veterinary College
University of Guelph
Guelph, Ontario, Canada

Sinai Yarus, M.Sc.
Department of Cellular
    Biochemistry
The Hebrew University - Hadassah
    Medical School
Jerusalem, Israel

Participants

H. Aarts
Molecular Biology
Coup Het Spelderholt
Spelderholt 9, Beekbergen 7361
DA, The Netherlands

G. Albers
Euribrid BV
P. O. Box 30, 5830 AA
Boxmeer, The Netherlands

J. A. Arthur
Hy-Line International
P. O. Box 310
Dallas Center, Iowa 50063

M. R. Bakst
Germplasm and Gamete Physiology
  Laboratory
USDA/ARS Bldg. 262
BARC-East
Beltsville, Maryland 20705

J. J. Bitgood
Department of Poultry Science
University of Wisconsin-Madison
1675 Observatory Dr.
Madison, Wisconsin 53706

S. E. Bloom
Poultry Science
Cornell University, Rice Hall
Ithaca, New York 14853

F. Bosch
Veterinary Medicine/Biochemistry
Autonomous University of Barcelona
Bellaterra, Barcelona 08193, Spain

R. A. Bosselman
Developmental Biology
Amgen Inc.
1840 De Havilland
Thousand Oaks, California 91320

C. L. Brazolot
Animal & Poultry Science
University of Guelph
Guelph, Ontario, N1G 2W1
Canada

J. A. Brumbaugh
Biological Sciences
University of Nebraska
348 Manter Hall
Lincoln, Nebraska 68588

W. H. Burke
Poultry Science
University of Georgia
120 Livestock Poultry Bldg.
Athens, Georgia

J. Burnside
Animal Science
University of Delaware
040 Townsend Hall
Newark, Delaware 19717

A. L. Cartwright
USDA-ARS Poultry Research Lab.
RD 2 Box 600
Georgetown, Delaware 19947

H. Y. Chen
Growth Biochemistry & Physiology
Merck Sharp & Dohme
P. O. Box 2000, R8OT-150
Rahway, New Jersey 07065

C. M. Chuong
Pathology
University of Southern California
2011 Zonal Ave., HMR 204
Los Angeles, California 90033

M.-E. Clark
Animal & Poultry Science
University of Guelph
Guelph, Ontario, N1G 2W1
Canada

G. Dambrine
Station Pathology Aviaire Parasitology
INRA, Nouzilly 37380
Monnaie, France

R. G. Deeley
Biochemistry
Queen's University, Botterell Hall
Kingston, Ontario, K7L 3N6
Canada

M. E. Delany
Poultry & Avian Science
Cornell University
401 Rice Hall
Ithaca, NY 14853, USA

F. Dieterlen-Lièvre
Institut d'Embryologie CNRS
49 bis, Ave de la Belle Gabrielle
Nogent Cedex 94736, France

M. Douaire
Animal Genetics
INRA-France
65 Rue De Saint Brieuc
Rennes, France

D. Dunon
Basel Institute for Immunology
487 Grenzacherstrasse
Basel, CH-4005, Switzerland

A. Elbrecht
Animal Biochemistry/Molecular
 Biology
Merck Sharp & Dohme
P. O. Box 2000
Rahway, New Jersey 07065

A. Emsley
ISA Babcock
P. O. Box 280
Ithaca, New York 14851-0280

E. R. Escapa
Avian Sciences
University of California-Davis
Davis, California 95616

R. J. Etches
Animal & Poultry Science
University of Guelph
Guelph, Ontario, N1G 2W1
Canada

D. I. Ewart
Wistar Institute
36th & Spruce Sts.
Philadelphia, Pennsylvania 19104

H. Eyal-Giladi
Zoology
Hadassah Hebrew University
Jerusalem 91904, Israel

M. J. Federspiel
ABL-Basic Research Program
NCI-Frederick Cancer Research
 Center
P. O. Box B/Bldg. 539
Frederick, Maryland 21702-1201

F. D. Flamant
Biologie Moleculaire
Ecole Normale Superieure
46 Allee d'Italie
Lyon Cedex 07 69364, France

R. A. Fraser
Physiology
University of Alberta
7-41 Medical Science Bldg.
Edmonton, Alberta, T6G 2H7
Canada

S. Fuerstenberg
Molecular Biology
Karolinska Institute
Box 60400
Stockholm 10401, Sweden

N. Fujihara
Animal Science
Kyuahu University, Hakozaki
Fukuoka 812, Japan

E. A. Garber
Animal Biochemistry
Merck Sharp & Dohme
P. O. Box 2000
Rahway, New Jersey 07065

J. S. Gavora
Centre for Food and Animal Research
Agriculture Canada
Ottawa, Ontario, K1A 0C6
Canada

A. M. Gibbins
Animal & Poultry Science
University of Guelph
Guelph, Ontario, N1G 2W1
Canada

C. Goddard
AFRC Inst. Animal Physiology &
 Genetics Research
Edinburgh Research Station
Roslin, Midlothian, EH25 9PS
United Kingdom

C. Grant
Biochemistry
Queen's University, Botterell Hall
Rm. 315
Kingston, Ontario, K7L 3N6
Canada

J. M. Grizzle
Reproduction Research
USDA-Meat Animal Research Center
 Box 166
Clay Center, Nebraska 68933

Y. Gruenbaum
Genetics
Hadassah Hebrew University
Givet Ram Campus
Jerusalem, 91904, Israel

T. C. Guido
Avian Sciences
University of California-Davis
3202 Meyer Hall
Davis, California 95616

T. I. Hargrove
Poultry Science
University of Maryland
College Park, Maryland 20742

J. Hillel
Genetics
The Hebrew University
P. O. Box 12
Rehovot 76-100, Israel

B. Hillgartner
Biochemistry
University of Iowa
Bowen Science Building
Iowa City, Iowa 52242

P. A. Hoodless
Cancer Research Lab.
Queen's University, Botterell Hall
Kingston, Ontario, K7L 3N6
Canada

S. H. Hughes
LBI-Basic Research Prog.
NCI/Frederick Cancer Research
 Institute
P. O. Box B/Bldg. 539
Frederick, Maryland 21701

N. J. Hutchison
Genetics AC-136
Fred Hutchinson Cancer Center
1124 Columbia Street
Seattle, Washington 98104

S. B. Jakowlew
Lab. Chemoprevention
National Inst. of Health-NCI
Bldg. 41
Bethesda, Maryland 20892

B. J. Kelly
Nicholas Turkey Breeding Farms
P. O. Box Y
Sonoma, California 95476

M. L. Kirby
Anatomy
Medical College of Georgia
Augusta, Georgia 30912-2000

U. Kuhnlein
Animal Science
McGill University, MacDonald
 College
Ste. Anne de Bellevue, Quebec
H9X 1C0, Canada

A. W. Kulenkamp
Shaver Poultry Breeding Farms
37 Randall Road
Cambridge, Ontario, N3C 1R8
Canada

S. J. Lamont
Animal Science
Iowa State University
201 Kildee Hall
Ames, Iowa 50011

A. Lichtler
Pediatrics
University of Connecticut
263 Farmington Ave.
Farmington, Connecticut 06030

G. Liu
Animal & Poultry Science
University of Guelph
Guelph, Ontario, N1G 2W1
Canada

K. Maruyama
Avian Physiology
USDA/ARS/LPSI, Bldg. 262
 BARC-East
Beltsville, Maryland 20705

M. P. Matise
NACS
University of Pittsburgh
Pittsburgh, Pennsylvania 15261

J. R. McCarrey
Genetics
Southwest Foundation for Biomedical
 Research
P. O. Box 28147
San Antonio, Texas 78228

W. T. McCormack
Howard Hughes Medical Institute
University of Michigan
1150 W Medical Center Dr.
Ann Arbor, Michigan 48109

J. C. McKay
Ross Breeders Ltd.
Roslin, Newbridge, Midlothian
United Kingdom

C. Miller
Zoology
University of Leicester
University Road
Leicester, LEI 7RH
United Kingdom

P. P. Minghetti
Biochemistry
University of California-Riverside
Riverside, California 92521

S. Mizuno
Agricultural Chemistry
Tohoku University
1-1 Tsutsumidori-Amamiyamachi
Sendai 980, Japan

U. Neumann
Poultry Clinic
Hannover Veterinary School
Bunteweg 17
3000 Hannover 71, FRG

A. W. Nordskog
Animal Genetics
Iowa State University
201 Kildee Hall
Ames, Iowa 50010

T. Ono
Faculty of Agriculture
Shinshu University
8304 Minami
Monowa, INA 399-45, Japan

J. N. Petitte
Poultry Science
North Carolina State University
Box 7608
Raleigh, North Carolina 27695-7608

C. A. Ricks
Research & Development
Embrex Inc.
P. O. Box 13989
Research Triangle Park, North
 Carolina 27709

T. Rigau
Reproduction
Veterinary School, UAB
Bellaterra, Barcelona 08193, Spain

C. I. Rosenblum
Growth Biochemistry & Physiology
Merck Sharp & Dohme
P. O. Box 2000/R8OT-150
Rahway, New Jersey 07065

B. Russell
Animal & Poultry Science
University of Guelph
Guelph, Ontario, N1G 2W1
Canada

A. K. Ryan
Cancer Research Laboratories
Queen's University, Botterell Hall
Kingston, Ontario, K7L 3N6
Canada

D. W. Salter
Microbiology & Public Health
Michigan State University
3606 Mt. Hope Rd.
USDA-ARS-RPRL
East Lansing, Michigan 48823

H. Sang
AFRC Inst. Animal Physiology &
  Genetics Research
Edinburgh Research Station
Roslin, Midlothian, EH25 9PS
United Kingdom

T. J. Schrader
Cancer Research Laboratories
Queen's University, Botterell Hall
Kingston, Ontario, K7L 3N6
Canada

D. L. Shaw
Animal & Poultry Science
University of Guelph
Guelph, Ontario, N1G 2W1
Canada

B. I. Sheldon
Animal Production
CSIRO, P. O. Box 184
North Ryde, Sydney, NSW 2113
Australia

K. Shimada
Animal Physiology
Nagoya University
Chikusa, Nagoya 46401, Japan

R. N. Shoffner
Poultry Science
University of Minnesota
St. Paul, Minnesota 55108

F. J. Shultz
Animal Breeding Consultants
P. O. Box 313
Sonoma, California 95476

R. M. Shuman
Gentra Systems, Inc.
3905 Annapolis Lane
Minneapolis, Minnesota 55447

M. B. Smiley
Institut de Selection Animale
Mauquerand
22800 Quintin, France

R. G. Somes
Nutritional Sciences
University of Connecticut
3624 Horsebarn Road
Storrs, Connecticut 06269-4017

W. E. Stumph
Chemistry
San Diego State University
San Diego, California 92182

L. J. Takemoto
Biology
Kansas State University
Ackert Hall
Manhattan, Kansas 68506

K. D. Taylor
Animal & Poultry Science
University of Guelph
Guelph, Ontario, N1G 2W1
Canada

C. B. Thompson
Howard Hughes Med Institute
University of Michigan
1150 W. Medical Center Dr.
Ann Arbor, Michigan 48109

P. Thoraval
Pathologie Aviaire
INRA, Domaine de l'Orprasiere
Monnaie 37380, France

M. H. Tixier-Boichard
Gentique Factorielle
Centre de Recherches INRA
78352 Jouy-en-Josas Cedex, France

R. H. Towner
Genetics
H & N International
3825-154 Ave NE
Redmond, Washington 98052

R. S. Tuan
Orthopaedic Surgery
Thomas Jefferson University
1015 Walnut Street
Philadelphia, Pennsylvania 19107

W. B. Upholt
Biostructure & Function
University of Connecticut Health
  Center
Farmington, Connecticut 06030

A. Valera
Biochem. & Biol. Molec.
Autonomous University
Facultie Veterinaria
Barcelona 08193, Spain

U. Vielkind
Animal & Poultry Science
University of Guelph
Guelph, Ontario, N1G 2W1
Canada

W. D. Wagner
Comparative Medicine
Bowman Gray School of Medicine
300 S. Hawthorne Rd.
Winston-Salem, North Carolina 27103

R. I. Walzem
Physiological Sciences
School of Veterinary Medicine
University of California
Davis, California 95616

Posters Presented at the Manipulation of the Avian Genome Colloquium

Abstracts of these posters were published in the Journal of Cellular Biochemistry, Supplement 15E, 1991, pp. 200–210

C1 100 The chicken genome contains many *ev*-gene loci.
H.J.M. Aarts, R.C. van der Hulst, G. Beuving & F.R. Leenstra "Het Spelderholt", Centre for Poultry Research and Information Services, Spelderholt 9, 7361 DA Beekbergen, The Netherlands.

C1 101 Affinity purification and characterization of turkey gonadotropin.
K.D. Taylor, C.E. Anderson-Langmuir & R.J. Etches, Dept. of Animal & Poultry Science, University of Guelph, Guelph, Ontario, Canada.

C1 102 Introduction of lipofected chicken blastodermal cells into the early chicken embryo.
C.L. Brazolot, J.N. Petitte*, M.E. Clark, R.J. Etches & A.M.V. Gibbins, Department of Animal & Poultry Science, University of Guelph, Guelph, Ontario. N1G 2W1, Canada, and *Department of Poultry Science, North Carolina State University, Raleigh, NC 27695, USA.

C1 103 A gene transfer vector for poultry which expresses mouse tyrosinase.
J. Brumbaugh, T. Frew, B. Whitaker, Biological Sciences, U of Nebraska, Lincoln, NE 68588, USA; H. Yamamoto, T. Takeuchi. Biological Sciences, Tohoku U, Sendai 980, Japan; S. Hughes. Frederick Cancer Research Facility, Frederick, MD 21701, USA; D. Salter, W. Payne, Depts Animal Science & Microbiology, Michigan State U and RPRL, E Lansing, MI 48823, USA.

C1 104 Developmental profile of growth hormone receptor gene expression in broiler chickens.
J. Burnside & L.A. Cogburn, Department of Animal Science & Agricultural Biochemistry, University of Delaware, Newark, DE 19717, USA.

C1 105 Gradients of homeoproteins in developing feather buds.
C.-M. Chuong, G. Oliver, S.A. Ting, B.G. Jegalian, H.M. Chen & E.M. De Robertis, Department of Pathology, University of Southern California, Los Angeles, CA 90033 and Department of Biological Chemistry, UCLA, Los Angeles, CA 90024, USA.

C1 106 The effects of a deficiency in rRNA gene copy number on progression of the early chick embryo through development.
M.E. Delany & S.E. Bloom, Department of Poultry and Avian Sciences, Cornell University, Ithaca, NY. 14853, USA.

C1 107 Mapping of single copy genes to chicken chromosomes using in situ hybridization.

M. Dominguez-Steglich & M. Schmid, Dept. of Human Genetics, University of Wurzburg, Federal Republic of Germany.

C1 108 Comparison of lipogenic enzymes and apoproteins messenger RNA from genetically lean and fat chickens.

M. Douaire, P. Langlois, N. Le Fur, F. Flamant, C. Mounier, J. Mallard, Laboratory of Animal Genetics, Institut National de la Recherche Agronomique, 35042 Rennes Cedex, France.

C1 109 Ultrastructural morphology of the pre-primitive streak chick embryo.

J.M. Watt, J.N. Pettite* & R.J. Etches. Department of Animal & Poultry Science, University of Guelph, Guelph, Ontario and * Department of Poultry Science, North Carolina State University, Raleigh, N.C., USA.

C1 110 Avian retrovirus vectors with internal promoter. Influence of 3' non coding sequences on gene transfer safety and efficiency.

F. Flamant, D. Aubert, J. Samarut. Laboratoire de Differentiation et Oncogenese virale. Ecole Normale Superieure, Lyon, France.

C1 111 The accessory reproductive fluids in the domestic male birds.

F. Noboru & K. Osamu, Department of Animal Science, Kyushu University 46-06, Hakozaki, Fukuoka 812, Japan.

C1 112 Avian cells expressing the murine Mx1 protein are resistant to influenza virus infection.

E.A. Garber[1], H. T. Chute[1], J.H. Condra[2], L. Gotlib[2], R.J. Colonno[2], E.O. Mills[3], J. Hancock[3], D. Hreniuk[1], & R.G. Smith[1], [1]Dept. of Animal Biochemistry & Molecular Biology, Merck, Sharp, and Dohme Research Laboratories, Rahway, NJ 07065; [2]Dept. of Virus and Cell Biology, Merck, Sharp, and Dohme Research Laboratories, West Point, PA 19486; [3]Hubbard Farms, Walpole, NH 03608, USA.

C1 113 The interspecific transfer of avian primordial germ cells.

T.C. Guido, U.K. Abbott, Department of Avian Sciences, University of California, Davis, CA 95616. John R. McCarrey, John Hopkins University, School of Hygiene and Public Health, Baltimore, MD 21205, USA.

C1 114 Attempts at production of transgenic chickens by manipulated primordial germ cell.

J.-Y. Han, K.S. Guise & R.N. Shoffner, Department of Animal Science, University of Minnesota, St. Paul, MN 55108, USA.

C1 115 Culture of the preovulatory avian embryo with consideration of cyropreservation.

T.L. Hargrove*, G.F. Gee+ and M.A. Ottinger*, *Department of Poultry Science, University of Maryland, College Park, MD 20742 and +Patuxent Wildlife Research Centre, Laurel, MD 20708, USA.

Cl 116 Isolation of a developmentally regulated protein binding upstream of the apoVLDLII gene from avian liver.
P.A. Hoodless & R.G. Deeley, Cancer Research Laboratories, Queen's University, Kingston, Ontario, Canada.

Cl 117 Gene mapping in chickens via fluorescent in situ hybridization to mitotic and meiotic chromosomes.
N.J. Hutchison & C. LeCiel, Genetics Department, F. Hutchinson Cancer Research Center, Seattle, WA 98104, USA.

Cl 118 Pattern of expression of transforming growth factor-84 in the developing chicken embryo.
S.B. Jakowlew, J. Cubert, M.B. Sporn & A.B. Roberts, Laboratory of Chemoprevention, National Cancer Institute, Bethesda, MD 20892, USA.

Cl 119 Chicken Thy-1; cDNA isolation and mRNA expression patterns.
P.L. Jeffrey, B.J. Dowsing & P.W. Gunning, Children's Medical Research Foundation, Camperdown, N.S.W. 2050, Australia.

Cl 120 Backtransplantation of cultured cardiac neural crest cells in early chick embryos rescues cardiovascular development.
Margaret L. Kirby, Donna H. Kumiski, Candace Rossignol, Department of Anatomy, Medical College of Georgia, Augusta, GA 30912-2000, USA.

Cl 121 The developmental stage-specific expression of myosin isoforms in turkey muscles.
K. Maruyama & N. Kanemaki, USDA, ARS, Avian Physiology Laboratory, Beltsville, MD 20705, USA.

Cl 122 Meiotic gene conversion in chickens and its implications for the introduction of foreign DNA into the avian germ line by homologous recombination.
W.T. McCormack & C.B. Thompson, Howard Hughes Medical Institute, Departments of Internal Medicine & Microbiology/Immunology, University of Michigan Medical Center, Ann Arbor, MI 48109, USA.

Cl 123 Hypervariable markers in the chicken genome.
C. Miller, M.W. Bruford & T. Burke. Department of Zoology, University of Leicester, University Road, Leicester LE1 7RH, U.K.

Cl 124 The development of the chicken immune system: investigations on the expression and possible immunological significance of alpha fetoprotein for T-cell dependent immune functions.
U. Neumann & L.D. Bacon, Clinic for Poultry, Hannover School of Veterinary Medicine, 3000 Hannover 71, Federal Republic of Germany, and Regional Poultry Research Laboratory, USDA, East Lansing, Michigan 48823, U.S.A.

Cl 125 Liposome-mediated gene transfer into chicken embryos in ovo.
C.I. Rosenblum & H.Y. Chen, Department of Growth Biochemistry & Physiology, Merck, Sharp and Dohme Research Laboratories, Rahway, NJ 07065, USA.

Cl 126 Developmental and tissue-specific protein interactions with a DNA element upstream of the avian apoVLDLII gene.
A.K. Ryan & R.G. Deeley, Cancer Research Laboratories, Queen's University, Kingston, Ontario, K7L 3K8, Canada.

Cl 127 Regulation of the chicken apoVLDLII gene: determinants of hormonal and tissue specificity.
T.J. Schrader, R.A. Burtch Wright & R.G. Deeley. Cancer Research Laboratories, Queen's University, Kingston, Ontario, K7L 3N6, Canada.

Cl 128 Mechanisms controlling the developmentally-regulated expression of genes encoding chicken U4 small nuclear RNA.
W.E. Stumph, Ihab W. Botros, & J.H. Miyake, Department of Chemistry & Molecular Biology Institute, San Diego State University, San Diego, CA 92182, USA.

Cl 129 Characterization of new genetic markers for poultry breeding by molecular genotyping of the major histocompatibility complex (MHC) of selected chicken lines.
P. Thoraval, A.-M. Chausse, F. Coudert & G. Dambrine, Station de Pathologie Aviaire et de Parasitologie, INRA, Nouzilly 37380 Monnaie, France.

Cl 130 Isolation of chicken cDNAS encoding homeodomain proteins and characterization of their temporal and spatial expression during embryonic limb development.
W.B. Upholt, R.A. Kosher, M. Barembaum, K.J. Blake, C.N.D. Coelho, Siew-Ging Gong, J. Paiva-Borduas, B.J. Rodgers & L. Sumoy, Departments of BioStructure and Function and Anatomy, University of Connecticut Health Center, Farmington, CT 06030, USA.

Cl 131 Turkey prolactin and growth hormone: molecular cloning, characterization and associated RFLPs.
D. Zadworny, C.N. Karatzas & U. Kuhnlein, Dept. of Animal Sci., McGill University, Ste-Anne de Bellevue, P.Q. H9X 1C0, Canada.

Cl 132 Characterization of RAV-related endogenous viral genes in six experimental strains of chicken.
M.H. Tixier-Boichard & Lucien Durand, Laboratoire de Genetique Factorielle, Centre de Recherches INRA, 78352 Jouy-en-Josas, Cedex, France.

Sponsors

We gratefully acknowledge support from:

Arbor Acres Farm, Inc.
Amgen, Inc.
Cobb Vantress, Inc.
DeKalb Genetics Foundation
Euribrid B. V.
Hubbard Farms
Hy-Line International
Lohmann Tierzucht
Maple Leaf Farms, Inc.
Merck Sharp & Dohme Research Laboratories
Natural Sciences and Engineering Research Council of Canada
Nicholas Turkey Breeding Farms
Ontario Chicken Producers' Marketing Board
Ontario Egg Producers' Marketing Board
Peterson Industries
Poultry Breeders' Union
Rhone-Merieux

This book is dedicated to the most important
people in our lives
- our families and our research group -
over whom we have shared
wonderment*, pride, laughter (lots), and tears.

---

* *Collins Concise Dictionary Plus* defines wonderment variously, but includes *rapt surprise* and *puzzled interest*. How apt!

Table of Contents

Chapter 1

A Genetic Approach to Physiology

Robert J. Etches

Summary. The genetic selection of poultry has relied on the skills of pragmatic breeders throughout human history. During the past 100 years, this pragmatism has been guided by an increasing body of information describing the physiological systems that have been altered by selection for morphology and phenotype. During the past several years, information has been obtained about the physiological systems that control the passage of genetic information from one generation to the next. This information includes the reproductive biology and embryology of the bird, the cytological structure and organization of genetic information in birds, and the physiology of the systems that are controlled by the genetic code. This information has facilitated the development of many techniques to assist the breeder in the selection of genetic variants from either existing populations or from combinations of genetic material from other sources. This information is likely to reduce the constraints of sexual reproduction on the recombination of genetic material that directs physiological processes from which benefit can be derived.

Historical Introduction

The avian genome has been manipulated for at least 3000 years, and specialized breeds and strains of domestic birds predate the recognition of genetics as a scientific discipline by at least 2000 years. It is possible that breeds of chickens were established by poultry breeders in China as early as 1400 B.C. (Darwin, 1896) and it is clear from the writings of Aristotle, Columella, and Pliny that Greek and Roman poultry breeders had created strains for the production of meat and for their ability to perform in the cockpit (Crawford, 1984). Darwin (1896) produced the first extensive catalog of poultry stocks around the world, and it is abundantly clear that many specialized and distinctive breeds had been developed in the Orient, in Europe, and in the Americas. The breeding and development of exhibition poultry, cage birds, and pigeons continues unabated today by many individuals who have little or no formal training in genetics but who continue to manipulate the avian genome in remarkable ways. For example, the Birmingham roller pigeon was created in the 1940s by Mr. Pensom, a bus driver from Birmingham, England, and a group of equally dedicated hobbyists from the area. These gentlemen selected a group of pigeons from a breed possessing a behavioral modification known as tumbling that was also described by Darwin (1896). Tumblers lose the ability to fly at about 6 months of age because the take-off hop initiates a backward roll in the air. The Birmingham group of amateur geneticists, however, selected birds who could fly and roll, and the descendants of these birds provide aerial acrobatics for thousands of devoted breeders of the performing roller.

When genetic selection was practiced prior to the development of a theoretical basis to guide the pragmatic breeder, progress relied entirely on the fortuitous "sports" and "freaks" that appeared from time to time. These animals were carefully nurtured by the breeder, who mated together phenotypes that most closely resembled the ideal animal. The ideal was originally defined in the mind of the breeder; subsequently, breed organizations formed to formalize the definition in writing. The remarkable diversity that ensued is a tribute to the diligence and stockmanship of generations of poultry enthusiasts. The legacy of mutations, strains, and breeds that we have inherited from these collectors is providing the tools to unravel the physiological processes involved in the replication, transcription, and translation of the message encoded in DNA.

Manipulation of the Avian Genome as Genetic Theory and Practice were Developed

The plethora of genetic variants in the poultry population played an important role in the understanding of inheritance from a Mendelian perspective and in the application of these principles in medicine and agriculture. For example, the first proof that Mendel's laws applied to animals as well as plants was presented to the Evolution Committee of the Royal Society by William Bateson, who described the inheritance of the rose- and single-comb genetic variants in the head furnishings of chickens (Bateson, 1902). While the rose comb had been known for some time as an esthetically pleasing genetic variant, it also became one of the first *quantitative trait loci* (QTL) to be investigated, when it was used in the selection of chickens whose comb resisted freezing in harsh northern climates. This goal was accomplished by applying the information published by Bateson and Punnett (1905) that demonstrated epistasis in animals for the first time. Their work identified the genotypes of single, rose, pea, and cushion combs as *rrpp*, *R-pp*, *rrP-*, and *R-P-*, respectively. The quantitative goal of increasing winter hardiness was achieved by incorporating the *P* and *R* alleles into a new breed, the Chantecler, whose small cushion comb reduced heat loss and resisted freezing (Cole, 1922).

The rose comb phenotype continued to be attractive to a number of poultry breeders and is a characteristic feature of the Wyandotte breed, which contributed significantly to the formation of modern broiler strains. It was again studied as a QTL when it was recognized that fertility in all rose-combed breeds was poor. Eventually, the source of infertility was attributed to a recessive effect of the dominant allele at the rose comb locus (Table 1). This

Table 1

Effect of Comb Genotype on Fertility in Chickens

Genotype of sire	Phenotype of sire	Fertility following natural mating	Fertility following artificial insemination
RR	Rose	78	44
Rr	Rose	91	84
rr	Single	92	82

Data from Crawford and Smyth (1964).

relationship between a single functional unit of inheritance and a quantitative trait was the prototype for the current investigations of the relationships between sequences of DNA that can be identified using the techniques of molecular biology and traits that are of commercial importance to the poultry industry (see Chapter 16).

The antiquarian geneticists who conserved and propagated the rose comb gene, with their keen eye and genetic skill, should be recognized not only for their contribution to the fundamental understanding of inheritance and the relationship between quantitative and qualitative traits, but also for the provision of the first autosomal locus to be placed on the genetic map of the chicken. In 1928, Serebrovsky and Petrov demonstrated autosomal linkage in the fowl for the first time when they reported that the rose comb gene was separated from the creeper gene by 9.1 crossover units. This map has been expanded during the past 63 years, and the most current edition (see Chapter 5) shows that markers identified using the tools of molecular biology, such as the *ev* loci, are being integrated into the map of morphological traits that was assembled during the first half of this century. At the same time, our knowledge of chromosome structure and function has expanded rapidly, and this body of information, together with recent advances in the field, have been summarized in this volume by Bloom et al. (see Chapter 4).

The shift in emphasis from morphologically defined traits to those that are physiologically defined and, more recently, to those that are defined by their base-pair sequences, has been the consequence of developments in the tools available to study inheritance. A good example of this change in emphasis can be traced in the development and use of the Sebright bantam. In about 1800, Sir John Sebright incorporated the rose comb, lacing, dwarfing, and hen-feathering genes into a single breed that today bears his name. In the eyes of Sir John and the devotees of the breed who have continued to perpetuate the Sebright bantam, the combination of mutations produces a work of art. Following the recognition that hen feathering (i.e., the female plumage structure of Sebright and Campine males) was inherited in a dominant Mendelian fashion, however, these breeds became the object of intense investigation by physiologists seeking an understanding of sexual differentiation (see review by Wilson et al., 1987). These investigations were stimulated by the apparent similarity between idiopathic femininization of genetic males in humans and the predictable femininization of the plumage of hen-feathered Sebright roosters. Many elegant, but indirect, studies conducted from 1920 to 1960 demonstrated that the gonadal secretions of the hen-feathered Sebright are normal and that the gene acts at the feather follicle to femininize the growing structure. Further definition of the aberrant physiological system that produces henfeathering awaited the development of an aromatase assay that was sufficiently sensitive to detect the massive increase in the ability of this enzyme to convert androgens to estrogen in the skin (Wilson et al., 1987). These studies showed that the growing feather was femininized because the activity of the androgen-binding cytochrome P-450 oxidase in males bearing the *Hf* allele is increased 20-fold in the heterozygote and 40-fold in the homozygote hen-feathered rooster. From the point of view of the physiologist who is concerned with enzyme function, therefore, the gene is co-dominant because aromatase levels in the heterozygote are intermediate between those of the homozygous parents. From the point of view of the breeder who is concerned with phenotype, however, the trait is dominant because the heterozygote produces sufficient aromatase to femininize the plumage of the heterozygote male.

Subsequent investigations have revealed the sequence of the mRNA and corresponding DNA that code for the extraglandular aromatase, and it would appear that the increase in activity of this enzyme in hen-feathered males is due to a retroviral promoter that has inserted into this gene (Matsumine et al., 1991). Furthermore, *in situ* hybridization has revealed that this gene is located on chromosome 1 (Tereba et al., 1991). This elegant work provides an excellent example of a morphological trait that can be related to a protein product that, in turn, can be related to a specific segment of the genome located on a specific chromosome in the chicken.

Unfortunately, our knowledge of the physiology and biochemistry of most traits of interest is far less comprehensive than our understanding of hen-feathering. In most cases, breeders manipulate the genome to alter physiological function with little, if any, understanding of the physiological systems that have been altered. Growth rate, for example, is easy to manipulate on a phenotypic basis, although the genetic basis of this massive physiological change is only beginning to be understood. The development of our understanding of the physiological systems that are manipulated by selection for the end product will be gained using the tools of molecular biology to explain morphology in terms of the base-pair sequences coding for the enzymes, hormones, and structural proteins that interact in a series of intertwined metabolic pathways. In this volume Dr. Goddard and his colleagues have assembled recent information concerning our knowledge of some of the hormonal mechanisms that regulate growth in poultry.

Our current inability to connect biochemical descriptions of genes with the trait we would like to manipulate is the major impediment to using transgenic systems. It could be of significant economic advantage, for example, to transfer the sex-linked dwarfing gene from the chicken to the turkey, because the dwarf broiler breeder hen produces more chicks using less feed (see review by Decuypere et al., 1991). If an extrapolation of the phenotypic action of this gene from the chicken to the turkey is justifiable, it is realistic to expect a reduction in feed consumption of about 10 – 15%. During the lifespan of a turkey breeder hen, this would equal 10 – 15 kg of feed. One of the major factors that prevents the execution of an experiment to test this extrapolation is the lack of information regarding the molecular biology of the gene and the gene product that limits growth in dwarfs. In a perfect world, the geneticist would have a genetic blueprint of the physiological functions governing the desired phenotype. The fields of genetics and physiology have evolved separately, however, and our current knowledge is a fragmented collection of facts drawn from a number of different perspectives. The objective of many individuals during the next decade will be the unification of the fragments into a useful description of the genetic code at the molecular level, the development of reliable predictions of the cellular consequences of altering the code, and, finally, the manipulation of the phenotypic expression of the genetic code in the whole animal. When this molecular understanding of genetics and the physiological consequences of genetic rearrangement is realized, manipulation of the avian genome can be undertaken with unprecedented precision.

Accessing the Avian Genome

The practicalities of manipulating the avian genome are inextricably linked to the morphology and physiology of cells in the germline, since they must be accessed if the genome is to be manipulated. In birds, the germline is believed to diverge from the embryonic tissues that give rise to the extraembryonic membranes and the embryonic endoderm, mesoderm, and ectoderm at about the time the egg is laid (Figure 1). At this time, the germline comprises less than a few hundred primordial germ cells, which can be identified in the anterior regions of the embryo after a few hours of incubation and in the embryonic vasculature in the second and third day of incubation. By the fourth day of incubation, they are migrating into the definitive gonad (Nieuwkoop and Sutasurya, 1979). These cells are good candidates in the search for a suitable cell type that can be obtained and manipulated *in vitro* to achieve the goal of gene insertion. Simkiss et al. (1990) have presented evidence indicating that it is possible to introduce DNA sequences into the genome via these cells, and Wentworth et al. (1989) have transferred primordial germ cells from chickens to quail. Furthermore, they have demonstrated that

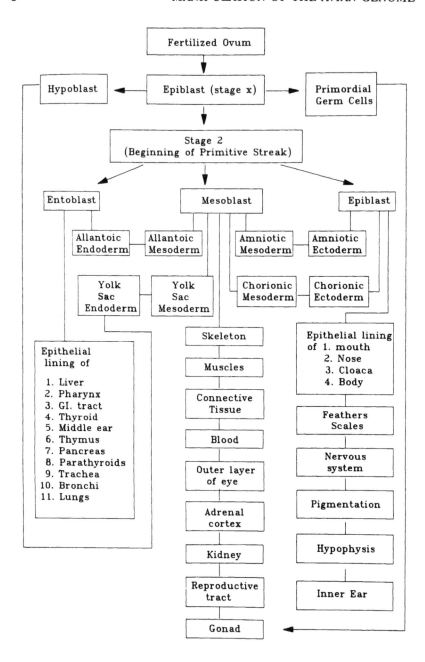

FIGURE 1.   A schematic diagram of development of the chick embryo indicating the derivation of tissues within the embryo and extraembryonic membranes. At the time of oviposition, the embryo is usually at stage X (E.G&K).

primordial germ cells can be introduced into the germline of a host and be transmitted to the next generation. In Chapter 6, Petitte et al. describe how blastodermal cells can be transferred between stage X (E.G&K) embryos and have shown that specific genes can be incorporated into these cells prior to their transfer to the recipient embryo. Dieterlen-Lievre and Le Douarin, in Chapter 7, describe elegant manipulations of embryos at later stages of development that have contributed greatly to our understanding of the interactions between cells and tissues as differentiation proceeds. Development of the embryo prior to stage X (E.G&K) and during the first few hours of incubation is described in Chapter 3 by Eyal-Giladi, who, in association with her colleagues, has charted the morphogenetic changes in the early chick embryo.

If we accept the basic organization of development that is outlined in Figure 1, it follows that a pluripotential cell type must exist in some early, but as yet undefined, stage of development. Although this cell type has not been identified, the chimeras produced by Petitte et al. (1990) were produced from a population of donor cells that may have contained a proportion of cells that are pluripotential. In mice, pluripotential cells from the early embryo have played a major role in the development of precise alteration of the murine genome. These embryonic stem cells (ES cells) have the potential to enter all of the embryonic tissues although, for the purposes of genetic engineering, their ability to enter the germline is their most important attribute. ES cells can also be maintained in culture for extended periods of time, during which their genome can be specifically altered by homologous recombination (see Chapter 19). Finally, if these cells are derived from a male embryo, they will masculinize a female host gonad and all of the germ cells produced by the mixed-sex chimera will be derived from the donor cell line. This preferential gametogenesis from the male ES cell-derived donor cells ensures transmission of the altered genome to the next generation (Evans et al., 1985).

In chickens, a pluripotential embryonic stem cell that is capable of replication without differentiation *in vitro* while retaining the ability to enter the germline following injection into a recipient embryo has not been identified. Additionally, we do not yet know what influence, if any, male primordial germ cells exert on a female host gonad and vice versa. It is clear, therefore, that although the use of chimeras to mediate genome manipulation has many theoretical advantages, its realization will depend on the development of embryonic stem cell lines and the compatibility of cells types in the developing gonad.

Manipulation of the genome in early embryos *in ovo* by direct injection of the newly fertilized egg has daunted even the most dexterous of those who wish to gain access to the avian genome. The volume of the yolk, the layers of shell, albumen, and vitelline membrane, and the fragility of the entire structure make this approach appear impractical. The albumen and shell can be avoided by using *in vitro* ovulation (Neher *et al.*, 1950), but the yolk that is recovered from this manipulation lacks the extravitelline membrane, which is normally deposited on the yolk in the upper regions of the infundibulum. The very elegant techniques developed by Perry (1988) and described in Chapter 8 by Sang et al., permit access to the genome of the newly fertilized egg *in vitro*, however, and this approach may provide a useful route to the chromosomes of birds.

Avian retroviruses have proved to be a useful biological vector for the insertion of genetic material into the genome of chickens. They can infect the cells of the early chick embryo, including those in the germline, and have been used on several occasions to modify the genome of domestic birds (Salter et al., 1987; Hippenmeyer et al., 1988; Bosselman et al., 1989a, 1989b; Lee and Shuman, 1990; Chen et al., 1990; Briskin et al., 1991). Although there may be some reservations about using retroviruses to deliver specific gene sequences into commercial breeding flocks, the technology for their use as vectors is well developed and will occupy a prominent position in the arsenal of poultry geneticists in the immediate future. Chapter 9 in this volume describes recent advances in this area.

The sperm cell is the natural vector for the introduction of DNA into the female pronucleus. There has been a great deal of speculation regarding the ability of sperm cells to introduce foreign DNA into the genome since the initial report of Lavitrano et al. (1989) describing sperm-mediated gene transfer into the mouse. The basic biology of the avian sperm and the egg are presented in Chapter 2, and the potential transfer of genetic material into chickens in association with the sperm cell is alluded to in Chapter 10.

The Usefulness of Accessing the Genome

A primary goal of manipulations of the avian genome is the creation of chickens and turkeys that are more suitable than existing stocks in the production of meat and eggs. The retail value of the poultry industry in the United States is approximately $16 billion. It is clear, therefore, that there are significant economic forces that stimulate the search for genetic control over the physiological processes that culminate in egg and meat production. These forces and the aspirations of the poultry industry as manipulation of the avian genome continues at the molecular level are presented in Chapter 18.

Manipulations of the avian genome that involve the Z and W chromosomes have potential interest in biology, medicine, and agriculture. In biology and medicine, the avian system of sex determination provides a useful mechanism to test generalized concepts of sexual differentiation in vertebrates because the female is the heterogametic sex. This difference shifts the timing of sex determination from the moment of fertilization as in mammals to the first meiotic division in the egg, when the homologous pairs of chromosomes separate into the female pronucleus and the first polar body. In birds, the first meiotic division begins about 6 h before ovulation and is completed at the time of ovulation. When the egg is released from the follicle, the female pronucleus is arrested in the second anaphase of meiosis, the first polar body contains the Z or W chromosome, and the sex of offspring has already been determined.

Broiler chickens and laying hens are hybrids from three or four pure lines, each of which is selected for specific attributes. Typically, males from lines 1 and 3 are bred to females from lines 2 and 4, respectively, and in the next generation, F_1 males from lines 1 and 2 are bred to F_1 females from lines 3 and 4 to produce the commercial hybrid. This hierarchial system of line crossing could provide an ideal scheme in which to implement genetic control over sex determination, since only one of the two sexes is required at each generation preceding the production bird. Sex determination would eliminate the production of chicks of the undesired sex and reduce the cost of incubation by half if the sex of the embryo were predetermined at the first meiotic division or if the femininizing effect of the W-bearing female gamete could be nullified by a sperm cell bearing an appropriate sex chromosomal consti- tution. Although realization of a breeding program that could deliver these goals has been hampered by both theoretical gaps in our understanding and the inability to manipulate the Z and W chromosomes, the recent studies of W-specific sequences, which are presented in Chapter 17, may open new frontiers to explore in the development of this technology. Since it has been recognized recently that triploid chickens are viable (Thorne et al., 1988), it may be possible to create sex-determining crosses that depend on the presence of extra copies of the genome if fertile triploids can be identified.

Traditionally, the poultry industry has relied on crossbreeding of pure lines to provide a relatively uniform bird with certain characteristics that are required by the marketplace. The genetic rearrangement that occurs as each generation is reproduced usually produces offspring that bear only a small proportion of the outstanding attributes of the most superior individuals in the parental population. Ideally, the breeder would like to manipulate the genome so that the best combination of genetic material that is available in any of the parental lines is transmitted as an exact copy to all of the offspring. This combination would be used until the breeder had identified a new com- bination of alleles in the genome that produced a more outstanding individual, which would be substituted in the dissemination of production stock. In order to implement such a scheme, it would be necessary to clone the chicken genome as a unit. Currently, we have little if any idea how this might be accomplished. It is possible, however, that the key to this approach lies in the relatively undifferentiated mass of cells in the early avian embryo. The ability to clone the chicken genome as a unit could give rise to an approach to poultry breeding that would resemble the strategy employed by some plant breeders, who propagate the preferred genotype by asexual means. Strawberry breeders, for example, make sexual crosses between parental plants that possess characters that the breeders would like to combine into a new stock. The cross produces several hundred or perhaps thousands of plants that are tested in field plots. Each of these plants is evaluated, and only the best one or two are kept. These plants, whose genotypes are closer to the ideal than any other plants existing in the past or present, are propagated asexually by runnering to ensure that this particular combination of genes is not lost or rearranged into a combination that is less productive. In the animal kingdom, cloning of entire genomes is the equivalent of asexual propagation of plants, and its appeal could be the ability to propagate outstanding genotypes in very large numbers. In the chicken, it may be possible to propagate large numbers of undifferentiated cells from the stage X (E.G&K) embryo *in vitro*, and return these cells to host embryos that would incorporate them into their germline. If the rate of transmission of the donor cell line in the chimera were sufficient and if large numbers of chimeras could be produced, a commercial breeding program could be envisaged in which the four pure lines are replaced by four genetically superior individuals.

Differentiation is still a process that is generally understood only at a morphological and descriptive level. Textbooks of development are atlases that chart the movement of cells and the growth of tissues with the passage of time. A molecular understanding of the developmental process that is comparable to the understanding of intermediary metabolism is only just emerging, because experimental paradigms to gain the information have been few. The development of techniques permitting targeted gene substitutions in the chicken would provide the tools to unravel the complexities of embryogenesis and aid in the development of a molecular description of the processes that direct the construction of a complex organism from a single cell. The chick embryo has enjoyed a preferred place in embryology because it is accessible and inexpensive. These attributes have contributed to the accumulation of a vast body of literature concerning the morphological changes in the embryo and the genetic factors that alter the developmental process during the past century. During the past 10 years, hundreds of genes coding for proteins that are presumed to be important in the regulation of interactions between cells during development have been isolated and combined with promoters that will drive their expression in novel ways within the embryo. Many transgenic mice have been formed following transfer of such gene constructs into embryos, and the accumulated information from these studies will provide the information that is required to chart a molecular description of development that parallels the existing morphological atlases (Rossant and Joyner, 1989).

A molecular understanding of the genetic processes that control physiological functions will facilitate the development of novel systems for the production of proteins and other products that are difficult to produce by conventional chemistry or using prokaryotic systems. In general, these would be complex proteins that are modified by posttranscriptional events. Since birds deposit large amounts of protein in their eggs, it is conceivable that modification of the genome could yield chickens capable of depositing large amounts of a desirable protein instead of, or in addition to, the normal egg white or egg yolk proteins. The ability to design genes that can direct protein synthesis and deposition to achieve these goals depends on our understanding of the regulation of genes controlling the production of the normal egg white and egg yolk components. The base-pair sequences for the genes coding for most of the major egg white and egg yolk proteins are now available (see review by Burley and Vadehra, 1989), and the interactions between the hormonal events that control reproduction and the expression of these genes is currently being elucidated. In Chapter 13, recent data concerning the structure of the regulatory sequences that control expression of the major yolk apolipoprotein and vitellogenin genes are presented.

In populations of domestic birds, disease control is still one of the major concerns. Public pressure to produce food at low cost and in the absence of vaccines, coccidiostats, and antibiotics will continue and intensify the search to find ways of producing chickens that are able to resist infectious and parasitic agents without intervention. In cases where the molecular biology of the genes and their products that confer resistance to disease are known, the techniques of gene transfer that are being developed using retroviruses, primordial germ cells, stem cells, and *in vitro* manipulations of embryos are likely to be employed. In other cases, the classical methods of gene introgression that rely on crossing and backcrossing into highly selected production stocks will continue to be used. DNA markers will be useful aids in the identification of those individuals that have inherited the largest proportion of DNA from the highly selected stock together with the gene of interest. This new approach to a classical breeding problem is developed in Chapter 16. The genetic control of disease and disease resistance in poultry is presented in Chapter 15, and the role of the major histocompatibility complex in the cell-mediated and humoral immune response to pathogens is presented in Chapter 12.

Conclusions

An understanding of the molecular biology controlling physiological processes in birds is evolving in many areas and for many different reasons. This information will be used to gain further information regarding the general principles that control developmental processes in vertebrates in general and, therefore, will serve as a general model in medicine and biology. In the industrial arena, birds may become useful bioreactors because they package and deposit large amounts of protein in eggs. In the poultry industry, directed alteration of the genetic code will produce chickens capable of performing physiological functions that are currently desirable but unavailable in the existing pool of genetic variation.

References

Bateson, W. (1902). Experiments with poultry. *Rep. Evolution Committee Roy. Soc.*, 1, 87–124.

Bateson, W., and Punnett, R. C. (1905). A suggestion as to the nature of the ''walnut'' comb in fowls. *Proc. Cambridge Phil. Soc.*, 13, 165–168.

Bosselman, R. A., Hsu, R., Boggs, T., Hu, S., Bruszewski, J., Ou, S., Kozar, L., Martin, F., Green, C., Jacobsen, F., Nicholson, M., Schulz, J. A., Semon, K. M., Richell, W., and Stewart R. G. (1989a). Germline transmission of exogenous genes in the chicken. *Science,* 243, 533–535.

Bosselman, R. A., Hsu, R., Boggs, T., Hu, S., Bruszewski, J., Ou, S., Souza, L., Kozar, L., Martin, F., Nicholson, M., Rishell, W., Schulz, J. A., Semon, K. M., and Stewart, R. G. (1989b). Replication defective vectors of reticuloendotheliosis virus transduce exogenous genes into somatic stem cells of the unincubated chicken embryo. *J. Virol.,* 63, 2680–2689.

Briskin, M. J., Hsu, R.-Y., Boggs, T., Schultz, J. A., Rishell, W., and Bosselman, R. A. (1991). Heritable retroviral transgenes are highly expressed in the chicken. *Proc. Natl. Acad. Sci.,* 88, 1736–1740.

Burley, R. V., and Vadehra, D. V. (1989). *The Avian Egg: Chemistry and Biology.* John Wiley & Sons, New York.

Chen, H. Y., Garber, E. A., Mills, E., Smith, J., Kopchick, J. J., DiLella, A. G., and Smith, R. G. (1990). Vectors, promoters, and expression of genes in chick embryos. *J. Reprod. Fertil., Suppl.,* 41, 173–182.

Cole, L. J. (1922). Chantecler poultry. *J. Heredity,* 13, 146–152.

Crawford, R. D. (1984). Domestic fowl. In *Evolution of Domestic Animals,* edited by I. L. Mason. Longman Group, Harlow, England.

Crawford, R. D., and Smyth, J. R., Jr. (1964). Studies on the relationship between fertility and the gene for rose comb in the domestic fowl. 1. The relationship between comb type and fertility. *Poultry Sci.,* 43, 1009–1017.

Darwin, C. H. (1896). *The Variation of Plants and Animals Under Domestication,* 2nd ed. Appleton, New York.

Decuypere, E., Huybrechts, L. M., Kuhn, E. R., Tixier-Boichard, M., and Merat, P. (1991). Physiological alterations associated with the chicken sex-linked dwarfing gene. *Crit. Rev. Poultry Biol.,* 3, 191–221.

Evans, M., Bradley, A., and Roberstson, E. (1985). EK cell contribution to chimeric mice: From tissue culture to sperm. In *Banbury Report 20: Manipulation of the Early Embryo.* Cold Spring Harbor Laboratory, Cold Springer Harbor, NY, pp. 93–102.

Hippenmeyer, P. J., Krivi, G. G., and Highkin, M. K. (1988). Transfer and expression of the bacterial NPT-II gene in chick embryos using a Schmidt-Ruppin retrovirus vector. *Nucleic Acids Res.,* 16, 7619–7632.

Lavitrano, M., Camaioni, A., Fazio, V. M., Dolci, S., Farace, M. G., and Spadafora, C. (1989). Sperm cells as vectors for introducing foreign DNA into eggs: Genetic transformation of mice. *Cell,* 57, 717–723.

Lee, M.-R., and Shuman, R. M. (1990). Transgenic quail produced by retrovirus vector infection transmit and express a foreign marker gene. In *Proceedings of the 4th World Congress on Genetics Applied to Livestock Production,* XVI, 107–110.

Matsumine, H., Herbst, M. A., Ou, S.-H. I., Wilson, J. D., and McPhaul, M. J. (1991). Aromatase mRNA in the extragonadal tissues of chickens with the henny feathering trait is derived from a distinctive promoter structure that contains a segment of a retroviral long terminal repeat. *J. Biol. Chem.,* 266:19900–19907.

Neher, B. N., Olsen, M. W., and Fraps, R. M. (1950). Ovulation of the excised ovum of the hen. *Poultry Sci.,* 29, 554–557.

Nieuwkoop, P. D., and Sutasurya, L. A. (1979). *Primordial Germ Cells of the Chordates.* Cambridge University Press Cambridge.

Perry, M. M. (1988). A complete culture system for the chick embryo. *Nature* (London), 331, 70–72.

Petitte, J. N., Clark, M. E., Liu, G., Verrinder Gibbins, A. M., and Etches, R. J. (1990). Production of somatic and germline chimeras in the chicken by transfer of early blastodermal cells. *Development,* 108, 185–189.

Rossant, J., and Joyner, A. L. (1989). Towards a molecular-genetic analysis of mammalian development. *Trends in Genetics,* 5, 277–283.

Salter, D. W., Smith, E. J., Hughes, J. S. H., Wright, S. E., and Crittenden, L. B. (1987). Transgenic chickens: Insertion of retroviral genes into the chicken germ line. *Virology,* 157, 236–240.

Serebrovsky, A. S., and Petrov, S. G. (1928). A case of close autosomal linkage in the fowl. *J. Heredity,* 19, 305–306.

Simkiss, K., Rowlett, K., Bumstead, N., and Freeman, B. M. (1989). Transfer of primordial germ cell DNA between embryos. *Protoplasma,* 151, 164–166.

Tereba, A., McPhaul, M. J., and Wilson, J. D. (1991). The gene for aromatase ($P450_{arom}$) in the chicken is located on the long arm of chromosome 1. *J. Heredity,* 82, 80–81.

Thorne, M. H., Collins, R. K., Sheldon, B. L., and Bobr, L. W. (1988). Morphology of the gonads and reproductive ducts of triploid chickens. In *Proceedings of the XVIII World's Poultry Science Association,* Nagoya, Japan, pp. 525–526.

Wentworth, B. C., Tsai, H., Hallett, J. H., Gonzales, D. S., and Rajcic-Spasojevic, G. (1989). Manipulation of avian primordial germ cells and gonadal differentiation. *Poultry Sci.,* 68, 999–1010.

Wilson, J. D., Leshin, M., and George, F. W. (1987). The Sebright bantam chicken and the genetic control of extraglandular aromatase. *Endocrine Rev.,* 8, 363–376.

Chapter 2

The Anatomy of Reproduction in Birds, with Emphasis on Poultry

Murray R. Bakst

Summary. The general strategy of reproduction in poultry differs considerably from that of other domesticated animals. If fertile, the oviparous hen lays an egg containing an embryo of about 60,000 cells. Fertilization had taken place about 26 h before oviposition with participating sperm that had been inseminated days or, in turkeys, weeks before fertilization. These sperm were subjected to an intense selection process within the vaginal segment of the oviduct and then stored in the sperm storage tubules (SSTs) localized in the uterovaginal junction (UVJ). It is this capacity to store, then slowly but continuously release sperm from these SSTs that assures an adequate number of sperm at the site of fertilization in the absence of repeated copulations or insemination.

Introduction

The approaches for insertion of genetic material into the avian genome vary and utilize both the male and female gametes and the fertilized ovum. To develop a full understanding and appreciation of this technology, a foundation in basic anatomy and physiology of avian reproduction is required. Toward this end, this chapter includes some aspects of selected areas of male and female reproductive anatomy and physiology and a brief explanation of artificial insemination technology. Since it is beyond the scope of this chapter to provide a comprehensive review of these areas, the reader is provided with an extended reference list covering both general and specific subject areas.

Anatomical Nomenclature

Often overlooked but extremely important in this type of descriptive work is the use of correct anatomical nomenclature. If unknown or if some doubt exists regarding an anatomical term, one should consult *Nomina Anatomica Avium* (Baumel et al., 1979). (A revised second edition will be available in 1993.) This text can serve the investigator in two ways: First, it provides standardized anatomical terms established by the International Committee on Avian Anatomical Nomenclature; and second, it is one of the more comprehensive accounts of avian anatomy available. In this chapter, general anatomical terms reflecting correct homologies and anatomical observations will

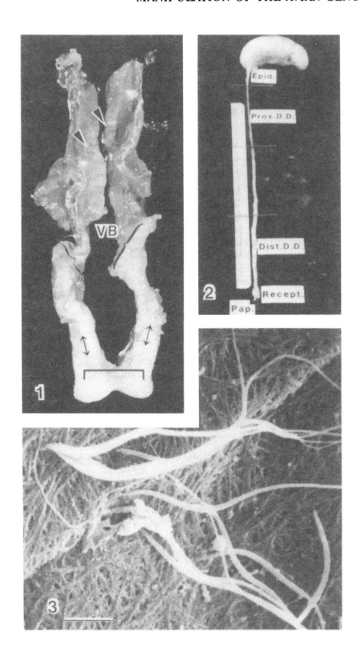

FIGURE 1. A latex-perfused preparation (cleared in glycerin) of the structures involved in the erection of the turkey phallus. The lighter perfusate highlights the lymph channels in the phallus (bracket), lymph folds (double arrows), and vascular bodies (VB). Also observed is a short section of the ductus deferens (arrowheads). (Bracket = 2 cm.)

FIGURE 2. The turkey testis and excurrent duct system, which includes the epididymal region and the proximal and distal regions of the ductus deferens, are observed adjacent to a 15-cm ruler. The distal end of the ductus deferens forms an oval-shaped dilation, the receptaculum. The papilla is a wartlike extension of the ductus deferens that projects into the central compartment of the cloaca, the urodeum.

FIGURE 3. Chicken sperm, which are filiform shaped and possess a characteristic curve of the head region, are observed on the perivitelline layer (PL) of a recently ovulated ovum. The PL is the fibrous reticulum enveloping the ovum at ovulation, which sperm must hydrolyze to penetrate and fertilize the ovum. (Bar = 6 μm.)

be used. For the correct Latin anatomical terms, refer to *Nomina Anatomica Avium.*

The Male Reproductive Tract

The genital systems of the male turkey (Figures 1 and 2) and chicken are nearly identical. The paired testes are located at the cranial end of the kidneys in the abdominal cavity. Sperm produced in the seminiferous tubules pass via the rete testis into the epididymal region, which is located in the hilar portion of the testes. From the rete, the semen passes through the efferent ducts, connecting tubules, and finally the epididymis before reaching the ductus deferens. Together the epididymal region and ductus deferens are commonly referred to as the excurrent duct system (Figure 2) (for reviews, see Hess et al., 1976; Bakst, 1980a; Van Krey, 1990).

Upon reaching the ductus deferens, chicken and turkey sperm do not appear to undergo any further maturational changes with respect to their motility or fertilizing capacity. Sperm removed from any level of the ductus deferens are highly motile and, if inseminated, will result in fertility levels as high as ejaculated sperm (Bakst and Cecil, 1981).

Unlike the mammal, which possesses true glandular accessory sex organs such as the ampullary gland, seminal vesicles, prostate, and bulbourethral glands, there are no discrete accessory sex glands or other such homologous structures associated with the avian male genital tract (Lake, 1981). Nevertheless, at the time of ejaculation a lymphlike fluid, commonly referred to as transparent fluid, is expelled from areas of the urodeum and the erectile tissue forming the phallus nonprotrudens. (The cloaca is divided into three compartments, the coprodeum, the urodeum, and the proctoderm, which is the most caudal.) Transparent fluid formation is better understood by examining the mechanism of phallic tumescence in poultry.

When sexually stimulated, there is a lymphlike fluid, actually a vascular transudate, produced within the dense capillary network formed within the *Corpora vascularia paracloacalia* (vascular bodies) (Figure 1). The vascular bodies are located in the wall of the cloaca just lateral to the distal end of the ductus deferens. The lymph formed during sexual stimulation flows from the capillaries to lymph capillaries and then into the larger lymph sinuses in the subcapsular region of the vascular body. The flow of this lymph continues into the more caudal lymph channels in the phallus (Figure 1). Tumescence results when the lymph channels within the phallus become engorged with lymph (for details, see Knight et al., 1984). It is this lymph fluid that, when under increased hydrostatic pressure during phallic tumescence, passes from the lymph channels to the surface of phallus. Alternatively, Fujihara et al. (1985) have suggested that transparent fluid is derived from the tissue around the papillae, wartlike projections of the ductus deferens, located in the urodeum. While the precise source of the transparent fluid needs to be resolved, there is no doubt that chicken and turkey semen is diluted at the time of insemination with a fluid very similar to blood plasma (Fujihara and Nishiyama, 1984).

The Female Reproductive Tract

In the female, only the left ovary and oviduct develop into functional organs (Figure 4). [More comprehensive reviews of the female genital system and the ovulatory cycle of the hen are found in Gilbert (1979) and Etches (1990), respectively.] With a hen in egg production, the largest of the megalecithal oocytes ovulates generally between 30 and 60 min after the previous oviposition. This ovulated ovum is grasped by the fimbriated region of the oviduct and is guided toward the ostium of the infundibulum (Figures 4 and 5). If sperm are present in the infundibulum, the ovum may be fertilized. Regardless of whether it is fertilized or not, the ovum continues in its progression through the five segments of the oviduct, where it accrues the albumen in the magnum, the shell membrane in the isthmus, and the shell in the uterus. The distal segment of the oviduct is the vagina, which serves as a conduit for the formed egg between the uterus and cloaca at oviposition (Figure 5).

There are no biological mechanisms in birds, such as a female mammal's estrus or heat period, that synchronize copulation with impending ovulation. Unlike most mammals, hens have the capacity to store sperm in their oviducts for prolonged periods of time. After copulation or artificial insemination, a relatively small percentage of the sperm deposited in the distal end of the vagina ascend to the uterovaginal junction (UVJ). Here, this select population of sperm enters discrete tubular invaginations of the surface epithelium, the sperm storage tubules (SSTs). In order to fertilize a near daily succession of ova, which amounts to four to seven ova per week, sperm are slowly but continuously released from the SSTs and ascend to the site of fertilization at the infundibulum.

In the chicken and turkey, optimal fertility is maintained with weekly inseminations of about $100 - 200 \times 10^6$ sperm. The actual frequency of insemination and the sperm number per insemination may vary based on such factors as season, past performance (flock fertility), and age of the hen.

The primary anatomical units responsible for prolonged oviducal sperm storage are the SSTs, which are collectively localized in a 7- to 10-mm band at the UVJ. Individual SSTs are tubular invaginations of the UVJ surface epithelium that collectively form a separate compartment within the UVJ mucosa (Figures 7 through 10). The SST epithelium is a simple, nonciliated, columnar-type characteractized by supranuclear lipid droplets, paucity of secretory granules in the apical cytoplasm, and a nucleus situated in the basal third of the cells. In contrast, the UVJ surface epithelium is pseudo-stratified and consists of alternating ciliated and secretory cells, both capable of secretory activity. These cells lack the supranuclear lipid droplet that characterizes the SST epithelium (Bakst, 1987).

It has been only in the last few years that we have begun to understand the mechanisms that regulate sperm selection in the vagina (reviewed by Wishart and Steele, 1990). For instance, dead sperm and particulate matter are not transported beyond the UVJ. Furthermore, if one inseminates testicular sperm or if ejaculated sperm are inseminated after neuraminadase treatment, subsequent fertility is very low. Yet if testicular sperm or neuraminidase-treated sperm are surgically inseminated anterior to the UVJ, hens lay fertile eggs. These kinds of observations have led to the suggestion that sperm surface glycoproteins containing terminal sialic acid residues may be necessary for the successful selection and transport of sperm through the vagina.

Our current understanding regarding the fate of sperm in the hen oviduct after insemination is far from complete (Bakst, 1989). The majority of sperm (about $80 - 85\%$) is excreted from the cloaca within 30 min of insemination (Howarth, 1971). However, a small select subpopulation of sperm, about $2 - 4 \times 10^6$, ascend to the UVJ and enter the SST (Brillard and Bakst, 1990). If there is no egg mass in the oviduct at the time of insemination, a yet smaller vanguard population of sperm bypasses the UVJ and populates the subepithelial tubular glands in the distal half of the infundibulum (Figures 11 and 12). Sperm residing in the SST are released slowly but continuously while the hen is in egg production. Fit sperm then ascend to the site of fertilization, while less fit or senescent sperm accrue in the vagina (Bakst, 1981). Whether fit sperm released from the SSTs require a period of capacitation prior to penetrating an ovum is not known.

There are situations that necessitate alternative methods of artificial insemination. It has been demonstrated that insemination of semen obtained from males of low fecundity or of semen that has been cryopreserved results

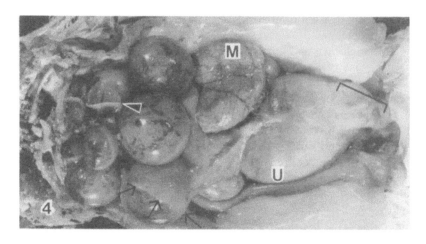

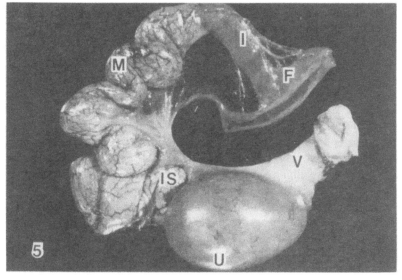

FIGURE 4. The mature left ovary and oviduct occupy a considerable amount of space in the abdominal cavity of the turkey. Yellow-yolk ovarian follicles of varying diameters and a post-ovulatory follicular sheath (arrowhead) are observed. The fimbriated region of the infundibulum (arrows) partially engulfs a follicle. Other segments of the oviduct that are apparent are the magnum (M), uterus (shell gland) (U), here containing a hard-shell egg, and the vagina (bracket), which is difficult to discern since it is tightly bound by connective tissue.

FIGURE 5. This excised turkey oviduct contains a hard-shell egg in the uterus (U). The fimbriated region (F) of the infundibulum (I) grasps and guides the ovum into the ostium of the oviduct. Sperm must be present in the infundibulum if the ovum is to be fertilized. In the course of the next 5 – 6 h, albumen derived from the magnum (M) and the shell membrane derived from the isthmus (IS) are deposited around the ovum. The egg mass resides in the uterus during the latter 20 h of egg formation. Here the calcareous shell is formed and, if it is fertile, the first cleavage furrow is formed and subsequent bilateral symmetry of the embryo is established. The vagina (V) is a conduit between the uterus and the cloaca.

FIGURE 6. A scanning electron micrograph of a fracture through the blastodisc (animal pole region) of a follicular oocyte. Observed are features of the blastodisc region that differentiate this region from the remaining ovum (vegetal region). These include columnar rather than cuboidal granulosa cells (G) overlying the blastodisc, white yolk (Y) spheres rather than yellow yolk spheres, and a microvillus border (arrows) not observed elsewhere in the mature oocyte. Granulosa cells are interconnected by cytoplasmic processes and are separated from the oocyte by the perivitelline layer (P). The basal surface of the granulosa cells is attached to a relatively thick basement membrane (B). (Bar = 20 μm.)

in low fertility. However, if the inseminations are made using intrauterine (nonsurgical) or intramagnal (surgical) techniques, fertility is moderately high (Ogasawara et al., 1966; Engel et al., 1991). By inseminating anterior to the UVJ, the sperm selection mechanism is bypassed and, if there is no egg mass anterior to the insemination site, a large percentage of the sperm reach the infundibulum, the site of fertilization. However, inseminations anterior to the UVJ and the consequent large numbers of sperm in the infundibulum may lead to an increase in early embryonic mortality, possibly due to pathological polyspermy.

Anatomy of the Ovum and Its Penetration by Spermatozoa

The hen ovum at ovulation consists of a 3- to 4-mm-diameter whitish disk, the blastodisc, which floats on the surface of a 3- to 4-cm-diameter yellow yolk mass. Enveloping the ovulated ovum is a single acellular investment, the perivitelline layer (PL) (Figure 3). The PL, which is often erroneously referred to as the vitelline membrane, is composed of fibers ranging between 0.4 and 0.7 μm in thickness. When first examined by transmission electron microscopy, investigators observed a dense but apparently open fibrous reticulum, which led some to speculate that sperm could maneuver between the fibers to reach the oolemma, the oocyte plasma membrane. However, with SEM a ground substance is observed occupying the spaces between the fibers.

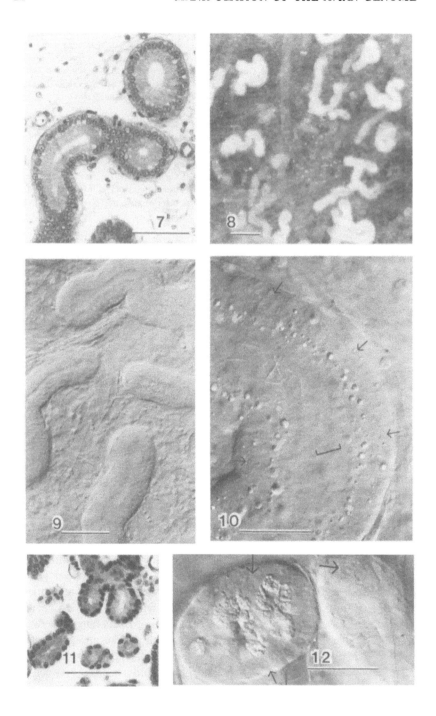

FIGURE 7. A paraffin section showing profiles of a sperm storage tubule (SST). The SST epithelium is simple columnar with a nucleus located in the basal aspect of the cell. The loose connective tissue surrounding the SST contains capillaries, fibroblasts, collagen fibers, and a few small lymphocytes. (Bar = 40 μm.)

FIGURE 8. Sperm storage tubules in fresh, unfixed mucosa isolated from the uterovaginal junction are observed using a stereomicroscope. (Bar = 200 μm.)

FIGURE 9. A squash preparation of fresh, unfixed mucosa from the uterovaginal junction and viewed by differential interference contrast microscopy (DIC). Compared to other modes of light microscopy, DIC offers better spatial and optical resolution of the SST and surrounding connective tissue elements. (Bar = 40 μm.)

FIGURE 10. The same type of preparation as shown in Figure 7, but at higher magnification, clearly reveals characteristics of the sperm storage tubule (SST) epithelium, including profiles of epithelial cell nuclei (arrows) and supranuclear lipid droplets. A few spermatozoa are observed in the lumen of the SST. Also observed here are terminal bars (in and around bracket), evidence of apical cell junctions. (Bar = 20 μm.)

FIGURE 11. A paraffin section showing profiles of a few subepithelial tubular glands located in the distal infundibulum. Nuclei are localized to the periphery of the glandular epithelium. (Bar = 40 μm.)

FIGURE 12. A squash preparation of fresh, unfixed mucosa from the infundibulum and viewed by differential interference contrast microscopy. Observed are secretory granules in the apical cytoplasm and profiles of epithelial cell nuclei (arrows) in a tubular gland. Individual glandular cells and their nuclei are observed in another tubular gland. (Bar = 20 μm.)

It is now accepted that chicken and turkey sperm must hydrolyze the PL in order to reach the oolema (Howarth, 1984; Bakst and Howarth, 1977a; Okamura and Nishiyama, 1978a).

The blastodisc of the large yellow yolk ovarian follicles is the only site where cell organelles are located. Such organelles include not only mitochondria, endoplasmic reticulum, golgi, and annulate lamellae, but the germinal vesicle, which is the female nucleus arrested in the diplotene stage of meiosis. Specific cytoplasmic inclusions found in the blastodisc include white yolk spheres, clear vacuoles, lipid droplets, and glycogen (Bakst and Howarth, 1977b). The oolemma forms a dense array of microvilli and more complex elaborations that occupy the perivitelline space, a narrow zone located between the underside of the PL and the oolemma (Figure 6). During oocyte maturation, the perivitelline space is also traversed by granulosa cell processes that articulate with the oolemma by way of gap junctions (Bakst, 1979).

The granulosa cells, which form a single layer around the follicular oocyte (Figure 6), consist of at least two morphologically distinct cell populations, one overlying the blastodisc and the second over the remaining yolky portion (vegetal region) of the oocyte (Bakst, 1979). Cytoplasmic extensions not only connect granulosa cells to the oocyte surface but also interconnect adjacent granulosa cells via gap junctions. The gonadotropin surge (luteinizing hormone) thought to be responsible for the resumption of meiosis in the pre-ovulatory oocyte, as manifested by the breakdown of the germinal vesicle, may also be responsible for the concomitant withdrawal of all granulosa cell cytoplasmic processes shortly before ovulation.

As mentioned previously, sperm must hydrolyze a path through the PL or penetrate a site already hydrolyzed by another sperm in order to reach the oolemma. Although the exact sequence of events has yet to be described, it appears that some of the sperm that penetrate the PL and make contact with the oolemma are initially grasped by the microvilli of the oolemma and then enveloped by the ovum (Howarth, 1984). Subsequent decondensation of sperm nuclei and male pronuclei formation has been described by Okamura and Nishiyama (1978b). Since there is no block to polyspermy triggered by the entry of the first sperm into the egg, physiological polyspermy is routinely observed. The actual block to pathological polyspermy is through the accretion of the outer perivitelline layer. This tertiary egg investment (the oolemma being the primary and the PL being the secondary) consists of oviducal secretions similar to albumen that are deposited around the ovum in transit through the distal infundibulum and proximal magnum. It is this layer, the outer PL, which sperm cannot penetrate (Bakst and Howarth, 1977a).

A relatively quick and simple technique developed by Wishart (1987) can be used to determine if inseminated sperm survive the rigors of sperm selection, transport, and storage in the oviduct. This procedure involves removing the PL of a fresh laid egg, staining the washed PL with a fluorescent nuclear stain, and then estimating the number of sperm per unit area embedded in the outer PL. If sperm were present at the site of fertilization, there should be sperm trapped in the outer PL. This procedure has been used to estimate both the length of the hen's fertile period (Wishart, 1987) and to estimate the number of sperm residing within the SST (Brillard and Bakst, 1990).

Artificial Insemination Technology

Artificial insemination (AI) is probably the single most powerful tool currently available for managing the genetic progress of livestock and poultry (for reviews, see Lake, 1983; Sexton, 1984; Bakst, 1990). The basis for this statement lies in the ability of the flock manager to select only those males possessing desirable genetic traits and characteristics for breeding purposes.

A successful AI program, or any research program requiring sperm, begins with proper semen collection. Semen collection is relatively rapid and easily accomplished (see Bahr and Bakst, 1987). The objective is to maximize the volume of semen obtained on a per-male basis with a minimal amount of contamination from fecal material, urates, blood, and/or an excessive amount of transparent fluid.

Unless inseminations are performed within 30 min of semen collection, a diluent should be added to the semen. The diluent, or extender as it is often called, is essentially a cell culture medium specifically designed for sperm. [See Bakst (1990) for diluent composition, suggested dilution rates, optimal storage times and conditions, and some commercial sources of prepared diluent.] Diluent serves two purposes: First, it dilutes the semen, thereby increasing the number of hens that can be inseminated on a per-male basis; and second, the diluent composition is such that it will extend the shelf life of sperm to be used for AI.

Regardless of which diluent is used, it is vitally important to mix the semen thoroughly with the diluent. Because of its high viscosity (neat turkey sperm concentration is about $8 - 10 \times 10^9$ sperm/mL), semen and diluent must be pipetted up and down for the full dispersal of the semen in the diluent.

Before using semen either for AI or for any other procedure, an estimate of semen quality, which includes such characteristics as sperm motility, morphology, percentage of live sperm (viability), and, in particular, sperm concentration, should be performed. Semen evaluation procedures applicable to livestock [reviewed by Hafez (1987)] are generally applicable to poultry semen.

Two common procedures used to estimate the viability of poultry sperm, the nigrosin/eosin stain and the ethidium bromide exclusion procedure (Bilgili et al., 1987; Bakst et al., 1991), rely on the same principles as the dye exclusion tests used to determine the viability of dispersed somatic cells. Certain dyes or stains, such as trypan blue, eosin, or the fluorescent dye ethidium bromide, are excluded by cells, including sperm, with intact plasma membranes. However, when the plasma membrane is damaged, these same dyes stain the nucleus. Such stained sperm are considered dead. Knowledge of the percentage of nonviable sperm is useful not only as a measure of semen quality but for the calculation of the total number of viable sperm for insemination or other procedures. At least one investigator and co-workers have been able to demonstrate significant positive correlations between certain poultry sperm characteristics and hen fecundity following insemination (Wishart and Palmer, 1986; Chaudhuri and Wishart, 1988).

With regard to morphology, chicken and turkey sperm are nearly indistinguishable by light and scanning electron microscopy (SEM) (Figure 3). However, closer examination reveals differences in lengths of their nuclei, midpieces, acrosomes, and consequently, overall length (Thurston and Hess, 1987). Compared to the sperm of most of the mammals examined, chicken and turkey sperm are rather simple. For example, chicken and turkey sperm acrosomes are homogenous in appearance and lack further regional specializations such as the acrosomal ridge, equatorial segment, or postacrosomal sheath.

With light microscopy, the task of defining sperm structures and more general morphological anomalies is demanding. Therefore, morphological assessments of sperm should be made by either phase-contrast or differential interference microscopy. Dilution of the semen should be high enough that individual sperm can be visualized under an oil-immersion objective lens. Probably the most common structural anomaly is bent sperm (Bakst, 1980b). Bent sperm are characterized by an acute bend in the area of the distal head or midpiece regions, which realigns the head nearly parallel with and adjacent to the tail. Even with their heads facing backward, bent sperm remain highly motile and are difficult to identify unless the semen is sufficiently diluted. If bent sperm are present in large numbers, there is a problem with the osmolarity of the diluent or the sperm are approaching senescence. Such semen should be discarded, since the higher the percentage of bent sperm, the lower the subsequent fertility (Saeki, 1960).

The plasmalemma is probably the most labile sperm organelle. Local distention of the plasmalemma overlying the midpiece precedes other sperm anomaly formation such as bent sperm and swelling of the mitochondria at the midpiece (Bakst, 1980b).

Conclusion

As forewarned at the outset, this chapter does not approach a comprehensive account of the anatomy and physiology of domestic poultry. However, the cited references should be sufficient to start an adequate search of the literature. Finally, if a text or report is not readily accessible at your institution, contact the National Agricultural Library (Beltsville, MD) for assistance.

References

Bahr, J., and Bakst, M. (1987). Poultry. In *Reproduction in Farm Animals,* 5th ed., edited by E. S. E. Hafez. Lea & Febiger, Philadelphia, pp. 379–497.

Bakst, M. R. (1979). Scanning electron microscopy of hen granulosa cells before and after ovulation. *Scanning Electron Microsc.,* III, 307–312.

Bakst, M. R. (1980a). Luminal topography of the male chicken and turkey excurrent duct system. *Scanning Electron Microsc.,* III, 419–426.

Bakst, M. R. (1980b). Fertilizing capacity and morphology of fowl and turkey spermatozoa in hypotonic extender. *J. Reprod. Fert.,* 60, 121–127.

Bakst, M. R. (1981). Sperm recovery from oviducts of turkeys at known intervals after insemination and oviposition. *J. Reprod. Fert.,* 62, 159–164.

Bakst, M. R. (1987). Anatomical basis of sperm-storage in the avian oviduct. *Scanning Microsc.,* 1, 1257–1266.

Bakst, M. R. (1989). Oviducal storage of spermatozoa in the turkey: Its relevance to artificial insemination technology. *Br. Poultry Sci.,* 30, 423–429.

Bakst, M. R. (1990). Preservation of avian cells. In *Poultry Breeding and Genetics.* edited by R. D. Crawford. Elsevier, New York, pp. 91–108.

Bakst, M. R., and Cecil, H. C. (1981). Changes in the characteristics of turkey ejaculated semen and ductus deferens semen with repeated ejaculations. *Reprod. Nutr. Develop.,* 21(6B), 1095–1103.

Bakst, M. R., Cecil, H. C., and Sexton, T. J. (1991). Modification of the ethidium bromide exclusion procedure for evaluation of turkey semen. *Poultry Sci.,* 70, 366–370.

Bakst, M. R., and Howarth, B. (1977a). Hydrolysis of the hen's perivitelline layer by cock sperm *in vitro. Biol. Reprod.,* 17, 370–379.

Bakst, M. R., and Howarth, B. (1977b). The fine structure of the hen's ovum at ovulation. *Biol. Reprod.,* 17, 361–369.

Baumel, J. J., King, A. S., Lucas, A. E., Brazile, J. E., and Evans, H. E. (1979). *Nomina Anatomica Avium.* Academic Press, New York.

Bilgili, S. F., Sexton, K. J., and Renden, J. A. (1987). Fluorometry of poultry semen: Influence of dilution and storage on chicken spermatozoal viability and fertility. *Poultry Sci.,* 66, 2032–2035.

Brillard, J. P., and Bakst, M. R. (1990). Quantification of spermatozoa in the sperm-storage tubules of turkey hens and the relation to sperm numbers in the perivitelline layer. *Biol. Reprod.,* 43, 271–275.

Chaudhuri, D., and Wishart, G. J. (1988). Predicting the fertilizing ability of avian semen: The development of an objective colormetric method for assessing the metabolic activity of fowl spermatozoa. *Br. Poultry Sci.,* 29, 837–845.

Engel, H. N., Froman, D. P., and Kirby, J. D. (1991). An improved procedure for the intramagnal insemination of the chicken. *Poultry Sci.,* 70, 1965–1969.

Etches, R. J. (1990). The ovulatory cycle of the hen. *Crit. Rev. Poultry Biol.,* 2, 293–318.

Fujihara, N., and Nishiyama, H. (1984). Addition to semen of a fluid derived from the cloacal region by male turkeys. *Poultry Sci.,* 63, 554–557.

Fujihara, N., Nishiyama, H., and Koga, O. (1985). The mechanisms of the ejection of a frothy fluid from the cloaca in the male turkey. *Poultry Sci.,* 64, 1377–1381.

Gilbert, A. B. (1979). Female genital organs. In *Form and Function in Birds,* Vol. 1, edited by A. S. King and J. McLelland. Academic Press, New York, pp. 237–360.

Hafez, E. S. E. (1987). Semen evaluation. In *Reproduction in Farm Animals,* edited by E. S. E. Hafez. Lea & Febiger, Philadelphia, pp. 455–480.

Hess, R. A., Thurston, R. J., and Biellier, H. V. (1976). Morphology of the epididymal region and ductus deferens of the turkey (*Meleagris gallopavo*). *J. Anat.,* 122, 241–252.

Howarth, B. (1971). Transport of spermatozoa in the reproductive tract of turkey hens. *Poultry Sci.,* 50, 84–89.

Howarth, B. (1984). Maturation of spermatozoa and mechanism of fertilization. In *Reproductive Biology of Poultry,* edited by F. J. Cunningham, P. E. Lake, and D. Hewitt. British Poultry Science Ltd., The Alden Press Ltd., Oxford, pp. 161–174.

Knight, C. E., Bakst, M. R., and Cecil, H. C. (1984). Anatomy of the *Corpus vasculare paracloacale* of the male turkey. *Poultry Sci.,* 63, 1883–1891.

Lake, P. E. (1981). Male genital organs. In *Form and Function in Birds,* Vol. 2, edited by A. S. King and J. McLelland. Academic Press, New York, pp. 1–61.

Lake, P. E. (1983). Factors affecting the fertility level in poultry, with special reference to artificial insemination. *World's Poultry Sci. J.,* 39, 106–117.

Ogasawara, F. X., Lorenz, F. W., and Bobr, L. W. (1966). Distribution of spermatozoa in the oviduct and fertility in domestic birds. III. Intra-uterine insemination of semen from low-fecundity cocks. *J. Reprod. Fert.,* 11, 33–41.

Okamura, F., and Nishiyama, H. (1978a). The passage of spermatozoa through the vitelline membrane in the domestic fowl, *Gallus gallus. Cell Tissue Res.,* 188, 497–508.

Okamura, F., and Nishiyama, H. (1878b). Penetration of spermatozoan into the ovum and transformation of the sperm nucleus into the male pronucleus in the domestic fowl, *Gallus gallus. Cell Tissue Res.,* 190, 89–98.

Saeki, Y. (1960). Crooked-neck spermatozoa in relation to low fertility in the artificial insemination of fowl. *Poultry Sci.,* 39, 1354–1361.

Sexton, T. J. (1984). Breeding by artificial insemination. In *Reproductive Biology of Poultry,* edited by F. J. Cunningham, P. E. Lake, and D. Hewitt. British Poultry Science Ltd., The Alden Press, Ltd., Oxford, pp. 175–182.

Thurston, R. J., and Hess, R. A. (1987). Ultrastructure of spermatozoa from domesticated birds: Comparative study of turkey, chicken, and guinea fowl. *Scanning Microsc.,* 1, 1829–1838.

Van Krey, H. (1990). Reproductive biology in relation to breeding and genetics. In *Poultry Breeding and Genetics,* edited by R. D. Crawford. Elsevier, New York, chap. 3.

Wishart, G. (1987). Regulation of the length of the fertile period in the domestic fowl by numbers of oviducal spermatozoa, as reflected by those trapped in laid eggs. *J. Reprod. Fert.,* 80, 493–498.

Wishart, G. J., and Palmer, F. H. (1986). Correlation of the fertilising ability of semen from individual male fowls with sperm motility and ATP content. *Br. Poultry Sci.,* 27, 97–102.

Wishart, G. J., and Steele, M. G. (1990). The influence of sperm surface characteristics on sperm function in the female reproductive tract. In *Proceedings of the Control of Fertility in Domestic Birds,* ed., Tours, France, July 1990, edited by J. P. Brillard. Les Colloques de l'INRA #54, pp. 101–112.

Chapter 3

Early Determination and Morphogenetic Processes in Birds

Hefzibah Eyal-Giladi

Summary. Orientation in the avian blastodisk is triggered by the force of gravity. Its manifestation is a gradual process of cell shedding from the lower layer, leading to the formation of an area pellucida. The marginal zone is the dominant source of a population of hypoblastic cells, which are responsible for the induction of the primitive streak in the epiblast. Posterior marginal zone cells have been demonstrated to move anteriorly into the forming hypoblast, while lateral marginal zone cells show a very limited movement. The marginal zone belt functions as a coordinating system, in which there is a very subtle equilibrium of inductive and inhibitory effects, which ensures the induction of a single primitive streak. The posterior marginal zone cells seem to gain their inductivity only after leaving the marginal zone belt and crossing Koller's sickle into the hypoblast.

Introduction

The developmental period of the chick covered by this chapter can be divided, as far as the environmental conditions are concerned, into two. The first 20 h of development are inside the hen (at $41 - 42°C$) and are referred to as the uterine period, while about 20 additional hours of incubation at $37°C$ are needed after the egg is laid to reach a definitive primitive streak stage (H&H stage 4; Hamburger and Hamilton, 1951).

The Embryo in the Uterus

The uterine period (E.G&K stages I – X) starts approximately 5.5 h after ovulation and fertilization in the upper section of the genital tract (Eyal-Giladi and Kochav, 1976; Kochav et al., 1980). During these early hours, the fertilized ovum is being gradually enveloped by several albuminous layers and the shell membranes.

Upon its entrance into the uterus the egg is still very flaccid, and continues to imbibe uterine fluid during the next 5 h. Shell deposition proceeds for most of the uterine period. The germinal disk, which includes the active cytoplasm (germinal plasm) of the zygote, starts to cleave around the time of entry of the egg into the uterus. The cleavage furrows appear somewhere in the central area of the disk (E.G&K stage I), and there is a remarkable variability of cleavage patterns. However, at the end of the cleavage process, which proceeds at a very rapid pace for 12 h, the entire germinal disk is divided into distinct cells. The resulting blastoderm, which is five to six cells in thickness (E.G&K stage VI), has a radial symmetric appearance. However, the blastoderm has already acquired a concealed labile bilateral symmetry, which has been gradually imprinted upon it during the first half of the uterine period. Kochav and Eyal-Giladi (1971) and Eyal-Giladi and Fabian (1980) have shown that it is gravity that dictates the anterio-posterior polarity of the blastoderm. As a result of the egg's rotation in the uterus, the blastoderm is stationed at an oblique position. The highest point of the blastoderm will then become the future posterior side, while the lowest point will become the anterior one. The first manifestation of the axis is the shedding of cells from the lower layers of the blastoderm at the future posterior side, a process that advances toward the anterior side and results in a local thinning out of the blastoderm. The shedding process, which takes about 8 h, continues from its posterior starting point anteriorly and, at the end, just before laying, the blastoderm is finally turned into an E.G&K stage X blastoderm, with a central circular area that is mainly one cell thick, the area pellucida, surrounded by a thicker belt, the area opaca. Once again the germ looks almost radially symmetric but, at the point of laying, the imprinting of bilateral symmetry is much more pronounced than it is at E.G&K stage VI. E.G&K stages VI – X are, therefore, those uterine stages during which the gradual establishment of the anterio-posterior axis of the blastoderm takes place. The determination of the axis can be manipulated during the first part of this period by changing the spatial position of the blastoderm. Toward the end of this period, and especially at E.G&K stages IX and X, the direction of the embryonic axis cannot be changed any longer by the application of a new gravitational vector (Eyal-Giladi and Fabian, 1980).

Laying is a very significant point in the development of the blastoderm, as all the parameters related to the uterine environmental conditions are abruptly changed. The laid egg is no longer bathed in uterine fluid, it enters a different gaseous environment (Arad et al., 1989) and requires a lower temperature, 37 – 38°C, to proceed with its development.

The Embryo in the Laid Egg

The newly laid egg is not always at a fixed developmental stage. Usually eggs laid by young hens are less developed than those laid by older hens. Young hens may sometimes lay eggs containing an E.G&K stage IX blastoderm, but usually the blastoderm has developed to E.G&K stage X. An early E.G&K stage X blastoderm does not reveal its polarity and looks radially symmetric; however, after 1 h of incubation at 37°C, the posterior side becomes conspicuous as a result of the appearance of a ridge that forms in the area pellucida near the area opaca. The ridge, which is parallel to the external area opaca belt and is limited to the posterior side of the blastoderm, is called

Koller's sickle due to its shape (Koller, 1882) and will be discussed in the next section.

The morphology of an E.G&K stage X blastoderm starts to change at incubation as a result of the gradual formation of a lower layer, the hypoblast. The hypoblast is formed of two cellular contributions (Vakaet, 1962; Eyal-Giladi and Kochav, 1976; Kochav et al., 1980). One component, which is already partially present at the time of laying, is in the form of isolated islands of cells attached to the lower surface of the blastoderm that have polyingressed from the area pellucida. The second component of the hypoblast forms as a coherent front of cells that gradually advances from the thickened ridge of Koller's sickle and moves in an anterior direction. After about 10 h of incubation, a confluent lower layer, the primary hypoblast, is formed that is believed to include both cell populations. At E.G&K stage XIII the central part of the blastoderm is thus two-layered: an upper epiblast and a lower hypoblast. Around this central area is the peripheral ring of the area pellucida, the marginal zone, which is continuous with the epiblastic layer. The marginal zone is again surrounded by the thicker belt of the area opaca.

The primary hypoblast has been shown by Waddington (1933) and Azar and Eyal-Giladi (1981) to be responsible for the induction of the primitive streak in the epiblastic layer above it. While all the embryonic tissue will be formed from derivatives of the epiblast, the cells of the primary hypoblast remain extraembryonic. The hypoblast cells first move laterally and anteriorly to form the lower layer of the germinal crescent (Azar and Eyal-Giladi, 1983), and later move on into the yolk sac.

Azar and Eyal-Giladi (1981), being aware of the dual origin of primary hypoblastic cells, performed a series of experiments aimed at clarifying which of the two populations is responsible for primitive streak induction. They have shown that, if the marginal zone as well as the hypoblast are removed from an E.G&K stage XIII blastoderm, the remaining epiblastic central disk regenerates a lower layer but is incapable of forming a primitive streak. In contrast, if the marginal zone is left *in situ* and the hypoblast alone is removed, not only will a lower layer form, but also a primitive streak. These and subsequent experiments described later led to the conclusion that hypoblastic cells derived from the marginal zone are those responsible for primitive streak induction in the epiblast. In order to further examine the above conclusion, we pursued several experimental tracks.

The experiments were based on the following hypotheses.

1. All the cells of the marginal zone are potentially inductive.
2. Marginal zone cells move into the growing hypoblast at the posterior side of the blastoderm via Koller's sickle.
3. Cells in the posterior marginal zone might have a higher inductive potential than other marginal zone cells, or, in other words, there may

be a gradient of potential inductivity in the marginal zone with a maximum at its posterior side and a minimum at the anterior side.

4. There is a temporal decline of the inductivity potential of the marginal zone, with a maximal inductivity at E.G&K stage X and a rapid gradual decrease.

5. A correlation exists between the potential inductivity and the migration potential of the marginal zone cells.

Khaner and Eyal-Giladi (1986, 1989) and Eyal-Giladi and Khaner (1989) tried to verify some of the above hypotheses concerning the contribution of the marginal zone to inductivity and to study the temporal and spatial potencies of this region of the blastoderm. The techniques implemented were: the rotation of the marginal zone at 90° to the axis of the central disk, explantation of a posterior fragment of the marginal zone to a lateral position in the marginal zone, or exchange of a lateral marginal zone fragment with a posterior fragment. The experimental results indicated very clearly that by changing the position of an E.G&K stage X posterior marginal zone fragment, the direction in which a primitive streak will be induced can be altered accordingly. However, if the posterior marginal zone fragment is deleted and not replaced, an axis will still be formed in a location lateral to the missing posterior fragment. This led to the following conclusions.

1. The marginal zone at E.G&K stage X is responsible for the capability of the hypoblast to induce a primitive streak.

2. The inductive potential of the marginal zone is introduced into the growing hypoblast by marginal zone cells that migrate into the hypoblast via Koller's sickle.

3. The inductive potential of the marginal zone is organized in the form of a gradient field, being maximal at the posterior side and gradually decreasing laterally toward the anterior side.

4. Inductivity can be experimentally quantified by determining the number of marginal zone cells that are capable of initiating an embryonic axis at an ectopic position.

5. There is a gradual decrease of the inductive potential of the marginal zone from E.G&K stage X to stage XIII.

6. The marginal zone with the highest inductive potential exerts an inhibitory effect on other marginal zone areas and prevents them from forming additional axes.

7. The blastoderm acts as an integrative system programmed to form a single embryonic field by the coordinated action of inductivity and inhibition. However, two conditions are required for this coordination: (a) cellular continuity within the marginal zone; and (b) the existence of only one area with "maximal" inductive potential.

The above conclusions are based on experimental data that did not, however, visually demonstrate the migration of marginal zone cells into a location in the hypoblast that would enable a direct interaction between the migrated cells and the competent epiblast above them to form a primitive streak. Therefore, we developed a method to demonstrate cell migration by labeling the posterior portion of the marginal zone.

A portion of the blastoderm consisting of a fragment of the posterior marginal zone attached to a section of the area opaca was cut out and incubated in a solution of Rhodamine-dextran-lysine. The fluorescent dye penetrated into the cells of the excised tissue, bound to proteins, and was unable to leak out of the cells subsequently. The stained tissue was washed and put back into the original blastoderm, which was further incubated at 37°C. Several blastoderms of each of E.G&K stages X, XI, and XII were operated upon. Various blastoderms of each group were fixed after reaching a different developmental stage, the most advanced being H&H stage 4. The material was then embedded in paraffin and serially sectioned, mounted on slides, and observed with a fluorescence microscope to detect dye-containing cells derived from the labeled tissue. Three-dimensional reconstructions of the blastoderm gave a dynamic overall picture of cell movements during hypoblast and primitive streak formation (H. Eyal-Giladi, A. Debby, and N. Harel, in press). During the stages of hypoblast formation there is an anteriorly directed movement of cells from Koller's sickle and the posterior marginal zone. The cells move initially in a narrow front and, on their way, push the islands of polyingressed cells anteriorly and laterally. The apparently uniform hypoblast thus proved to contain a median strip of labeled cells originating in the posterior marginal zone that was surrounded anteriorly and laterally by non-labeled cells. The cell movements within the hypoblast continued also during H&H stages 2 – 4, when labeled cells gradually occupied most of the hypoblast. The above observations support the conclusion of previous work (Azar and Eyal-Giladi, 1979) that, during the entire period of induction of the primitive streak, the epiblast is in direct contact with cells originating from the posterior marginal zone and that these cells induce the primitive streak in the epiblast.

Another very interesting, unexpected point has emerged from the analysis of the above experimental material (H. Eyal-Giladi, A. Debby, and N. Harel, in press). The posterior marginal zone fragments were cut out from the blastoderm in such a way that they included not only the marginal zone fragment plus Koller's sickle but also a narrow strip of the area pellucida anterior to Koller's sickle. Consequently, some future epiblastic cells were also labeled in this preparation. This strip of epiblastic cells seems to have stayed in place until E.G&K stage XIII, namely, during the entire period of the anterior growth of the hypoblast. It had, therefore, been constantly under the influence of the posterior marginal zone cells moving forward into the hypoblast. At H&H stages 2 – 4, this strip of epiblastic cells started to stretch longitudinally and gradually formed the entire definitive primitive streak. The process of gastrulation via the primitive streak starts between H&H stages 3 and 4, and seems to proceed anteriorly from the posterior sections of the primitive streak. At H&H stage 4, most of the labeled cells that were found in the primitive streak at H&H stage 3 had migrated to the mesoblastic layer. Only the anterior part of the primitive streak (Hensen's node) still contained a concentrated population of labeled cells, which had, however, started to move into the mesoblast.

We have also tried to determine when the cells from the posterior marginal zone acquire their inductivity and whether acquisition of inductivity is correlated to their migration (H. Eyal-Giladi, T. Levin, and O. Avner, unpublished results). The following experimental system was used: Sheets of competent epiblast (capable of reacting to primitive streak inductive stimuli) from the central disk of E.G&K stage XIII embryos were explanted onto a vitelline membrane with their lower side up. The entire surface of the central disk was covered either with longitudinal strips of posterior marginal zone (experimental group) or with similar strips of a stage XIII hypoblast (control). In the control embryos with hypoblastic strips the primitive streak developed in the epiblast, whereas the primitive streak did not develop in the experimental series under the influence of the marginal zone strips. The conclusion, therefore, was that in order to become inductive, the posterior marginal zone cells have to migrate from their position in the marginal zone and enter the hypoblast.

The Origin of Primordial Germ Cells

Another subject that is still puzzling is the ontogenesis of the germline in birds. The dilemma as to whether the primordial germ cells (PGCs) arise from a localized fraction of the uncleaved germinal plasm of the zygote, or as a result of inductive processes, is not yet solved. It was already established by Swift (1914) that the PGCs are localized in the germinal crescent at H&H stages 4 and later. The germinal crescent is situated extraembryonically at the extreme anterior side of the area pellucida. It was not clear, however, where the PGCs came from. In the late 1960s and early 1970s, several researchers (Rogulska, 1968; Fargeix, 1969; Eyal-Giladi et al., 1976) tried to transect early blastoderms, culture the fragments, and look for the presence of PGCs in the cultures. The results did not supply any clear answer and triggered differently designed experiments. Ginsburg and Eyal-Giladi (1987, 1989) demonstrated that the PGCs originate mainly from the central part of the blastoderm and that their appearance is totally disconnected from the existence of an embryonic axis. Previously, independent experiments by Sutasurya et al. (1983) and Eyal-Giladi et al. (1981) determined which of the two germ layers of an E.G&K stage XIII blastoderm give rise to PGCs and showed that the PGCs are epiblastic. This result caused Ginsburg and Eyal-Giladi (1986) to make a systematic study of the temporal, spatial, and quantitative aspects of PGC migration from the epiblast into the germinal crescent during the period from E.G&K stages XII E.G&K to H&H 10.

From the outcome of the experiments the following scenario was drawn: The PGCs begin to translocate from the epiblast at E.G&K stages XII – XIII, and land directly on the hypoblast with which they establish close contact. At later stages, an intermediate layer of mesoderm spreads centrifugally from the primitive streak and invades the space between the epiblast and the hypoblast. The PGCs, therefore, have to land on the dorsal side of the mesodermal wings and move to the germinal crescent, either by ameboidal movement within the lateral mesoderm or by translocation to the hypoblastic layer and then by passive transport into the germinal crescent by the anteriorly directed morphogenetic movements. The experiments also supported Vakaet's assumption (Vakaet, 1962) that the hypoblast, after completing its anteriorly directed movement, is in effect the main component of the germinal crescent (Azar and Eyal-Giladi, 1983). The hypoblast, therefore, probably has three very important tasks concerning PGC translocation: (1) It attracts the PGCs; (2) it serves as a substrate for the anterior spreading of the mesodermal wings, which carry the PGCs that leave the epiblast at relatively late stages; (3) it functions as the main carrier into the germinal crescent of the PGCs attached to it.

Efforts to trace either an earlier existence of PGCs or a localized fraction of the uncleaved germinal plasm that contributes to PGC formation have been unsuccessful. Several attempts have been made to approach this problem in avian and murine embryos by immunological methods (Hahnel and Eddy, 1982, 1986; Pardanaud et al., 1987; Urven et al., 1988; Ginsburg et al., 1989; Loveless et al., 1990). To date, these experiments have provided ambiguous information regarding the origin of PGCs within the epiblast.

References

Arad, Z., Eylath, U., Ginsburg, M., and Eyal-Giladi, H. (1989). Changes in uterine fluid composition and acid-base status during shell formation in the chicken. *Am. J. Physiol.*, 257, R 732–737.

Azar, Y., and Eyal-Giladi, H. (1979). Marginal zone cells — The primitive streak inducing component of the primary hypoblast in the chick. *J. Embryol. Exp. Morphol.*, 52, 79–88.

Azar, Y., and Eyal-Giladi, H. (1981). Interaction of epiblast and hypoblast in the formation of the primitive streak and the embryonic axis in chick, as revealed by hypoblast rotation experiments. *J. Embryol. Exp. Morphol.*, 61, 133–144.

Azar, Y., and Eyal-Giladi, H. (1983). The retention of primary hypoblastic cells underneath the developing primitive streak allows for their prolonged inductive influence. *J. Embryol. Exp. Morphol.*, 77, 143–151.

Eyal-Giladi, H., and Fabian, B. (1980). Axis determination in uterine chick blastodiscs under changing spatial positions during the sensitive period for polarity. *Develop. Biol.,* 77, 228–232.

Eyal-Giladi, H., and Khaner, O. (1989). The chick's marginal zone and primitive streak formation. II. Quantification of the marginal zone's potencies — temporal and spatial aspects. *Develop. Biol.,* 134, 215–221.

Eyal-Giladi, H., and Kochav, S. (1976). From cleavage to primitive streak formation: A complementary normal table and a new look at the first stages of the development of the chick. I. General morphology. *Develop. Biol.,* 49, 321–337.

Eyal-Giladi, H., Ginsburg, M., and Farbarov, A. (1981). Avian primordial germ cells are of epiblastic origin. *J. Embryol. Exp. Morphol.,* 65, 139–147.

Eyal-Giladi, H., Kochav, S., and Menashi, M. K. (1976). On the origin of primoridal germ cells in the chick embryo, *Differentiation,* 6, 13–16.

Eyal-Giladi, H., Debby, A., and Harel, N., The posterior section of the chick's area pellucida and its involvement in hypoblast and primitive streak formation. *Development* (in press).

Fargeix, N. (1969). Les cellule germinales du Canard chez des embryons normaux et des embryones de regulation. Etude des jeunes stades du development. *J. Embryol. Exp. Morphol.,* 22, 477–503.

Ginsburg, M., and Eyal-Giladi, H. (1986). Temporal and spatial aspects of the gradual migration of primordial germ cells from the epiblast into the germinal crescent in the avian embryo. *J. Embryol. Exp. Morphol.,* 95, 53–71.

Ginsburg, M., and Eyal-Giladi, H. (1987). Primordial germ cells of the young chick blastoderm originate from the central zone of the area pellucida irrespective of the embryo-forming process. *Development,* 101, 209–219.

Ginsburg, M., and Eyal-Giladi, H. (1989). Primordial germ cell development in cultures of dispersed central disks of stage X chick blastoderms. *Gamete Res.,* 23, 421–428.

Ginsburg, M., Hochman, J., and Eyal-Giladi, H. (1989). Immunohistochemical analysis of the segregation process of the quail germ cell lineage. *Int. J. Dev. Biol.,* 33, 389–395.

Hahnel, A. C., and Eddy, E. M. (1982). Three monoclonal antibodies against cell surface components on early mouse embryos. *J. Cell Biol.,* 95, 156a.

Hahnel, A. C., and Eddy, E. M. (1986). Cell surface markers of mouse primordial germ cells defined by two monoclonal antibodies. *Gamete Res.,* 15, 25–34.

Hamburger, V., and Hamilton, H. L. (1951). A series of normal stages in the development of the chick. *J. Morphol.,* 88, 49–92.

Khaner, O., and Eyal-Giladi, H. (1986). The embryo-forming potency of the posterior marginal zone in stages X through XII of the chick. *Develop. Biol.,* 115, 275–281.

Khaner, O., and Eyal-Giladi, H. (1989). The chick's marginal zone and primitive streak formation. I. Coordinative effect of induction and inhibition. *Develop. Biol.,* 134, 206–214.

Kochav, S., and Eyal-Giladi, H. (1971). Bilateral symmetry in chick embryo, determination by gravity. *Science,* 171, 1027–1029.

Kochav, S., Ginsburg, M., and Eyal-Giladi, H. (1980). From cleavage to primitive streak formation: A complementary normal table and a new look at the first stages of the development of the chick. II. Microscopic anatomy and cell population dynamics. *Develop. Biol.,* 79, 296–308.

Koller, C. (1882). Untersuchungen uber die Blatterbildung im Hunerkeim. *Arch. Mikr. Anat.,* 20, 174–211.

Loveless, W., Bellairs, R., Thorpe, S. J., Page, M., and Feizi, T. (1990). Developmental patterning of the carbohydrate antigen FC10.2 during early embryogenesis in the chick. *Development,* 108, 97–106.

Pardanaud, L., Buck, C., and Dieterlen-Lievre, F. (1987). Early germ cell segregation and distribution in the quail blastodisc. *Cell Differentiation,* 22, 47–60.

Rogulska, T. (1968). Primordial germ cells in normal and transected duck blastoderms. *J. Embryol. Exp. Morphol.,* 20, 247–260.

Sutasurya, L. A., Yasugi, S., and Mizuno, T. (1983). Appearance of primordial germ cells in young chick blastoderms cultured in vitro. *Devel. Growth Differ.*, 25, 517–521.

Swift, C. H. (1914). Origin and early history of the primordial germ cells in the chick. *Am. J. Anat.*, 15, 483–516.

Urven, L. E., Erickson, C. A., Abbot, U. K., and McCarrey, J. R. (1988). Analysis of germ line development in the chick embryo using an anti mouse EC cell antibody. *Development*, 103, 299–304.

Vakaet, L. (1962). Some new data concerning the formation of the definitive endoblast in the chick embryo. *J. Embryol. Exp. Morphol.*, 10, 3857.

Waddington, C. H. (1933). Induction by the endoderm in birds. *Roux' Arch. Dev. Biol.*, 128, 502–521.

Chapter 4

Constant and Variable Features of Avian Chromosomes

Stephen E. Bloom, Mary E. Delany, and Donna E. Muscarella

Summary. Chromosome complements in avian species include both constant and variable features that must be considered in studies to map genes, identify sites of integration of foreign DNA, and track donor cells to tissue sites in chimeras. Detailed cytogenetic studies in the domestic chicken are restricted primarily to the first 10 chromosome pairs, each of which is morphologically distinct. High-resolution banding permits analyses of yet smaller chromosomes (e.g., chromosomes 1 – 15). The primary constitutive features of the avian genome have been revealed by G banding for structural landmarks, C banding for constitutive heterochromatin, RBG banding for DNA replication patterns, and *in situ* hybridization for localizing highly repeated sequences to centromeric and telomeric regions. Variable features of the chromosome complement and of gene expression in selected repeated gene families provide convenient metaphase and interphase phenotypes for studies *in vitro* and *in vivo*. A significant amount of genomic variability has been observed among early chick embryos (1 – 15% aberration frequency among genetic lines), including haploidy, triploidy, trisomy, and mosaicism. The inadvertent selection of an aberrant embryo as a source of cells for chimeric production could yield surprising or confusing results if cells are not karyotyped. Variability at the ribosomal RNA gene cluster generates polymorphic nucleolar (PNU) patterns in interphase cells. Polymorphic cells have one macro- and one micronucleolus per cell (*Pp*). This phenotype is easily diagnosed in cytological preparations. We have developed chicken genetic strains with defined PNU patterns. A particular nucleolar variant is expressed in the embryo and at all other stages including adult. We have detected the PNU phenotype in stage X embryos, in embryonic tissues representing ectoderm, mesoderm, and endoderm, and in feather pulp cells from chickens. Heterozygous PNU cells should be easy to track to tissue sites when transferred to nonpolymorphic recipient embryos. While development of homozygous PNU embryos (*pp*) is arrested at early primitive streak formation, heterozygotes (*Pp*) develop normally. Studies with rDNA variants should be useful for investigating the relative contributions of maternal ribosomes versus embryo-derived rRNA in early developmental processes.

Introduction

The domestic chicken has proven to be a valuable model system for investigations in biology and medicine. There is renewed excitement about working with avian species since there are many unanswered questions and even unexplored areas that can now be addressed more effectively with the newer tools of molecular biology. New studies are likely to expand greatly our knowledge of the organization of genes in avian chromosomes, provide new insights into evolutionary forces that have and continue to mold the genome, and identify genes and gene complexes that play central roles in particular developmental processes and cellular functions. A better understanding (and utilization) of the specific roles of genes and gene mutations in disease formation is likely to be achieved, affecting approaches toward controlling diseases in humans and animals alike.

In this chapter we review knowledge of major constant (i.e., constitutive) features and some noteworthy variable features of avian genomes, especially in *Gallus domesticus*. Our goal is to provide a broad perspective and stimulus for studies directed toward (1) mapping genes of interest, (2) improving knowledge of genetic organization in chromosomes, and (3) advancing knowledge concerning the genetic control of developmental processes. We also discuss a model system for modifying the genome that uses *in vivo* mechanisms that can bring about small to large changes in gene copies within multigene families.

Constant Features of the Avian Genome

Karyotypes

The presence of numerous very small chromosomes or microchromosomes (MICS) complicates the study and understanding of gene and chromosome organization and behavior in avian genomes. For this reason, it is essential to determine and differentiate the more or less constant (i.e., constitutive) from the more variable features in avian chromosome complements. Constant features to be discussed include (1) chromosome complement (karyotype) for a species, (2) cell and chromosome DNA contents, (3) DNA replication pattern, (4) higher-order chromosome architecture, and (5) content and locations of repetitive DNA sequences. In addition, sites for cleavage by restriction enzymes, particularly rare cutters, may provide landmarks throughout the genome; they can be used to generate DNA fragments useful for mapping genes across kilo- and megabase stretches of DNA.

Chromosome complements for many avian species have been described (Bloom, 1969; Ray-Chaudhuri, 1973; Takagi and Sasaki, 1974; Bitgood and Shoffner, 1990). The favorite organisms for genetic studies include domestic bird species such as the chicken, turkey, quail, and duck. These species have high diploid chromosome numbers, and karyotypes that include numerous MICS (Table 1). There is general agreement that the modal chromosome count represents the true chromosome number for a species. For chickens, a mode of 78 is obtained in counts of somatic cells, and 39 bivalents and synaptonemal complexes are detected in meiotic preparations (Ohno, 1961; Kaebling and Fechheimer, 1983). Distinct sex chromosome heteromorphism (ZZ, male, and ZW, female) is displayed in the complements of all

Table 1
Constant Features of the Avian Genome: Karyotype

Common name	Modal diploid number[a]	Sex chromosomes[b]
Chicken	78	5Z, 9W
Turkey	80	4Z, 7W
Chinese ringneck pheasant	82	4Z, 7W
Japanese quail	78	4Z, 6W
Mallard duck	78	4Z, 7W
Falconiformes:		
Lanner falcon	52	Unknown
Kestrel	52	1Z, 10W
Buzzard	68	4Z, 15W
Parakeet	58	5Z, 9W
Ratitae:		
Ostrich	80	No heteromorphism
Cassowary	80	No heteromorphism
Emu	80	No heteromorphism
Rhea	82	6Z, 6W (slight size difference)

[a] Chromosome numbers represent modes obtained from counts in somatic cells. Numbers for Ratitae are based on limited data and should be regarded as estimates. (*Sources*: Bloom, 1969; Takagi and Sasaki, 1974.)

[b] Size designations for the W chromosome are based on closest match to other chromosomes in the karyotype.

gallinaceous bird species studied. The Z sex chromosome is large and metacentric, and the W is much smaller but also metacentric (Table 1). Quite a different chromosome plan is found among the Falconiformes. Here, low chromosome numbers are found (e.g., $2n = 52$), and the Z chromosome can be quite large. Still a different chromosome plan is found among members of the flightless birds, the Ratites. Although these species have high chromosome numbers, as in the Gallinaceous birds, some species show no sex chromosome heteromorphism. Only one species (*rhea*) demonstrates a slight size dimorphism in chromosome pair no. 6 in females (Takagi et al., 1972). It is of great interest to determine mechanisms underlying evolution of these different karyotype plans in birds. It would be of interest to determine conserved versus broken linkage arrangements of genes in avian lineages and to determine which gene groups were retained in other forms leading to mammalia.

Since avian chromosomes are generally considered to be small, it is useful to compare them with chromosomes from other groups such as mammals. For example, the first six pairs of chromosomes in the chicken are in the same size range as human chromosomes (Figure 1). However, chicken chromosomes 7 through 39 are much smaller. Examination of the idiogram of the chicken reveals also the graded nature of the reductions in size from larger to smaller chromosomes (Figure 1). The size ratios of largest to smallest

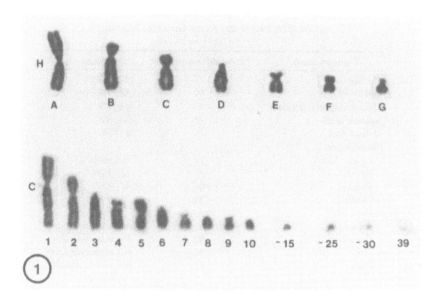

FIGURE 1. Comparison of chromosomes from human (H) and chicken (C) lymphocytes photographed and printed at the identical magnification (largest chromosomes are about 8 μm). Chromosomes from each of the major groups are shown for the human. Chicken chromosomes are in size order, including the Z sex chromosome in position 5. The size of the smallest human chromosome (containing about 50 Mb of DNA) is intermediate between chicken chromosomes 6 and 7.

chromosomes in the karyotypes are about 5:1 for the human and 23:1 for the chicken.

The larger pairs of chicken chromosomes (i.e., pairs 1–5) have often been referred to as macrochromosomes (MACS) to distinguish these larger, morphologically distinct chromosomes from the smaller, dotlike microchromosomes (MICS). However, chromosome sizes decrease gradually in the karyotype, and so these terms are used somewhat arbitrarily. We have used a system established in an earlier review of chicken chromosomes (Bloom, 1981). Chromosome pairs 1 – 5 are considered MACS, pairs 6 – 10 are intermediate (larger MICS), and pairs 11 – 39 are MICS.

Higher-Order Chromosome Architecture

Chromatin fibers in mammalian chromosomes are organized into regions of high versus low packing densities. This differential packing results in a banded appearance along the chromosome, especially visible in meiotic chromosomes (the classical chromomeres) (Nokkala and Nokkala, 1986). Enzyme treatments reveal such bands in mitotic chromosomes (e.g., G bands) (Hsu, 1974).

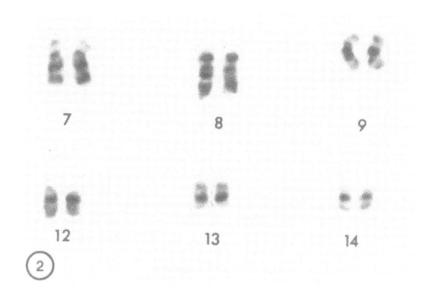

FIGURE 2.   Avian microchromosomes display G bands upon treatment with trypsin or other enzymes. Banding on several homologous pairs of microchromosomes is illustrated.

Trypsin and/or urea treatments of mitotic chromosome preparations from birds reveals a longitudinal banding pattern on the macrochromosomes (MACS) (Takagi and Sasaki, 1974; Stock et al., 1974). In addition, bands can be seen on many of the larger MICS (Figure 2). These findings indicate a chromomere type of organization for the chromatin fiber in avian MACS and MICS.

DNA Replication Pattern in Chromosomes

The presence of numerous small chromosomes in avian genomes calls into question whether they are replicated consistently and faithfully over multiple rounds of cell division. Minute chromosomes in plants (B chromosomes) and in certain tumor cells can be replicated and distributed unevenly or out of synchrony with the rest of the genome (Mark, 1967; Barker and Hsu, 1979).

We have used bromodeoxyuridine (BrdU) labeling of replicating DNA to study replication of MICS over multiple cell cycles. Labeling of cells *in vitro* or *in vivo* with BrdU for two cell cycles followed by Hoechst staining or a fluorescence-plus-Giemsa procedure yields a characteristic differential staining of sister chromatids (SCD) on all chromosomes at metaphase (Bloom and Hsu, 1975; Bloom, 1982). That is, all chromosomes have one dark and one light chromatid reflecting single versus double substitution of DNA strands by BrdU. Application of this technique to avian cells, including the chicken, shows the expected SCD pattern on both MACS and MICS. Thus, DNA is replicated in all chromosomes of the avian genome in a semiconservative fashion, and replication of all chromosomes is completed during any given S phase of the cell cycle.

It is also possible to track the consistency of chromosome replication by observing the chromosome phenotypes in endoreduplicated (ENDO) chromosomes. Cells with ENDO chromosomes have passed through two rounds of DNA synthesis with no intervening mitosis. They subsequently reenter the cell cycle. Thus, the products of two rounds of DNA synthesis can been visualized in a metaphase cell as replicated chromosomes lying adjacent to each other (Figure 3). Studies of chicken cells with ENDO chromosomes from tissue cultures show that both MACS and MICS are replicated consistently over multiple cell cycles. That is, the replication products have identical chromosomal morphologies (Figure 3). Thus, avian MACS and MICS are replicated strictly and consistently in the course of cell divisions.

Content and Locations of Repetitive DNA Sequences

Constitutive Heterochromatin

Knowledge of the content and locations of repetitive DNA sequences may help to guide gene mapping studies. Repetitive DNA sequences are found in abundance in mammalian genomes (Table 2). Highly repeated DNA makes up some 36% of the human genome; this DNA is located at the centromeres of all 23 chromosomes pairs (Lewin, 1980). This localization can be shown very clearly using C banding; the dark-stained material at centromeres is referred to as constitutive heterochromatin. This type of heterochromatin is highly resistant to alkali and salt extractions, and therefore stains deeply with Giemsa dye. The chicken genome has much less (12%) highly repetitive DNA; most is located at the centromeres of MICS (Figure 4) (Stefos and Arrighi, 1974; Lewin, 1980). One arm of the W sex chromosome is rich in constitutive heterochromatin. Recently, Tone et al. (1982, 1984) detected and characterized W-specific repetitive DNA sequences. These sequences probably account for the bright fluorescence observed after staining with quinacrine mustard or 4'-6-diamidino-2-phenylindole (DAPI). In fact, a distinct quinacrine-positive W body can be seen in interphase nuclei of female cells (Bloom and Macera, 1974). Some MICS appear totally heterochromatic. In addition, the tips of the Z sex chromosomes are C-band positive (Figure 4). A counterstain-enhanced fluorescence technique (CDD) has recently been employed to detect primarily G-C–rich and A-T–rich areas of chromosomes (Schweizer, 1980; Auer et al., 1987). G-C rich heterochromatic blocks were detected at the end of the short arm of chromosome 1, the end of the long arm of chromosome 2, and on the nucleolar organizer microchromosome (about 16th or 17th in size). Less prominent G-C–rich blocks were found at the end of the Z chromosome and on one arm of the W chromosome. A-T–rich areas of the W were also identified (ends and centromere). No A-T–rich blocks

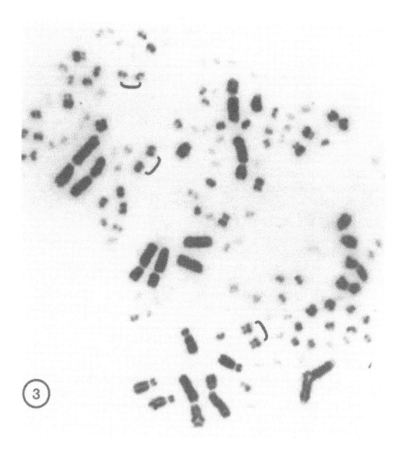

FIGURE 3. A cultured cell that has undergone endoreduplication. The products of two rounds of replication (two S periods with no intervening mitosis) are visualized as morphologically identical chromosomes lying adjacent to each other. Microchromosomes (see brackets to denote examples) demonstrate consistent and faithful replication over the two periods of DNA synthesis.

were found among the MICS. Thus, it would appear that C-band–positive regions are G-C rich, except for the W chromosome where there are also A-T–rich blocks.

Ribosomal RNA Genes

Vertebrates typically have numerous copies of the 18S + 28S ribosomal RNA genes (rDNA) in order to support substantial and changing levels of protein synthesis. Humans and other mammals generally have their rDNA distributed on multiple chromosomes (Goodpasture and Bloom, 1975). Chickens and other avian species have either just one site or a limited number (Table 2). The entire rDNA cluster in chickens consists of 145 repeats (290 in a diploid cell), and it is located on a microchromosome approximately 16th in size in

Table 2

Content and Localization of Repetitive Sequences in the Avian and Other Vertebrate
Genomes

Type of DNA	Percent of genome	Localization
I. Repetitive sequences (intermediate and highly repeated)[a]		
Chicken	12	(1) Most of W sex chromosome
		(2) Centromeres of microchromosomes[b]
Human	36	All centromeres
II. Telomeric sequences[c] (TTAGGG)$_n$)		
Avian species		All telomeres
G. domesticus		
Vireo bellii		
Passer domesticus		
Buteo jamaicensis		
Pisces, Amphibia, Reptilia, Mammalia		All telomeres
III. Ribosomal DNA[d] (18S, 5.8S, 28S)		
Chicken: 290 repeats/cell (40 kb repeat)	0.5	Microchromosome 16
Human: 560 repeats/cell (44 kb repeat)	0.4	Chromosomes 13, 14, 15, 21, 22

[a] Data from Bloom, 1974; Stefos and Arrighi, 1974; Lewin, 1980.
[b] Exact number of microchromosomes with repeated DNA not known. *In situ* hybridization
    reveals about 15 labeled chromosomes. However, C banding shows dark staining on most if
    not all microchromosomes. Macrochromosomes may also contain some repeated DNA including
    the Z sex chromosome as indicated indirectly by C banding. The primary location of the
    intermediate and highly repeated DNA is at centromeres but also in the arms of the W
    chromosome and possibly the Z chromosome.
[c] Data from Meyne et al., 1989.
[d] Data from Lewin, 1980; Muscarella et al., 1985.

the karyotype (Bloom and Bacon, 1985; Muscarella et al., 1985). The col-
lective transcriptional activity from the rDNA cluster results in the formation
of a nucleolus. Diploid chicken cells thus have two nucleoli per cell. The
rDNA site can be localized on metaphase chromosome preparations using
silver nitrate (Ag banding), which has a high affinity for nucleolar proteins
associated with the rDNA transcription complex. Silver stains much of the
long-arm region of the MIC (Bloom and Bacon, 1985), thus indicating the
location of the NOR.

An intriguing feature of the MIC containing the rDNA is the presence
on this chromosome of two other gene complexes, the major histocompatibility
complex (MHC) and the β subunit of a G protein (Bloom and Bacon, 1985;
Guillemot et al., 1989). This linkage group has not been conserved in evolution and is therefore not found in mammalian genomes.

Telomeres

Special families of repeated sequences occur at telomeres of chromosomes.

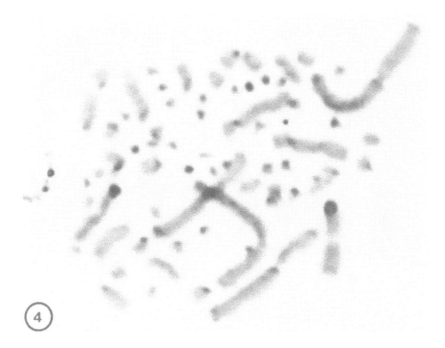

FIGURE 4. A C-banded metaphase from a male (ZZ) chicken demonstrating the localization of constitutive heterochromatin. Major sites include the centromeric regions of microchromosomes and the ends of the Z sex chromosomes. The W sex chromosome in females is also heterochromatic and stains darkly with C banding.

Recently, the sequence $(TTAGGG)_n$, found at human telomeres, was used in *in situ* hybridization studies (Meyne et al., 1989). Using fluorescence reporting, this sequence was found to hybridize to the telomeres of members of all vertebrate classes including avian (Table 2). Both MACS and MICS showed hybridization in chickens and three other avian species tested. These results provide additional evidence to support the view that avian MICS are simply small versions of larger chromosomes.

We have recently demonstrated separation of large DNA fragments from the chicken genome by pulse field gel electrophoresis (PFGE) following digestion with rare cutter enzymes. The human telomeric sequence hybridizes to numerous DNA fragments generated with either *Not*I or *Mlu*I in the size range of 50 kb to 2.2 Mb (Figure 5).

Sizes and DNA Contents of Individual Chromosomes

The chicken genome contains approximately 2.5 pg of DNA per diploid cell, or 2.4×10^9 base pairs (bp). This represents about 37% of the human genome. To determine the complete sequence of the haploid complement of the chicken genome, the order of approximately 1.2×10^9 bp would need to be

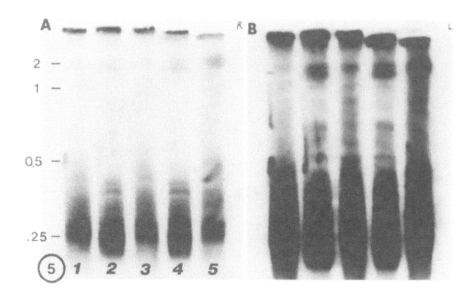

FIGURE 5. Pulse field gel electrophoretic separation of chicken DNA cut with rare cutter enzymes and then blotted to detect hybridization of the human telomeric sequence (TTAGGG)$_n$ to large fragments in the chicken genome. DNA was cut with *Not*I (lanes 1, 3, 5) or *Mlu*I (lanes 2, 4) restriction enzymes and run on a CHEF DR II pulse field apparatus for 24 h. A and B represent shorter and longer exposures of film, respectively, from the same filter. The ethidium bromide stained gel is shown in Figure 7. Sizes of fragments are in megabases (largest fragment resolved was actually 2.2 Mb).

established, a task that would be laborious using current technology. Sequencing of specific genes of interest to various investigators has been achieved, but it is likely to be a long time before extensive sequencing is performed. What is perhaps more important at this time is to improve on the rather meager genetic map of the chicken and other birds. Mapping in chickens and other birds is facilitated to some extent by the fact that most of the genome is packaged into the larger chromosomes. In chickens, some 55% of the genome is packed into the first five chromosomes (including the Z sex chromosome as number 5). The next 20% of the genome resides in chromosomes 6 through 10 (Table 3). Chromosomes 11 through 39 contain about 25% of the genome.

Estimates of the bp contents of individual chromosomes indicates a considerable range in values (Table 3). Although, MICS are small in overall size (0.2 to 1.0 μm), they are estimated to contain from 23 to 7 Mb of DNA (chromosomes 11 to 39). The actual DNA packing ratios for avian chromosomes are not known; it is possible that these values of chromosome sizes or masses, which were calculated from photographic emulsions, are overestimates. However, recent attempts at separating uncut chicken DNA by PFGE have not yielded more than two large bands at the top of gels even

Table 3

Sizes and DNA Contents of Individual Chromosomes from the Chicken

Chromosome number	Size in μm mitosis[a]	Size in μm meiosis[b]	Percent of the haploid genome[c]	Estimated DNA in megabases[d]
1	6.2	30.2	20.5	246
2	4.6	25.2	12.8	154
3	3.2	20.5	9.0	108
4	2.7	17.3	7.1	85
5 (Z)	2.6	15.0	7.1	85
6	1.8	13.5	5.8	70
7	1.5	11.0	3.2	38
8	1.3	9.2	3.2	38
9	1.0	7.8	1.9	23
W	1.1	nd	1.9	23
10	1.0	7.4	1.9	23
16		nd	1.3	16
19		nd	1.26	15
29		nd	0.87	10
39		1.3	0.54	7
11–39	1.0–0.2	7.1–1.3	27.6	331

[a]   B cells *in vivo.*
[b]   From Ford and Woollam, 1964; male pachytenes.
[c]   Based on weights of chromosomes from cutouts of photomicrographs of mitotic cells.
[d]   Calculated by multiplying percentage of haploid genome for each chromosome times 1.2×10^9 base pairs (haploid genome size).

after extended runs that successfully separate large yeast chromosomes (e.g., 5.7 Mb in *S. pombe*) (Figure 6). These results suggest strongly that even the smallest of MICS may still be too large for separation with instruments and technologies currently available.

Production of Large DNA Fragments for Mapping

While it is not yet possible to separate intact avian chromosomes by existing electrophoretic technologies (Figure 6), it is relatively straightforward to produce large DNA fragments. Digestion with the restriction enzymes *Not*I (GCGGCCGC recognition sequence) or *Mlu*I (ACGCGT recognition sequence) followed by PFGE allows for the separation of fragments in the size range of about 40 kb to more than 2.2 Mb (Figure 7). Particularly distinct fragments, seen after ethidium bromide staining, are produced following *Mlu*I digestion and PFGE for 24 h (Figure 7). Thus, it should be possible to perform mapping of genes over kilo- and megabase stretches of DNA in the avian genome. Hybridization with a chicken ribosomal RNA gene probe detects a fragment of about 40 kb, which represents the size of one rDNA repeat (Figure

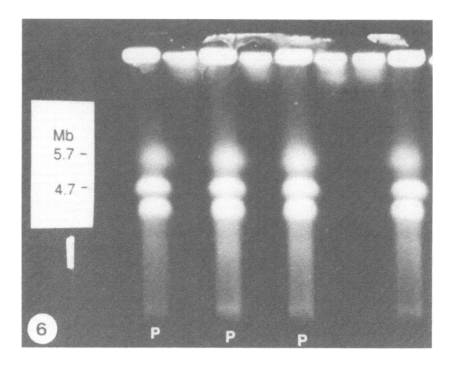

FIGURE 6. Pulse field gel separation of uncut chicken DNA from MSB-1 tumor cells. An 0.8% agarose gel was run for 6 days, 12°C, 50 V, with a switch time of 60 min on a CHEF DR II apparatus. *S. pombe* chromosomes (lanes marked P) were easily separated with significant migration of the large, 5.7-Mb chromosome. Uncut chicken DNA displayed minimal migration as shown in the lanes adjacent to *S. pombe*. The chicken ribosomal DNA sequences were detected at the top of the gel using a ³²P-labeled rDNA fragment (data not shown). The rDNA is located on chicken chromosome 16.

8). This result is one of the first concerning the mapping of genes to particular large fragments of chicken DNA separated by PFGE.

Variable Features of the Avian Genome

Sex Chromosomes

In most avian species, karyotypes of males and females differ by the presence of a heteromorphic pair of chromosomes. Males are ZZ and females ZW (Bloom, 1974). Generally, the Z is a MAC and the W is a MIC. In the chicken, the W chromosome is similar in size and morphology to the ninth largest pair, which are metacentrics. The W can sometimes be distinguished based on its slightly submetacentric appearance and by the fact that it is

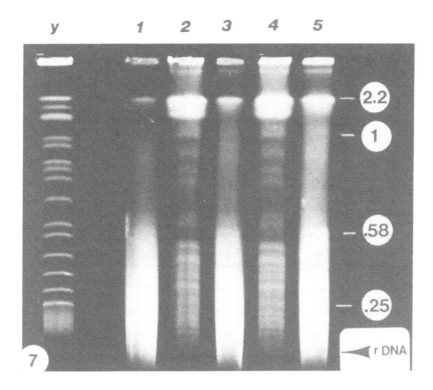

FIGURE 7. Pulse field gel electrophoretic separation of chicken DNA fragments produced with rare cutter enzymes. A clear, reproducible pattern of fragments revealed with ethidium bromide fluorescence is obtained with *Mlu*I. MSB-I chicken DNA was cut with either *Not*I (lanes 1, 3, 5) or *Mlu*I (lanes 2, 4) restriction enzymes and run on the CHEF system for 24 h. Uncut DNA from *S. cerevisiae* was included (lane Y) to allow sizing of fragments. All 15 yeast chromosomes were resolved; sizes in megabases are 2.2, 1.6, 1.1, 1.0, 0.95, 0.85, 0.80, 0.77, 0.70, 0.63, 0.58, 0.46, 0.37, 0.29, and 0.25. Position of rDNA genes is shown with arrowhead (see also Figure 8).

negatively heteropycnotic, i.e., it decondenses more than other chromosomes at metaphase.

There is much interest in identifying genes on the Z and W (and autosomes) that are essential for normal male versus female gonadal development. Parallel studies with mammals will help reveal the common genetic strategies used in vertebrates to achieve appropriate sexual development. One hypothesis involving a common mechanism for sex development in birds and mammals, put forward by Mittwoch (1971), may be directly testable, namely, that genes on the Y or W chromosome regulate the growth rate of the gonadal rudiment. In this model, early proliferation of medullary cells of the developing gonad results in the differentiation of a testis in the XY embryo. In birds, similar

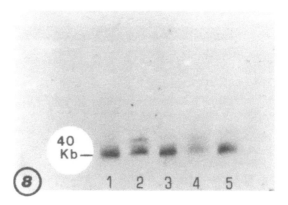

FIGURE 8. Southern blot of gel shown in Figure 7 using a ^{32}P-labeled chicken ribosomal DNA probe. A 40-kb band is detected near the bottom of the gel (see arrowhead in Figure 7 for position). This is about the size of one rDNA repeat. Less prominent bands represent the products of partial digestion.

stimulation would occur in the cortical region of the gonad, driving the ZW embryo toward female development.

Several genes in mammals have been identified and studied for their potential roles in primary sex determination. These have included the H-Y antigen, testis-determining factor (TDF), and SRY, located on the Y chromosome (Page et al., 1987; Berta et al., 1990). Similar sequences probably exist in avian species. Indeed, a human Y-specific sequence (probe pDP 10071) hybridizes readily in Southern blots to chicken sequences in males and females (Page et al., 1987; see also Chapter 17).

Deviations from the Normal Chromosome Complement

Spontaneous abnormalities in chromosome number and also morphology occur with surprisingly high frequency in many vertebrate species. In humans, 5% of all conceptions involve a chromosomal abnormality, either in the number of chromosomes (heteroploidy) or in chromosome structure (translocations, deletions, etc). The incidence of heteroploidy has been assessed in great detail in various strains of chickens (Table 4) (Bloom, 1981). While many strains have about 3–4% of fertile eggs with chromosomal abnormalities, some are much higher, approaching 15% of all fertile eggs. These higher frequencies have been detected in rapid growth lines, including broilers.

The gain or loss of one chromosome for a particular pair (aneuploidy) or deviation by a whole set of chromosomes (haploidy, triploidy) is usually associated with reduced developmental potential of the embryo. Most such embryos die by 4 to 5 days of incubation (Bloom, 1972). However, triploid zygotes can sometimes develop to hatching and even to adult stages in chickens (and other lower vertebrates) (Fankhauser and Humphrey, 1959; Ohno et al.,

Table 4

Spontaneous Chromosomal Abnormalities in
Chicken Embryos

Genetic strain or breed	Proportion of fertile eggs with chromosome aberrations (%)
Cornell obese	0.9
Cornell K-resistant	1.0
Araucana	2.7
Jungle fowl	2.8
Commercial egg line B	3.3
Cornell C-resistant	3.5
Commercial inbred B	5.4
Virginia LWS	6.2
Broiler stock	11.9
Virginia HWS	14.5

1963; Abdel-Hameed and Shoffner, 1971). Triploid individuals have been detected in some commercial egg-laying stocks. They are sterile intersexes. Certain genetic backgrounds may support triploid development, while many do not. There is also evidence that production of triploids and haploids is under genetic control (Bloom, 1972). Thus, heteroploidy is a significant cause of embryo lethality and can also contribute to reduced reproductive performance of poultry flocks.

With renewed interest in producing chimeric chickens using very early embryonic cells, it will be important to consider the chromosome status of the donor and recipient. Use of heteroploid cells would either cause arrested development of chimeras or contribute to teratogenesis. On the other hand, the deliberate use of a particular trisomy in a chimera would be useful as a model to understand the mechanisms of trisomy-induced teratogenic development in vertebrates.

Variants of Use in Experimental Embryology

In the chicken there are several cellular markers that have or could be used to track donor cells in chimeric embryos for the purpose of understanding developmental processes (Table 5). The heterochromatin of quail cells and the ZZ and ZW sex chromosomes have both been used in such studies (Sundick et al., 1973; Le Douarin, 1973). Two other cellular phenotypes could be used. The W body in interphase cells presents a readily detectable phenotype. This structure, representing constitutive heterochromatin, is quite prominent in cells of female turkey embryos and neonates (Bloom and Macera, 1974). It is very likely that other species will show a prominent W body as well.

In chickens, there are just two nucleoli in interphase cells, and these are very distinctive in embryonic cells. We have recently defined genetic strains of chickens having either a monomorphic or polymorphic nucleolar phenotype

Table 5
Cellular Markers for Tracking Donor Cells in
Chimeras

Species	Cell marker or genotype	Stage in mitosis
Chicken	ZZ, ZW chromosomes	Metaphase
Chicken	Polymorphic nucleoli	Interphase
Turkey	Sex chromatin or W body	Interphase
Quail	Heterochromatin	Interphase

(PNU) in interphase cells (Delany et al., 1991). The PNU phenotype is typified by the presence of one large and one small nucleolus within a given cell nucleus (monomorphics have two equal-sized nucleoli). The PNU phenotype results from reduced numbers of gene copies within the rDNA cluster, and therefore represents a heritable condition. The PNU phenotype has been detected in embryonic tissues derived from all three embryonic layers (i.e., ectoderm, endoderm, mesoderm). Using the PNU strain, it should be possible to make chicken × chicken (histocompatible) chimeras and track PNU donor cells to tissue sites in non-PNU (or monomorphic) recipient embryos.

Generating and Utilizing Variations in Multigene Families

Model Systems and Multigene Families

It is now recognized that many essential functions in cells are encoded by genes within multigene families (MGFs). Examples include the immunoglobulin superfamily (including the MHC), actins, homeobox, and histone MGFs. We have been particularly interested in the 18S and 28S rRNA genes because they play a central role in regulating protein synthesis in cells. This regulation is achieved at multiple levels, including recruitment of rRNA genes for transcription, RNA processing, and in translation (Nomura et al., 1984).

Studies with model genetic systems such as yeast and *Drosophila melanogaster* reveal some of the fascinating features of rDNA and variations in this complex. It is now appreciated that rDNA clusters are somewhat dynamic entities, with reductions and additions in copy numbers in clusters as a result of unequal sister-strand crossing over in mitosis and unequal recombination in meiosis. There are specific genes that "detect" and regulate rDNA copy number in a cluster. The rDNA cluster has been a useful model in population genetics to detect rates and mechanisms of evolution of repeated gene families (Frankham, 1982).

Developmental and physiologic effects of variations in rDNA clusters have also been studied. Selection for rapid versus slow emergence from the larval stage in *D. melanogaster* was shown to be accompanied by changes in repeated elements of the rDNA spacer regions (Cluster et al., 1987). These studies suggest roles for altered copy numbers and structure of rDNA in modulating aspects of development.

Induction of rDNA Variation in Chickens

We have developed a system to induce extensive alterations in rDNA copy numbers (Delany et al., 1991). Chickens trisomic for the rDNA/MHC microchromosome are used to generate rDNA variants in progeny. We initially detected the trisomy-induced alterations in the rDNA complex by studying nucleolar sizes in cells. Some 18% of progeny, derived from trisomic parents with three equal-sized nucleoli, demonstrate dramatically altered nucleoli phenotypes. The mechanism we favor for this *de novo* production of nucleolar polymorphisms is trisomy-mediated out-of-register pairing (ORP) in the rDNA cluster followed by crossing over during meiosis. Trivalent pairing may enhance the opportunity for ORP. Both reduced and expanded complexes would be generated by such a mechanism. We have so far isolated and studied heterozygotes having one normal and one reduced complex in the cells.

Recovery and Utilization of rDNA Variants

A new genetic line, PNU, was bred containing birds homozygous, *PP*, for nucleolar sizes (equal-sized nucleoli) and birds that are heterozygotes, *Pp*, having one larger and one smaller nucleolus in cells (Delany et al., 1991). From *in situ* hybridization studies we calculated approximate copy numbers for the various genotypes, i.e., 145 copies of the rDNA repeat for a normal or wild-type cluster and about 30 rDNA repeats for the reduced or mutant cluster. In each generation in the breeding of the PNU line we have crossed *Pp* birds to normals (*PP*) from a control strain (MFO). Thus, each *Pp* individual has one wild-type and one mutant rDNA cluster. The crosses of *PP* × *Pp* birds generates a 1:1 ratio of *PP:Pp*, indicating a Mendelian type of transmittance of the normal versus the reduced rDNA clusters or "blocks".

The PNU genetic line will be extremely useful for investigating the biological effects of rDNA modulations in chickens. We are particularly interested in determining the minimum number of rDNA repeats needed to support embryonic development in the chick and when embryo rRNA and ribosomes are first needed to support development beyond the early cleavage stages. A cross of *Pp* × *Pp* birds is expected to generate a 1:2:1 ratio of *PP:Pp:pp* birds in the F_1. Our studies of newly hatched chicks reveal only the presence of *PP* and *Pp* chicks, and these two genotypes appear to have equal viability. The *pp* individuals are detected only among very early embryos. In fact, development of these mutant embryos is arrested during initial primitive streak formation. Thus, rDNA copy numbers of 290 and 175 repeats per cell in wild types (*PP*) and heterozygotes (*Pp*), respectively, represent levels that confer equal developmental potential to embryos. However, a copy number of about 60 rDNA repeats per diploid cell confers little or no support for development. Survival of *pp* embryos to the primitive streak stage may be due to the contribution of ribosomes from the oocyte and/or the capability of 60 rDNA repeats to support development to this stage but no further.

We are also interested in the potential biological effects of heterozygous conditions for rDNA complexes and nucleoli. Polymorphisms in nucleoli in various chicken populations are known to occur, but their bases have not been defined. Alterations in gene copy numbers and gene spacer size as well as differential recruitment of rDNA genes may be involved. It will be of interest to determine if heterozygosity in nucleolar phenotypes is associated with enhanced performance of poultry stocks.

The system we have developed for rDNA may have broad applications for the modification of the avian genome. That is, we showed that rDNA copy numbers can be altered via normal cellular processes, e.g., unequal crossing over. We found that the extent of this process can be increased by having a trisomic condition (allows for more slippage enhancing out-of-register pairing) or even in diploids where the two homologous rDNA-containing chromosomes have different sized rDNA clusters (Delany et al., 1991). This latter situation can lead to still further changes in rDNA clusters over multiple generations of breeding.

Areas for Future Investigations

This is a truly exciting era in biology where long-standing and newer questions are being addressed and answered with powerful methodologies. Avian species have been and will continue to be an integral part of basic studies in biology due to the many attractive features offered and the impressive base of information on which to build. There are several areas that stand out that should be both exciting and beneficial to explore. First, it will be essential to expand knowledge of gene locations on chromosomes for genes and multigene families of high interest (homeobox, immunoglobulin, P450, and DNA repair genes, etc.). Such information would expand knowledge of genetic organization, allow one to detect useful genetic variants, and help in the effort to target foreign DNA sequences to specific areas of the genome. In addition, knowledge of gene locations will further the analysis of disease-causing genes and aid in the development of preventative strategies.

There is presently only rudimentary knowledge of genetic organization of avian genomes and the forces that have molded them. We know little about the mechanisms and significance of locating some genes and families of DNA in microchromosomes versus macrochromosomes. One speculation is that locating genes on MICS is one way of enhancing genetic stability; i.e., such genes or small gene blocks would have comparatively low rates of crossing over and would be smaller targets for endogenous and natural exogenous mutagens that might enter the cell nucleus.

The chick embryo is one of the most powerful vertebrate models to probe the mysteries of cell and tissue formation and differentiation in specific cell lineages. It will be of interest to identify and understand the roles of specific genes and gene families in very early developmental events, in differentiation of gonads toward male versus female directions, and in events leading to tissue-specific functions in key systems such as muscle, cartilage, bone, nerves, etc.

The outcome of new research will be a much more detailed understanding of the genes and their actions in molding the organism and its various functions. Such knowledge can have important applications in agriculture and medicine, leading to the improvement of quality of life for humans and animals alike.

Acknowledgments

We thank T. C. Hsu, S. A. Latt (deceased), H. L. Robinson, V. M. Vogt, D. C. Page, M. M. Miller, and C. Goodpasture for stimulating discussions and various collaborations over the years. The various research discussed in this review was supported by grants from the NIEHS (ES03499), USDA (NY157433), and Cornell Biotechnology Program. We thank Diane Colf for editorial assistance.

References

Abdel-Hameed, F., and Shoffner, R. N. (1971). Intersexes and sex determination in chickens. *Science,* 172, 962–964.

Auer, H., Mayr, B., Lambrou, M., and Schleger, W. (1987). An extended chicken karyotype, including the NOR chromosome. *Cytogenet. Cell Genet.,* 45, 218–221.

Barker, P. E., and Hsu, T. C. (1979). Double minutes in human carcinoma cell lines with special reference to breast tumors. *J. Natl. Cancer Inst.,* 62, 257–262.

Berta, P., Hawkins, J. R., Sinclair, A. H., Taylor, A., Griffiths, B. L., Goodfellow, P. N., and Fellous, M. (1990). Genetic evidence equating SRY and the testis-determining factor. *Nature,* 348, 448–450.

Bitgood, J. J., and Shoffner, R. N. (1990). Cytology and cytogenetics. In: *Poultry Breeding and Genetics,* edited by R. D. Crawford. Elsevier, New York, pp. 401–427.

Bloom, S. E. (1969). A current list of chromosome numbers and karyotype variations in the avian subclass *Carinatae. J. Heredity,* 60, 217–220.

Bloom, S. E. (1972). Chromosome abnormalities in chicken embryos: Types, frequencies, and phenotypic effects. *Chromosoma,* 37, 309–326.

Bloom, S. E. (1974). Current knowledge about the avian W chromosome. *BioScience*, 24, 340–344.

Bloom, S. E. (1981). Detection of normal and aberrant chromosomes in chicken embryos and in tumor cells. *Poultry Sci.*, 60(7), 1355–1361.

Bloom, S. E. (1982). Avian and aquatic systems for *in vivo detection of sister chromatid exchange*. In: *Sister Chromatid Exchange*, edited by A. A. Sandberg. Alan R. Liss, New York, pp. 249–277.

Bloom, S. E., and Bacon, L. D. (1985). Linkage of the major histocompatibility (B) complex and the nucleolar organizer in the chicken: Assignment to a microchromosome. *J. Heredity*, 76, 146–154.

Bloom, S. E., and Hsu, T. C. (1975). Differential fluorescence of sister chromatids in chicken embryos exposed to 5-bromodeoxyuridine. *Chromosoma*, 51, 261–267.

Bloom, S. E., and Macera, M. (1974). Fluorescence detection of the W-body in turkey interphase nuclei. *Avian Chromosomes Newsl.*, 3, 1–7.

Cluster, P. D., Marinkovic, D., Allard, R. W., and Ayala, F. J. (1987). Correlations between developmental rates, enzyme activities, ribosomal DNA spacer-length phenotypes, and adaptation in *Drosophila melanogaster*. *Proc. Natl. Acad. Sci. USA*, 84, 610–614.

Delany, M. E., Muscarella, D. E., and Bloom, S. E. (1991). Formation of nucleolar polymorphisms in trisomic chickens and subsequent microevolution of rRNA gene clusters in diploids. *J. Heredity*, 82:213–220.

Ford, E. H. R., and Woollam, D. H. M. (1964). Testicular chromosomes of *Gallus domesticus*. *Chromosoma* (Berlin), 15, 568–578.

Frankham, R. (1982). Contributions of *Drosophila* research to quantitative genetics and animal breeding. In: *Proceedings, 2nd World Congress on Genetics Applied to Livestock Production*, vol. 5, pp. 43–56.

Fankhauser, G., and Humphrey, R. R. (1959). The origin of spontaneous heteroploids in the progeny of diploid, triploid, and tetraploid axolotl females. *J. Exp. Zool.*, 142, 379–421.

Goodpasture, C., and Bloom, S. E. (1975). Visualization of nucleolar organizer regions in mammalian chromosomes using silver staining. *Chromosoma*, 53, 37–50.

Guillemot, F., Billault, A., and Auffray, C. (1989). Physical linkage of a guanine nucleotide-binding protein-related gene to the chicken major histocompatibility complex. *Proc. Natl. Acad. Sci. USA*, 86, 4594–4598.

Hsu, T. C. (1974). Longitudinal differentiation of chromosomes. *Ann. Rev. Genetics*, 7, 153–176.

Kaebling, M., and Fechheimer, N. S. (1983). Synaptonemal complexes and the chromosome complement of domestic fowl, *Gallus domesticus*. *Cytogenet. Cell Genet.*, 35, 87–92.

Le Douarin, N. (1973). A biological cell labeling technique and its use in experimental embryology. *Dev. Biol.*, 30, 217–222.

Lewin, B. (1980). *Gene Expression 2*, 2nd ed., John Wiley & Sons, New York.

Mark, J. (1967). Double minutes — a chromosomal aberration in Rous sarcomas in mice. *Hereditas*, 57, 1–22.

Meyne, J., Ratliff, R. L., and Moyzis, R. K. (1989). Conservation of the human telomere sequence $(TTAGGG)_n$ among vertebrates. *Proc. Natl. Acad. Sci. USA*, 86, 7049–7053.

Mittwoch, U. (1971). Sex determination in birds and mammals. *Nature*, 231, 432–434.

Muscarella, D. E., Vogt, V. M., and Bloom, S. E. (1985). The ribosomal RNA gene cluster in aneuploid chickens: Evidence for increased gene dosage and regulation of gene expression. *J. Cell Biol.*, 101, 1749–1756.

Nokkala, S., and Nokkala, C. 1986. Coiled internal structure of chromonema within chromosomes suggesting hierarchical coil model for chromosome structure. *Hereditas*, 104, 29–40.

Nomura, M., Gourse, R. L., and Baughman, G. (1984). Regulation of the synthesis of ribosomes and ribosomal components. *Ann. Rev. Biochem.*, 53, 75–117.

Ohno, S. (1961). Sex chromosomes and microchromosomes of *Gallus domesticus*. *Chromosoma*, 11, 484–498.

Ohno, S., Kittrell, W. A., Christian, L. C., Stenius, C., and Witt, G. A. (1963). An adult triploid chicken (*Gallus domesticus*) with a left ovotestis. *Cytogenetics*, 2, 42–49.

Page, D. C., Mosher, R., Simpson, E. M., Fisher, E. M. C., Mardon, G., Pollack, J., McGielivray, B., de la Chapelle, A., and Brown, L. G. (1987). The sex-determining region of the human Y chromosome encodes a finger protein. *Cell*, 51, 1091–1104.

Ray-Chaudhuri, R. (1973). Cytotaxonomy and chromosome evolution in birds. In: *Cytotaxonomy and Vertebrate Evolution*, edited by A. B. Chiarelli and E. Capanna. Academic Press, New York, pp. 425–481.

Schweizer, D. (1980). Simultaneous fluorescent staining of R bands an specific heterochromatic region (DA/DAPI-bands) in human chromosomes. *Cytogenet. Cell Genet.*, 27, 190–193.

Stefos, K., and Arrighi, F. E. (1974). Repetitive DNA of *Gallus domesticus* and its cytological location. *Exp. Cell Res.*, 83, 9–14.

Stock, A. D., Arrighi, F. E., and Stefos, K. (1974). Chromosome homology in birds: Banding patterns of the chromosomes of the domestic chicken, ring-necked dove and domestic pigeon. *Cytogenet. Cell Genet.*, 13, 410–418.

Sundick, R. S., Bloom, S. E., and Kite, J. H., Jr. (1973). Parabiosis between obese (OS) and normal strain chicken embryos. *Clin. Exp. Immunol.*, 14, 437–442.

Takagi, N., and Sasaki, M. (1974). A phylogenetic study of bird karyotypes. *Chromosoma* (Berlin), 46, 91–120.

Takagi, N., Itoh, M., and Sasaki, M. (1972). Chromosome studies in four species of Ratitae (aves). *Chromosoma* (Berlin), 36, 281–291.

Tone, M., Nakano, N., Takao, E., Narisawa, S., and Mizuno, S. (1982). Demonstration of W chromosome-specific repetitive DNA sequences in the domestic fowl, *Gallus domesticus*. *Chromosoma* (Berlin), 86, 551–569.

Tone, M., Sakaki, Y., Hashiguchi, T., and Mizuno, S. (1984). Genus specificity and extensive methylation of the W chromosome-specific repetitive DNA sequences from the domestic fowl, *G. gallus domesticus*. *Chromosoma* (Berlin), 89, 228–237.

Chapter 5

The Genetic Map of the Chicken and Availability of Genetically Diverse Stocks

J. James Bitgood

Summary. The chicken gene map has been under development since 1902, and a number of updates have been published since 1936. The majority of loci that have been assigned to linkage groups code for morphological traits, but some biochemical loci have been mapped. The most extensive chromosome maps are of the Z (sex) chromosome and chromosome 1. A few loci have been assigned to chromosomes 2, 6, and 17, and there are three major linkage groups that have yet to be assigned to a chromosome. Each of these known linkage groups and chromosome maps is discussed in this review. New technologies are identifying more loci, and these technologies have great promise for increasing our knowledge of the genetic maps of all avian species. The improved map will be of benefit to commercial breeders, avian genetic researchers, and other members of the biological research community.

Introduction

It has been almost 90 years since Bateson (1902) presented his first studies on inheritance in chickens. These studies demonstrated that Mendel's laws of inheritance apply to animals as well as plants. They might also be considered to be the first steps toward constructing a chicken gene map. Efforts to develop this map have been phasic, and it appears that the techniques of molecular biology are inspiring a new phase at the present time.

The chicken has a number of favorable attributes that should make it a good model for gene mapping, as enumerated below:

1. The chicken has a short generation time when compared to other species of economic importance, such as cattle or swine. However, when compared to species traditionally used for theoretical genetics, such as the fruit fly and the laboratory mouse, the generation time is lengthy.
2. The chicken has a small physical size, making it easy to handle and examine and resulting in a relatively small space requirement.
3. A large number of progeny can be generated in a short period of time because the chicken is oviparous. The ability to hatch a number of full sibs at the same time, and maintain them in the same or comparative environments during hatching, brooding, rearing, and adult periods of life, has made genetic studies in poultry informative.
4. Artificial insemination allows specific matings that are structured.

5. Identification of genetic variants at all stages of development is facilitated because the embryo is accessible as soon as the egg is laid.
6. A large number of genetic markers are available, although relatively few of these have been categorized.
7. Nucleated red blood cells can be used to obtain large amounts of DNA, eliminating the need to culture cells for DNA extraction.

In spite of these favorable characteristics and the economic importance of poultry, the chicken has not been recognized by many researchers and granting agencies as a good species for genetic studies, and the gene map is not well developed. With its experimental advantages, the chicken should be a good species to use as a vertebrate model with which to develop new mapping procedures.

The current chicken gene map has been developed primarily using classical test crosses. Molecular and biochemical techniques will aid the mapping effort considerably (see Chapter 19). Some reference populations are being developed for use in gene mapping. Results using these new procedures are just beginning to be seen. A large number of genes have been cloned, and many new single genes have been reported. A large number of these genes, however, have been reported at facilities where the research focus does not include gene mapping. These studies have provided a much more detailed understanding of the action of the particular genes, although they may provide no information regarding their location. As the techniques for handling the genes become more refined and more accessible to nonspecialized laboratories, the knowledge that is being generated can be applied to assist in developing the gene map. There are some cases in which this is beginning to happen, and additional allocations can be anticipated as more anchor loci are identified.

Cytogenetic principles have proven valuable in assigning genes to specific chromosome regions. Chromosome rearrangement break points segregate in a Mendelian manner, and gene loci can be mapped against these break points. Also, when the loci being tested lie in the interstitial segment (that region between the break point and the centromere), they can be mapped against the centromere, further refining the location of these loci. Single crossovers in the interstitial region will place the recombinants in unbalanced (duplication/ deficient) chromosomes. These unbalanced chromosomes create a lethal condition in the zygote, and the recombinants will not be recovered (Burnham, 1956). As fewer recombinants are recovered, the loci appear to be more tightly linked. For example, this procedure was used to map pea comb and blue egg, and recovered recombinants were reduced from the expected 4% to about 0.8% (Bitgood et al., 1980). A test of this type allows placement of the centromere in relation to the traits in question.

Although the emerging tools of molecular biology will probably yield additional information regarding the genome of poultry, the disappearance of genetically characterized stocks will reduce the opportunity to define their genetic diversity. More specialized lines may be lost over the next 5 to 10 years, further reducing the availability of diverse, well-characterized material with which to work in the future. Many laboratories work with eggs or chicks provided by commercial facilities, but there is little diversity within or between the lines from which these samples are derived. There are many lines of chickens that have been specifically selected based on various characteristics, be it metabolic characteristics, production characteristics, specific single-gene mutations, or in the case of the University of Minnesota and Ohio State University, chromosome rearrangements (Somes, 1988). Ohio State University has disposed of the one inversion and 12 translocation lines that were induced by Fechheimer and his colleagues. These unique lines were dropped due to a lack of funding (Fechheimer, personal communications). The University of Minnesota has most of their lines saved as frozen semen samples (Shoffner, personal communications). The lead time to recover these rearrangements for use is a minimum of 6 months, and would be available only in the heterozygous state for another generation or two.

Several poultry qualitative genetic researchers, as well as some highly regarded poultry quantitative geneticists who recognized the value of characterized genetic lines, have retired or will retire in the near future. Some of their lines have been distributed to other researchers for various purposes, but their interests do not usually include placement of the genes on the map. Lines that have been characterized for behavioral characteristics over a number of generations may not be available in the future. There does not seem to be much interest in replacing these highly productive scientists who created these specialized lines with poultry geneticists who will maintain them. Human resources, in the form of the poultry scientists with an interest in preserving these resources, who are equipped with the tools of molecular biology and with an appropriate budget, are not emerging in the traditional land-grant colleges and research institutes.

Identification and use of molecular markers that are closely linked to current morphological markers will help to maintain those mutations that cannot be perpetuated in pure breeding lines. These include a number of lethal mutations. Currently, potential carriers need to be progeny tested each generation. If closely linked markers were available, it would be much easier to maintain the lines, allowing more flexibility in utilization of resources. Every individual that hatched would not need to be reared. This would also provide additional markers to assist in mapping newly identified mutations. Several participating laboratories, each using selected, nonredundant markers, could analyze DNA samples from carriers of newly found mutations for mapping purposes. One test mating at one facility could be conducted, and appropriate blood samples could then be provided to these participating laboratories for analysis.

To maximize mapping efforts, scientists using various tools that are available to study avian genomes must collaborate and share data to the maximum extent possible. A central repository or data bank for the information that is being generated is needed. Because of the volume and negative nature of much of the data, it may not be possible to publish these data in paper format, and computerized networks may be required to disseminate this information among workers in the field.

The first chicken gene assigned to a chromosome was the barring gene, shown by Spillman (1908) to be sex linked. The first linkage map was published by Hutt (1936), followed by updated publications by Hutt and Lamoreux (1940), Hutt (1949, 1960, 1964), Etches and Hawes (1973), Somes (1973, 1978, 1987), and most recently by Bitgood and Somes (1990) and Somes and Bitgood (1990). Abbott and Yee (1975) presented a review of chicken genetic tests published to that time. This included a listing of negative linkage relationships. Bloom (1979), Etches and Hawes (1979), and Somes (1979) presented lists of different types of linkage associations noted to that date. Somes (1988) has published a directory of poultry genetic stocks, which includes a brief description of many mutations and a gene map. Bitgood and Somes (1990) and Somes and Bitgood (1990) have listings of loci that show independent segregation when tested against other loci and several chromosome rearrangement break points.

This chapter will review the development of the current chicken genetic map over the last 15 or 20 years. While limited with respect to the chromosomes mapped, the chicken gene map as it exists contains a series of anchor loci that can be used for further studies. In most cases, the inheritance and characterization of these loci is fairly well documented. There have been some recent assignments of genes to specific chromosomes, but these have not yet been mapped in relation to other loci on the linear map. These assignments will be noted.

The Z Chromosome

Until recently, the Z chromosome was the best mapped of the chicken chromosomes. Spillman (1908) first suggested the sex-linked nature of the inheritance of barring (B). This is impressive when one considers that only six years earlier, Bateson (1902) had reported his first studies on inheritance in the chicken. To comprehend the new science of genetics, recognize sex linkage, critically examine the evidence available to support this new concept of inheritance based on sex of the parent, and then publish this new information so rapidly was a credit to Spillman.

Probably the biggest change that has been made recently to the published map of the Z chromosome is its orientation (Figure 1). The map has been inverted (Somes, 1988) so that the cytogenetic and the classical genetic data

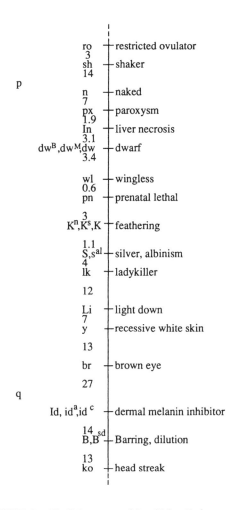

FIGURE 1. The linkage map of the chicken Z chromosome.

are in agreement, as suggested by Bitgood (1986). A series of studies using several translocations involving the Z chromosome has led to this reorientation, and they are discussed below.

Cytological studies involving the Z chromosome were conducted using the NM 7659 t(Z;1) translocation described by Kaelbling and Fecheimer (1983), who showed that the breakpoint of this translocation was on the short arm. The gene coding for silver plumage, S, was tested against this breakpoint and no recombinants were recovered, indicating that these markers are closely linked (J. J. Bitgood, unpublished results). By contrast, Bitgood (1985b) reported that dermal melanin inhibitor (*Id*) segregated independently from this break point, suggesting that *Id* is on the opposite arm from NM 7659

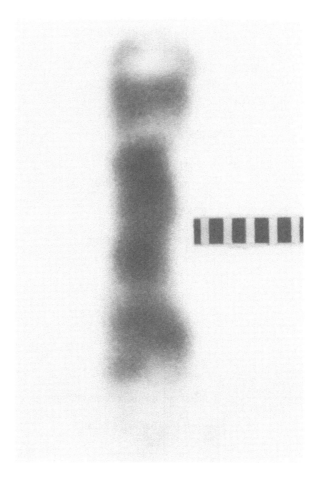

FIGURE 2.   The G-banded chicken Z chromosome, showing the G-band light-staining region on the long arm. The dotted line indicates the placement of the centromere.

t(Z;1). Bitgood et al. (1980) tested the breakpoint of another translocation, NM 7092 t(Z:1) (Zartman, 1973), in a test with S, and found close linkage (6.3 centimorgans, or cM). The normal Z chromosome shows a large G-band light-staining region, a good marker region, at the end of one of the arms (Figure 2). When chromosomes from birds carrying the NM 7092 t(Z;1) were G-banded, the large G-band light-staining region was shown to be on the intact long arm (Bitgood, 1980).

Wang and Shoffner (1974) had earlier shown that the MN t(Z;3) breakpoint was on the arm opposite to this same G-band light-staining region, i.e., on the same arm as the NM 7092 t(Z;1) breakpoint. Bitgood (1985b) tested this breakpoint against *S* and found no recombinants. This breakpoint was later tested against both *S* and *K* obtained from a Light Brahma line, and again no recombinants were recovered (Bitgood, 1986). This indicated that the breakpoint of MN t(Z;3) and the loci for *S* and *K* were on the short arm. This finding complemented two other reports. Kaelbling and Fechheimer (1983) studied the synaptonemal complex formed in the OH t(Z;micro) and found that the breakpoint was on the short arm. Telloni et al. (1976) had earlier reported that this breakpoint is linked to *K* by 23 cM.

There have been several other assignments of loci to the short arm of the Z chromosome. Bitgood and Whitley (1985) found linkage between pop eye (*pop*) and the breakpoint of NM 7092 t(Z;1) (Zartman, 1973), although a precise distance was not established. The linear order of *pop* and the NM 7092 t(Z;1) breakpoint in relation to other loci on this arm has not yet been tested. Congenital baldness (*ba*) was mapped 7.6 cM from *S* (Somes, 1989), but the direction was not established.

The finding by Bacon et al. (1988) that the integration site for the endogenous retroviral locus, *ev21*, and *K* are intimately related is of concern to the breeding industry. *K* is a dominant sex-linked allele that causes a slow rate of feather growth, noticeable at the time the chicks hatch. The recessive fast feathering allele, k^+, is the wild type. *K* is used by commercial poultry breeders to create autosexing lines of birds. If the dam line is *K* and the sire line is k^+, all the male chicks will be slow feathering and are easily identified at time of hatch by the much shorter primary flight feathers. The finding by Bacon et al. (1988) of a close association between *K* and *ev21* suggested that slow feathering may be a result of retroviral-induced insertional mutagenesis. While the ability to autosex the chicks at hatch time is an economic benefit, the breeders are concerned because *ev21* encodes an infectious virus that can be congenitally transmitted. This compromises the immune response against exogenous avian leukosis virus infection of progeny from *ev21* dams (Smith and Crittenden, 1988). Smith and colleagues at the USDA Avian Disease and Oncology Laboratory at East Lansing, Michigan, are analyzing the molecular arrangement of this region of the Z chromosome (Levin and Smith, 1990). It is unfortunate that stocks carrying the delayed feathering gene, K^N, and the slow feathering K^S alleles reported by Somes (1969) and McGibbon (1977), respectively, have been lost, because these alleles could have added a new dimension to these studies.

In summary, the markers that have been mapped recently to the short arm of the Z chromosome are MN t(Z;3), NM 7092 t(Z;1), NM 7659 t(Z;1), OH t(Z;micro), silver (*S*), late feathering (*K*), pop eye (*pop*), and congenital baldness (*ba*).

Recent studies of the genetic markers on the long arm of the Z chromosome have revealed that the breakpoint for the MN t(Z;1) (Wang et al., 1982) is linked to *B* by 22.3 cM. G-banding studies showed the break was very close y--(40)--Id--(14)--B

y--(17)--MN Z;1--(19)--B

FIGURE 3.   Diagram of the results of two matings testing the linkage relationships of recessive white skin (*y*), the MNt(Z;1) rearrangement break point, melanin (*Id*), and the barred feather pattern (*B*).

to, or in the large G-band light-staining region on the long arm (Bitgood, 1980). A study of synaptonemal complexes by Solari et al. (1988) confirmed that this breakpoint is on the long arm. As this is the only breakpoint that has been reported on the long arm, it is a unique resource and should be maintained.

In three-point test crosses, Bitgood (1988) established the relationships between recessive white skin (*y*), *Id*, and *B*, and also between *y*, MN t(Z;1), and *B* (Figure 3).

In summary, the loci that have been mapped to the long arm of the Z chromosome are MN t(Z;1), the large G-band light-staining region, barring (*B*), inhibitor of dermal melanin (*Id*), and recessive white skin (*y*). The placement of *y* on this arm is tentative. Of the loci on this arm that have been studied, *y* is closest to the centromere and may be on the short arm. The relationship of *Id* to the breakpoint of MN t(Z;1) is unknown. The actual linear relationship between these two markers has not been tested. If *Id* and *y* are in the interstitial segment, a mating using these loci can be constructed as described earlier to test them against the breakpoint. Due to the length of the Z chromosome (over 120 cM) shown in Figure 1, independent assortment between many loci on the two arms will be seen in test matings.

Chromosome 1

The current linkage map of chromosome 1 is shown in Figure 4. Intensive investigation aimed at developing this linkage map is underway in several laboratories. Zartman (1973) first assigned the pea comb locus (*P*) to this chromosome. He irradiated semen from a Cornish male carrying *P* and semen from a single-comb White Leghorn male. Using artificial insemination on chromosomally normal hens, two chromosome translocations involving the Z chromosome and chromosome 1 were recovered. Linkage tests then showed that the breakpoints of these translocations were linked with *P*. These translocations have been used by other investigators and have played a key role in delineating the chromosome 1 linkage map.

Bitgood et al. (1980) verified the assignment of *P* to chromosome 1, and also mapped the closely linked blue egg gene (*O*). This study indicated that these loci were both near the centromere on the short arm of chromosome 1. Bruckner and Hutt (1939) found only two recombinants among 35 individuals

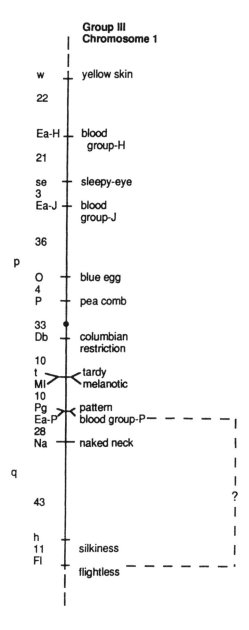

FIGURE 4. The linkage map of chicken chromosome 1. The assignment of the region marked by the question mark needs to be verified.

P--(33)(28)-⊢-Db------(17)----------Lg
Db--(10)--Ml--(12)--Lg
P-----------(46)------------Ml

FIGURE 5. Summary of results of several tests of linkage that involve overlapping regions on chromosome 1. The direction of the centromere has not been confirmed.

in their initial study that discovered linkage between the O and P loci. A later study refined the map distance between P and O to about 4 cM (Bitgood et al., 1983). Crawford (1986) calculated the distance to be 2.4 cM using 136 individuals. With close distances such as this, a large number of individuals needs to be examined to obtain sufficient recombinants for an accurate estimate of the map distance. As only sexually mature females can be studied to evaluate sex-limited traits such as blue egg, a good test is expensive and time-consuming.

Bitgood (1985a) used the NM 7659 t(Z;1) breakpoint to show that O was more distal than P from the centromere. The order is O, P, centromere, and the breakpoint. This seems to be the first nonacrocentric centromere mapped in any economically important animal species.

A study (Bitgood et al., 1987) involving more than 1000 chicks examined the linkage relationship between P and the tardy feathering gene (t) (Warren, 1933). Their results verified the finding of McGibbon and Monro, as cited by Etches and Hawes (1973), of a distance of about 41 cM between these loci. Bitgood et al. (1987) also examined the order of the P and t loci relative to the centromere using the MN t(Z;1). If the breakpoint is between P and t, linkage should have been seen between the breakpoint and t, since P and the breakpoint show linkage (Bitgood et al., 1980). No linkage was seen, so it appears that the order is MN t(Z;1), P, and t (Figure 4).

Bitgood et al. (1991) confirmed the report of Briles et al. (1967) showing that the loci for erythrocyte alloantigen P (Ea-P) and naked neck (Na) are linked on chromosome 1. A total of 467 individuals has been examined, and linkage of about 29 cM has been calculated. This same report suggests that Ea-I and P may be linked by 33 cM. Also, t may be linked with Ea-D by 38 cM, and Ea-D and the NM 7092 t(Z;1) breakpoint may be linked by 31 cM. No linkage was noted between either P and NM 7092 t(Z;1), or P and Ea-D. Based on the analysis of a three-point test cross, the probable order is NM 7092 t(Z;1), Ea-D, and t. Based on the previously discussed linkage of P and t, t is probably the closest of the three markers to the centromere.

Within chromosome 1, the linkage relationships between P, dark brown columbian restriction (Db), melanotic (Ml), and patterning (Pg) can be consolidated (Figure 5). Hertwig (1933) showed linkage of about 33 cM between marbling (ma) and P. Washburn and Smyth (1967) reported that the map distance between Db and P is 28 cM, and since Carefoot (1988) and Smyth (personal communication) have indicated that *ma* was most likely *Db* in an *S* background, the original map distance of 33 cM has been confirmed. Moore and Smyth (1972a) reported that 17 cM separated *Db* and autosomal barring (*Ab*). Carefoot (1985, 1986) stated that *Ab*, lacing (*Lg*), and pencilling (*Pg*) are all the result of the same gene, but expression depends on other background genes. He suggested retention of the symbol *Pg* for pattern. Carefoot (1987a, 1987b, 1990) also reported that *Db* and *Ml* are linked by 10 cM (1987a), and *P* and *Ml* are linked by 46 cM. *Ml* and *Lg* (now *Pg*) are linked by 10 cM (Moore and Smyth, 1972b) and this relationship was confirmed by Campo and Alvarez (1991), who reported linkage of 12 cM. In summary, the genetic map of this section of chromosome 1 is shown in Figure 5.

The assignment of the complex illustrated in Figure 5 to chromosome 1 is tentative and based on the suggestion by Hertwig (1933) that *ma* is linked to naked neck (*Na*) by approximately 46 cM, a long distance for linkage mapping. Because of the long distance between *ma* and *Na*, the complex is shown on the long arm on currently published maps. However, recent investigations of the linkage between *Na* and *Ea-P* and several markers, including several chromosome rearrangements, have not confirmed linkage of either locus to this chromosome (Bitgood, unpublished data; Shoffner, personal communication). For example, no evidence for linkage was found in 87 individuals from a cross designed to investigate the map distance of 38 cM between *t* and *Na* inferred from the map.

Crawford (1986) reported a plumage color mutation closely linked to *P* by 0.3 cM and lying between *P* and the centromere. Initially thought to be *Ml*, it was later determined that this was most likely a different recessive mutation and was named charcoal (*cha*) (Carefoot, 1990).

Other approaches are contributing to the gene map of chromosome 1. Shaw et al. (1990) used *in situ* hybridization to assign the chicken growth hormone gene to the long arm of chromosome 1. No studies of linkage with other mapped loci have been conducted. More recently, Tereba et al. (1991) used *in situ* hybridization to show that the henny feathering gene (*Hf*) is apparently located on the proximal one-third of the long arm of chromosome 1. An earlier study (Somes et al., 1984) failed to find any linkage of *Hf* with *P*, sleepy eye (*se*), or *Na*, all reportedly on chromosome 1 (Figure 4).

Ponce de Leon (personal communication) and Hutchison (personal communication) also used *in situ* hybridization techniques for mapping studies. Working independently, they found that the *ev1* locus is on the short arm, close to the centromere. A separate study, using a combination of classical mapping and molecular techniques, suggests that *ev1* is within 10 cM of *P* (Bitgood and Crittenden, personal observations). These results are all different from those reported by Tereba and Astrin (1980), which indicated that the *ev1* locus is near the middle of the long arm. Clearly, additional data are required to confirm the location of *ev1*.

Using *in situ* hybridization, Hutchison and LeCiel (1991) have reported that the red blood cell-specific histone H5 gene and the 21-gene histone cluster are located approximately in the center of the short arm of chromosome 1.

Chromosome 2

While several loci have been mapped on the Z chromosome and chromosome 1, only a few loci have been assigned to other chromosomes. Langhorst and Fecheimer (1985) reported that the shankless mutation segregated with a pericentric inversion of chromosome 2, indicating that these markers are linked. Subsequent studies in three different laboratories have shown that this linkage is very close, since no recombinants have been recovered (Fechheimer, personal communication; Shoffner, personal communication; Bitgood, personal observations).

Several genes have been located on chromosome 2 by *in situ* hybridization. Shaw et al. (1989) reported that the β-actin gene was either on chromosome 2 or on one of the microchromosomes in the size range 9 – 12. Since two sites were labeled, they may be closely related members of the actin gene family. Hutchison and LeCiel (1991) reported that ovalbumin and two closely linked genes, *X* and *Y*, were on the long arm of chromosome 2, confirming the results of Hughes et al. (1979), who had suggested previously that ovalbumin was on either chromosome 2 or 3. Durand and Merat (1982) reported that the loci for ovoglobulin G_3 and ovalbumin were linked by 7 cM and, therefore, the gene encoding ovoglobulin G_3 must be on chromosome 2. The linear order in relation to other markers is not known. Other loci reported to be on chromosome 2 are *ev2* (Tereba et al., 1981), two oncogenes (Symonds et al., 1984, 1986), and possibly the β-globin genes (Hughes et al., 1979). No linear relationships have been reported among any of these genes.

Chromosome 6

Kao (1973) used somatic cell hybridization techniques to map to chromosome 6 a gene involved in adenine synthesis that was later identified as phosphoribosyl pyrophosphate amidotransferase (*PPAT*) (Palmer and Jones, 1986). The loci for phosphoglucomutase (*PGM-2*), serum albumin (*Alb*), and vitamin D-binding protein (*Gc*) were also mapped to chromosome 6 using chicken and Chinese hamster somatic cell hybrids (Palmer and Jones, 1986). Since the linear order cannot be determined by this method, the relative locations of these genes are not available.

Chromosome 17

The nuclear organizer region (NOR) and the major histocompatibility complex (MHC) have been assigned to chromosome 17 (Bloom and Bacon, 1985;

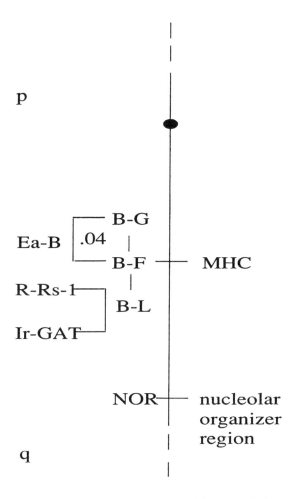

FIGURE 6. The linkage map of the chromosome containing the nucleolar organizer region and the major histocompatibility complex of the chicken. This is presumably chromosome 17.

Auer et al., 1987) (Figure 6). The NOR and MHC are discussed in Chapters 4 and 12, respectively.

Linkage Groups Unassigned to Chromosomes

Linkage groups I, II, and IV, as published by Hutt (1964), remain similar (Figure 7). Another allele, dun (I^D) has been identified at the *I* locus in linkage group II (Ziehl and Hollander, 1987), and a second mutant allele has been identified at the duplex locus (*D*) in linkage group IV (Somes, 1986).

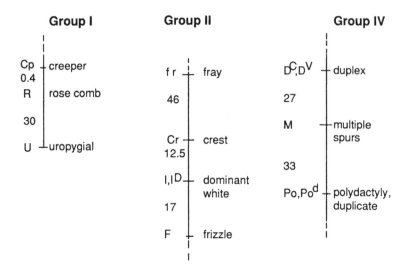

FIGURE 7. Maps of linkage groups I, II, and IV of the chicken. These have not yet been assigned to a specific chromosome.

The Future of the Map

In the last comprehensive listing of chicken genes coding for morphological traits, blood groups, enzymes, and structural proteins, Somes (1980) listed more than 250 genes that had been reported since the early 1900s. Bulfield (1990) listed about 125 cloned genes, although the relationship between these cloned genes and previously reported mutant phenotypes has not been established in most cases.

Benefits of a more complete gene map range from use at the DNA level for a more complete understanding of gene action and interactions to the use by commercial breeders in marker-assisted selection (MAS). Once procedures are simplified and marked regions of chromosomes are identified, it will be much easier to utilize MAS.

Reference mapping populations are being created by crossing diverse lines and then backcrossing to one or both parental lines (Crittenden and Abplanalp, personal communications). The parental lines are expected to have a high degree of DNA polymorphism. Blood samples or DNA samples are being preserved for future genetic analysis. We are preserving blood samples from a large test cross using a series of markers reportedly on chromosome 1. Grandparent, parent, and test cross progeny are being sampled. As DNA markers are identified, these samples can be screened to determine if linkages can be established without the need to conduct further laborious and time-consuming matings.

Identification of molecular markers that are closely linked to known morphological traits will allow rapid selection of carriers at hatch. Fewer resources will then be required to maintain mutations in populations, particularly in the case of lethal mutations. The development of these types of markers will also assist in mapping the genome.

There are a number of laboratories that are cloning and investigating chicken genes. As these clones become available, procedures such as *in situ* hybridization will allow physical mapping of these genes. Other procedures will be needed to place these genes in the genetic map, which is more difficult if there are no polymorphisms at that locus.

It must also be kept in mind that linkage information found in other species, including humans, can provide direction for mapping avian genomes and vice versa, since segments of the genome appear to be conserved throughout evolution. For example, Palmer and Jones (1986) indicated that the linkage group on chicken chromosome 6 (*PPAT*, *PGM-2*, *Alb*, and *Gc*) is similar to a linkage group found in mammals. *PGM-1* and *Alb* are on chromosome 5 in the mouse, and *PPAT*, *PGM-2*, *Alb*, and *Gc* are on chromosome 4 in humans. Therefore, this syntenic group has been conserved for over 300 million years.

References

Abbott, U. K., and Yee, G. W. (1975). Avian genetics. In: *Handbook of Genetics*, edited by R. C. King, Plenum Press, New York, pp. 151–200.

Auer, H., Mayr, B., Lambrou, M., and Schleger, W. (1987). An extended chicken karyotype including the NOR chromosome. *Cytogenet. Cell Genet.*, 45, 218–221.

Bacon, L. D., Smith, E., Crittenden, L. B., and Havenstein, G. B. (1988). Association of the slow-feathering (*K*) and an endogenous viral (*ev21*) gene on the Z chromosome of chickens. *Poultry Sci.*, 67, 191–197.

Bateson, W. (1902). Experiments with poultry. *Rep. Evolution Committee Roy. Soc.*, 1, 87–124.

Bitgood, J. J. (1980). Gametogenesis and zygote formation in domestic fowl (*Gallus domesticus*) with chromosome rearrangements. Ph.D. thesis, University of Minnesota, St. Paul, MN.

Bitgood, J. J. (1985a). Locating pea comb and blue egg in relation to the centromere of chromosome 1 in the chicken. *Poultry Sci.*, 64, 1411–1414.

Bitgood, J. J. (1985b). Additional linkage relationships within the Z chromosome of the chicken. *Poultry Sci.*, 64, 2234–2238.

Bitgood, J. J. (1986). Refining the linkage map of the chicken Z chromosome. In: *Proceedings, 3rd World Congress on Genetics Applied to Animal Production. X. Breeding Programs for Swine, Poultry and Fish*, pp. 289–292.

Bitgood, J. J. (1988). Linear relationship of the loci for barring, dermal melanin inhibitor, and recessive white skin on the chicken Z chromosome. *Poultry Sci.*, 67, 530–533.

Bitgood, J. J., Kendall, R. L., Briles, R. W., and Briles, W. E. (1991). Erythrocyte alloantigen loci *Ea-D* and *Ea-I* map to chromosome 1 in the chicken. *Animal Genet.*, 22, 449–454.

Bitgood, J. J., Kiorpes, C. A., and Arias, J. A. (1987). Tardy feathering locus (*t*) located on chromosome 1 in the chicken. *J. Heredity*, 78, 329–330.

Bitgood, J. J., Otis, J. S., and Shoffner, R. N. (1983). Refined linkage value for pea comb and blue egg: Lack of effect of pea comb, blue egg, and naked neck on age at first egg in the domestic fowl. *Poultry Sci.*, 59, 1686–1693.

Bitgood, J. J., Shoffner, R. N., Otis, J. S., and Briles, W. E. (1980). Mapping of the genes for pea comb, blue egg, barring, silver, and blood groups A, E, H, and P in the domestic fowl. *Poultry Sci.*, 59, 1686–1693.

Bitgood, J. J., and Somes, R. G., Jr. (1990). Linkage relationships and gene mapping. In: *Poultry Breeding and Genetics*, edited by R. D. Crawford. Elsevier Science Publishers, Amsterdam, pp. 469–495.

Bitgood, J. J., and Whitley, R. D. (1985). Pop-eye: An inherited Z-linked keratoglobus in the chicken. *J. Heredity*, 77, 123–125.

Bloom, S. E. (1979). Linkage relationships: Chicken. In: *Inbred and Genetically Defined Strains of Laboratory Animals. Part 2. Hamster, Guinea Pig, Rabbit, and Chicken*, edited by P. L. Altman, Federation of American Societies for Experimental Biology, Bethesda, MD, pp. 620–622.

Bloom, S. E., and Bacon, L. D. (1985). Linkage of the major histocompatibility (*B*) complex and the nucleolar organizer in the chicken. *J. Heredity*, 76, 146–154.

Briles, W. E., Briles, R. W., and Shoffner, R. N. (1967). Tests for linkage between ten blood group loci and the loci of seven morphological traits. *Poultry Sci.*, 46, 1791 (abstract).

Bruckner, J. H., and Hutt, F. B. (1939). Linkage of pea comb and blue egg in the fowl. *Science*, 90, 88–89.

Bulfield, G. (1990). Molecular genetics. In: *Poultry Breeding and Genetics*, edited by R. D. Crawford. Elsevier Science Publishers, Amsterdam, pp. 543–584.

Burnham, C. R. (1956). Chromosomal interchanges in plants. *Botanical Rev.*, 22, 419–552.

Campo, J. L., and Alvarez, C. (1991). Further study on the plumage pattern of the Blue Andalusian breed. *Poultry Sci.*, 70, 1–5.

Carefoot, W. C. (1985). Effect of the eumelanin restrictor *Db* on plumage pattern phenotypes of the domestic fowl. *Br. Poultry Sci.*, 26, 409–412.

Carefoot, W. C. (1986). Pencilled and double-laced plumage pattern phenotypes in the domestic fowl. *Br. Poultry Sci.*, 27, 431–433.

Carefoot, W. C. (1987a). Test for linkage between the eumelanin restrictor (*Db*) and the eumelanin extension (*Ml*) genes in the domestic fowl. *Br. Poultry Sci.*, 28, 69–73.

Carefoot, W. C. (1987b). Relative positions of the loci of the peacomb (*P*), eumelanin restrictor (*Db*), eumelanin extension (*Ml*) and plumage pattern (*Pg*) genes of the domestic fowl. *Br. Poultry Sci.*, 28, 347–350.

Carefoot, W. C. (1988). Inheritance of the spangled plumage pattern and the marbled chick down phenotypes of the Silver-Spangled Hamburgh bantam. *Br. Poultry Sci.*, 29, 785–790.

Carefoot, W. C. (1990). Test for linkage between the eumelanin dilution blue (*Bl*), the extended black (*E*) allele at the E-locus and the linked pea comb (*P*) and eumelanin extension (*Ml*) genes in the domestic fowl. *Br. Poultry Sci.*, 31, 465–472.

Crawford, R. D. (1986). Linkage between pea comb and melanotic plumage loci in chickens. *Poultry Sci.*, 65, 1859–1862.

Durand, L., and Merat, P. (1982). Frequence et comparaison des performances pour las genotypes a quatre loci affectant des proteines de l'oeuf chez la poule. *Annales de Genetiques et Selectionne d'Animales*, 14, 49–66.

Etches, R. J., and Hawes, R. O. (1973). A summary of linkage relationships and a revised linkage map of the chicken. *Can. J. Genet. Cytol.*, 15, 553–570.

Etches, R. J., and Hawes, R. O. (1979). Linkage relationships: Chicken. In: *Inbred and Genetically Defined Strains of Laboratory Animals. Part 2. Hamster, Guinea Pig, Rabbit, and Chicken*, edited by P. L. Altman. Federation of American Societies for Experimental Biology, Bethesda, MD, pp. 628–637.

Hertwig, P. (1933). Geschlechtsgebundene und autosomale Koppelungen bei Huhnem. *Verhandlungen der Deutschen Zoologischen Gesellschaft*, 112–118.

Hughes, S., Stubblefield, E., Payvar, F., Engel, J., Dodgson, J., Spector, D., Cordell, B., Schimkel, R., and Varmus, H. (1979). Gene localization by fractionation: Globin genes are on at least two chromosomes and three estrogen-inducible genes are on three chromosomes. *Proc. Natl. Acad. Sci. USA*, 76, 1348–1352.

Hutchison, N. J., and LeCiel, C. (1991). Gene mapping in chickens via fluorescent in situ hybridization to mitotic and meiotic chromosomes. *J. Cell. Biochem.* (Suppl. 15E), 205.

Hutt, F. B. (1936). Genetics of the fowl. VI. A tentative chromosome map. *Neue Forschungen in Tierzucht und Abstammungslehre (Duerst Festschrift)*, 105–112.

Hutt, F. B. (1949). *Genetics of the Fowl*. McGraw-Hill Book Co., New York.

Hutt, F. B. (1960). New loci in the sex chromosome of the fowl. *Heredity*, 15, 97–110.

Hutt, F. B. (1964). *Animal Genetics*. Ronald Press Co., New York.

Hutt, F. B., and Lamoreux, W. F. (1940). Genetics of the fowl. *J. Heredity*, 31, 231–235.

Kaelbling, M., and Fechheimer, N. S. (1983). Synaptonemal complex analysis of chromosome rearrangements in domestic fowl, *Gallus domesticus. Cytogenet. Cell Genet.* 36, 567–572.

Kao, F. (1973). Identification of chick chromosomes in cell hybrids formed between chick erythrocytes and adenine requiring mutants of Chinese hamster cells. *Proc. Natl. Acad. Sci. USA*, 70, 2893–2898.

Langhorst, L. J., and Fechheimer, N. S. (1985). Shankless, a new mutation on chromosome 2 in the chicken. *J. Heredity*, 76, 182–186.

Levin, I., and Smith, E. J. (1990). Molecular analysis of endogenous virus *ev21*-slow feathering complex of chickens. 1. Cloning of proviral-cell junction fragment and unoccupied site. *Poultry Sci.*, 69, 2017–2026.

McGibbon, W. H. (1977). A sex-linked mutation affecting rate of feathering in chickens. *Poultry Sci.*, 56, 872–875.

Moore, J. W., and Smyth, J. R., Jr. (1972a). Genetic factors associated with the plumage pattern of the barred Fayoumi. *Poultry Sci.*, 51, 1149–1156.

Moore, J. W., and Smyth, J. R., Jr. (1972b). Inheritance of the silver-laced Wyandotte plumage pattern. *J. Heredity*, 63, 179–184.

Olsen, M. W. (1965). Twelve year summary of selection for parthenogenesis in Beltsville Small White turkeys. *Br. Poultry Sci.*, 6, 1–6.

Palmer, D. K., and Jones, C. (1986). Gene mapping in chicken-Chinese hamster somatic cell hybrids. *J. Heredity*, 77, 106–108.

Shaw, E. M., Guise, K. S., and Shoffner, R. N. (1989). Chromosomal localization of chicken sequences homologous to the β-actin gene by in situ hybridization. *J. Heredity*, 80, 475–478.

Shaw, E. M., Shoffner, R. N., Foster, D. N., and Guise, K. S. (1990). Mapping growth hormone gene on chicken chromosomes by in situ hybridization. *Poultry Sci.* 69 (Suppl. 1), 122.

Smith, E. J., and Crittenden, L. B. (1988). Genetic cellular resistance to subgroup E avian leukosis virus in slow-feathering dams reduces congenital transmission of an endogenous retrovirus encoded at locus *ev21. Poultry Sci.*, 67, 1668–1673.

Solari, A. J., Fechheimer, N. S., and Bitgood, J. J. (1988). Pairing of ZW gonosomes and the localized recombination nodule in two Z-autosome translocations in *Gallus domesticus. Cytogenet. Cell Genet.*, 48, 130–136.

Somes, R. G., Jr. (1969). Delayed feathering, a third allele at the *K* locus of the domestic fowl. *J. Heredity*, 60, 281–286.

Somes, R. G., Jr. (1973). Linkage relationships in domestic fowl. *J. Heredity*, 64, 217–221.

Somes, R. G., Jr. (1978). New linkage groups and revised chromosome map of the domestic fowl. *J. Heredity*, 69, 401–403.

Somes, R. G., Jr. (1979). Linkage relationships: Chicken. In: *Inbred and Genetically Defined Strains of Laboratory Animals. Part 2. Hamster, Guinea Pig, Rabbit, and Chicken*, edited by P. L. Altman. Federation of American Societies for Experimental Biology, Bethesda, MD, pp. 622–628.

Somes, R. G., Jr. (1980). Alphabetical list of the genes of domestic fowl. *J. Heredity*, 71, 168–174.

Somes, R. G., Jr. (1986). Multiple alleles at the duplex comb locus of the domestic chicken. *Poultry Sci.*, 65 (Suppl. 1), 128.

Somes, R. G., Jr. (1987). Linked loci of the chicken — *Gallus gallus (G. domesticus)*. In: *Genetic Maps 1987. A Compilation of Linkage and Reconstruction Maps of Genetically Studied Organisms*, Vol. 4, edited by S. J. O'Brien. Cold Spring Harbor Press, Cold Spring Harbor, NY, pp. 422–429.

Somes, R. G., Jr. (1988). International Registry of Poultry Genetic Stocks. *Storrs Agricultural Experiment Station Bulletin 476.*

Somes, R. G., Jr. (1989). Hereditary congenital baldness: A sex-linked trait in the domestic fowl. *J. Heredity*, 80, 493–496.

Somes, R. G., Jr., and Bitgood, J. J. (1990). Gene map of the chicken (*Gallus gallus* or *G. domesticus*). In: *Genetic Maps — Locus Maps of Complex Organisms*, Vol. 5, edited by S. J. O'Brien. Cold Spring Harbor Press, Cold Spring Harbor, NY, pp. 4.169–4.175.

Somes, R. G., Jr., George, F. W., Baron, J., Noble, J. F., and Wilson, J. D. (1984). Inheritance of the henny-feathering trait of the Sebright bantam chicken. *J. Heredity*, 75, 99–102.

Spillman, W. J. (1908). Spurious allelomorphism results of some recent investigations. *Am. Naturalist*, 42, 610–615.

Symonds, G., Stubblefield, E., Guyaux, M., and Bishop, J. M. (1984). Cellular oncogenes (c-*erb*-A and c-*erb*-B) located on different chicken chromosomes can be transduced into the same retroviral genome. *Molecular and Cellular Biol.*, 4, 1627–1630.

Symonds, G., Quintrell, N., Stubblefield, E., and Bishop, J. M. (1986). Dispersed chromosomal localization of the proto-oncogenes transduced into the genome of Mill Hill 2 or E26 leukemia virus. *J. Virol.*, 59, 172–175.

Telloni, R. V., Jaap, R. G., and Fechheimer, N. S. (1976). Cytogenetic and phenotypic effects of a chromosomal rearrangement involving the Z-chromosome and a micro-chromosome in the chicken. *Poultry Sci.*, 55, 1886–1896.

Tereba, A., and Astrin, S. M. (1980). Chromosomal localization of *ev-1*, a frequently occurring endogenous retrovirus locus in White Leghorn chickens, by in situ hybridization. *J. Virol.*, 35, 888–894.

Tereba, A., Crittenden, L. B., and Astrin, S. M. (1981). Chromosomal localization of 3 endogenous retrovirus loci associated with virus production in White Leghorn chickens. *J. Virol.*, 39, 282–289.

Tereba, A., McPhaul, M. J., and Wilson, J. D. (1991). The gene for aromatase ($P450_{arom}$) in the chicken is located on the long arm of chromosome 1. *J. Heredity*, 82, 80–81.

Wang, N., and Shoffner, R. N. (1974). Trypsin G- and C-banding for interchange analysis and sex identification in the chicken. *Chromosoma* (Berlin), 47, 61–69.

Wang, N., Shoffner, R. N., Otis, J. S., and Cheng, K. M. (1982). The induction of chromosomal structural changes in male chickens by the alkylating agents triethylene melamine and ethyl methanesulfonate. *Mutation Res.*, 96, 53–66.

Warren, D. C. (1933). Retarded feathering in the fowl. A new factor affecting manner of feathering. *J. Heredity,* 24, 430–434.

Washburn, K. W., and Smyth, J. R., Jr. (1967). A gene for partial feather achromatosis in the fowl. *J. Heredity,* 58, 131–134.

Zartman, D. L. (1973). Location of the pea comb gene. *Poultry Sci.,* 52, 1455–1462.

Ziehl, M. A., and Hollander, W. F. (1987). Dun, a new plumage-color mutant at the *I*-locus in the fowl *(Gallus domesticus). Iowa State J. Res.,* 62, 337–342.

Chapter 6

Accessing the Genome of the Chicken Using Germline Chimeras

James N. Petitte, Cynthia L. Brazolot, Mary Ellen Clark,
Guodong Liu, Ann M. Verrinder Gibbins, and Robert J. Etches

Summary. Somatic and germline chimeras can be formed in the chicken by transferring stage X (E.G&K) blastodermal cells recovered from the unincubated embryo into recipients at the same stage of development. The frequency of somatic chimerism appears to be dependent on the genotype, the source of cells within the donor embryo, and the degree of developmental synchronization between the donor and the recipient. Survival of chimeric embryos during development appears to be related to the genotype of the donor/recipient combination. Dispersed blastodermal cells can be cultured, transfected, or frozen prior to their introduction into the recipient and, therefore, are potential candidates for experiments and techniques that require access to the germline. Applications of these techniques could include the production of transgenic chickens and the crypopreservation of the genome. Data supporting the feasibility of each of the manipulations required to achieve these applications are presented.

Introduction

During the past few years, considerable interest has emerged regarding manipulation of the avian genome as a tool for experimental biologists and as a means of facilitating the genetic improvement of commercial poultry. Much of the attraction in developing transgenic birds lies in the goal of modifying the genome in a specific, site-directed manner so that the function of targeted changes to the genome may be more predictable. At the present time, however, mice remain the only species in which site-directed modifications of the genome have been successful. This has come about through the ability to establish pluripotent embryonic stem cell lines (Evans and Kaufman, 1981; Martin, 1981), the ability to develop constructs capable of undergoing homologous recombination coupled with selection *in vitro* for the recombinant of interest (Thomas and Capecchi, 1987), and the ability to inject embryonic stem cells into the blastocyst to produce germline chimeras (Bradley et al., 1984) (see the discussion in Chapter 19). In this chapter we will discuss the application of a similar technical scheme for the production of transgenic chickens to facilitate specific modifications to the avian genome, and describe current work on the production of avian germline chimeras.

Primordial Germ Cells and Germline Chimeras

Primordial germ cells (PGCs) are the progenitors of egg and sperm cells and, therefore, are potential target cells to mediate the introduction of desired modifications to the genome. In birds, PGCs arise from the epiblast and migrate to the hypoblast during the first hours of incubation (Eyal-Giladi et al., 1981; Sutasurya et al., 1983; Urven et al., 1988; see also Chapter 3). During the formation of the primitive streak and the onset of gastrulation, PGCs move anteriorly with the hypoblast and reside in the extraembryonic area referred to as the germinal crescent (Swift, 1914; Ginsburg and Eyal-Giladi, 1986). Concomitant with the formation of the extraembryonic and embryonic vasculature, PGCs become localized in vascular elements and can be found in blood samples up to about 65 h of incubation. Thereafter, the number of circulating PGCs declines rapidly as the cells begin actively to migrate to the presumptive gonad due to chemotaxic attraction (Singh and Meyer, 1967; Swartz and Domm, 1972). This developmental history of PGCs immediately suggests that the isolation and genetic modification of these cells or their precursors, coupled with their return to a host embryo, could facilitate the insertion of novel genes into the germline.

Primordial germ cells appear to have a strong attraction to the gonadal anlage, and PGCs from quail, turkey, and even mouse will respond to the chemotaxic stimulus produced by a chick embryo (Reynaud, 1969; Rogulska et al., 1971; Didier and Fargeix, 1976). Hence, it is not surprising that several intraspecific embryonic chimeras have been developed by the transfer of primordial germ cells. In this regard, Reynaud (1969, 1976) must be credited with demonstrating the extragonadal origin of PGCs as well as the first successful inter- and intraspecific transfer of PGCs *in ovo*. Initially, this was accomplished by injecting a mixture of cells from the germinal crescent, which contained PGCs, into the extraembryonic vasculature of recipient chick embryos at 3–5 days of incubation that had been sterilized by exposure to ultraviolet irradiation. Likewise, Shuman (1981) transferred PGCs isolated from the germinal crescent of chicks to chick embryos of a similar stage and, using a chromosomal marker, identified donor cells within the gonad of the host. Recently, an endogenous retroviral marker was used to demonstrate the presence of viral DNA sequences in gonadal tissue after the injection of donor embryonic blood into virus-free chick embryos (Simkiss et al., 1989). These reports, coupled with studies in which quail/chick chimeras were constructed surgically to examine the interrelationship between germ cells and gonadal differentiation (Tachinante, 1974; Didier and Fargeix, 1976; Didier 1977; Dieterlen-Lievre et al., 1985; Hajji et al., 1988), indicate that exogenous PGCs can readily colonize the embryonic gonad.

Although the experiments described above provide convincing evidence that exogenous PGCs will colonize the gonad of a host embryo, it has been relatively difficult to demonstrate that exogenous PGCs will give rise to functional gametes within a chimeric gonad. Reciprocal injections of PGCs from White Wyandottes and Rhode Island Reds were conducted by Reynaud (1976) in an attempt to demonstrate that functional gametes could be derived from the donor-derived PGCs. This combination of breeds provided feather pigmentation markers at the autosomal color (*C*) locus [White Wyandottes can be recessive white (*cc*) and Rhode Island Reds are colored (*CC*)]; the sex-linked silver locus [White Wyandottes can be silver (*SS* or *S⁻*) and Rhode Island Reds are gold (s^+s^+ or s^{+-})] and a morphological marker at the rose comb locus [White Wyandottes are rose combed (*RR*) and Rhode Island Reds are single combed (r^+r^+)]. Following injection of PGCs from Rhode Island Red embryos into White Wyandotte recipients sterilized by exposure to ultraviolet irradiation, two phenotypically White Wyandotte females hatched. These hens were reared to sexual maturity, mated to Rhode Island Red males, and two single-combed chicks with brown down were obtained. Although the *in ovo* manipulations employed by Reynaud (1976) resulted in poor egg production and fertility of the recipient, the two phenotypically Rhode Island Red offspring indicate that functional germ cells can be derived from transferred PGCs.

Recently, blood containing PGCs from Dwarf White Leghorn (DWL) embryos, which were homozygous dominant for *I* at the locus for "dominant white," was transferred into normal, nonirradiated Barred Plymouth Rock (BPR) embryos at 53, 72, or 96 h of incubation (Petitte et al., 1991). Since Barred Plymouth Rocks are homozygous for the recessive allele (*i*) of "dominant white," the presence of the Dwarf White Leghorn genotype in the gametes was tested by breeding the putative chimeras with Barred Plymouth Rocks. Incorporation of the Dwarf White Leghorn PGCs into the Barred Plymouth Rock recipient would be indicated in this cross by white BPR × DWL chicks at hatch, but none were observed among the 3117 offspring from 59 putative chimeras. The absence of BPR × DWL chicks may have been due to the transfer of insufficient numbers of PGCs (although the estimate of the mean number of PGCs that were transferred was 3.5) and/or to abnormal development of the donor germ cells in the gonad of the recipient (Petitte et al., 1991).

The transfer of PGCs between individuals marked by single gene differences in plumage color has been reported in the quail (Wentworth et al., 1989). These chimeric quail produced functional donor-derived gametes, but the details of the techniques used to prepare the donor PGCs and the recipients were not reported. The production of the donor-derived gametes by the recipient declined after the onset of sexual maturity; the extent of this decline and associated changes in gonadal morphology were not reported.

Interspecific transfer of tissue containing PGCs has been used to demonstrate that exogenous PGCs can colonize the gonad of the recipient

(Tachinante, 1974; Didier and Fargeix, 1976; Didier 1977; Dieterlen-Lievre et al., 1985; Hajji et al., 1988), and the transfer of isolated turkey PGCs into chicken embryos that were compromised by exposure to ultraviolet light produced adult males containing spermatozoa that morphologically resembled those of turkeys (Reynaud, 1976). Despite their normal appearance, however, these spermatozoa were incapable of fertilizing either chicken or turkey ova (Reynaud, 1976). Again, the technical difficulty of sterilizing the recipient embryo may have altered gonadal development sufficiently to render the spermatozoa infertile, although it is also possible that maturation of turkey spermatozoa in a chicken testis cannot be completed because essential physiological interactions within and between cells might be absent in the chimeric testis. Attempts have also been made to populate the sterile gonad of the quail-chicken hybrid with PGCs from quail (Wentworth et al., 1989), but these attempts have not been successful to date. It would appear, therefore, that while it may be possible to transfer functional primordial germ cells both within and between gallinaceous species, the technology of accomplishing this feat requires further development. Methods must be developed to concentrate PGCs isolated from blood or the germinal crescent, and for sterilization of the host embryo without compromising its development.

Blastodermal Cell Transfer and Germline Chimeras

An alternative approach to accessing the genome of chickens is to manipulate the progenitors of the cells that eventually give rise to the PGCs. The location of these cells during early development is unknown, but they are presumed to reside in the stage X (E.G&K; Eyal-Giladi and Kochav, 1976) embryo (Ginsburg and Eyal-Giladi, 1987). In a series of experiments designed to assess this approach, chimeras were produced that contained donor-derived cells in both the germline and somatic tissues (Petitte et al., 1990).

The development of the chick embryo from fertilization to early primitive streak formation has been classified into a series of well-defined stages by Eyal-Giladi and Kochav (1976). Using this scheme, the freshly laid egg often contains a stage X (E.G&K) embryo. At this stage of development, the blastoderm is composed of about 40,000–60,000 cells, which can be divided into two distinct regions, the area opaca, which is a peripheral ring of cells attached to the yolk, and the area pellucida, which encompasses the center of the embryo and is distinguished from the area opaca by its transparency when viewed from above. Within the first few hours of incubation [stages XI–XIII (E.G&K)], the area pellucida differentiates into two distinct layers: the epiblast, which gives rise to the embryo proper; and the hypoblast, which will contribute to the germinal crescent and eventually give rise to extraembryonic endoderm. Several observations suggested that manipulation of the stage X (E.G&K) embryo could yield both somatic and germline chimeras.

First, PGCs or their precursors cannot be identified with certainty at stage X (E.G&K). Second, the area pellucida has yet to differentiate into the epiblast and hypoblast. Furthermore, several studies *in vitro* suggested that differentiation of the stage X (E.G&K) embryo is not irreversibly fixed. For example, disaggregated cells from the central region of the area pellucida can differentiate into epithelial and mesenchymal tissues and often organize into domelike structures resembling embryoid bodies of murine teratocarcinoma lines (Sanders and Dickau, 1981; Mitrani and Eyal-Giladi, 1982; J. N. Petitte and R. J. Etches, unpublished observations). Lastly, Marzullo (1970) reported the production of chimeric embryos after the transfer of pieces of the unincubated blastoderm from one embryo to another. In this case, feather color markers were used and the embryos were evaluated at 15 days of incubation.

In order to evaluate chimerism after blastodermal cell transfer, stage X (E.G&K) embryos from a line of Barred Plymouth Rocks were isolated, dissociated into a single-cell suspension, and injected into the subgerminal cavity of embryos from a line of Dwarf White Leghorns. Several somatically chimeric embryos, which died during incubation, were distinguished by patches of black down characteristic of the Barred Plymouth Rock chick among the yellow down of a Dwarf White Leghorn. Of the embryos that survived to hatch, a male was phenotypically chimeric with respect to feather color. This BPR/DWL chimera was mated to several Barred Plymouth Rock hens to test for germline transmission of the Barred Plymouth Rock phenotype. Of the 719 chicks produced, two were phenotypically Barred Plymouth Rocks, demonstrating that cells capable of incorporation into the germline had been transferred. This conclusion was verified using fingerprinting of blood and sperm DNA (Petitte et al., 1990). Taken together, these data indicated that blastodermal cells taken from stage X (E.G&K) embryos and transferred to recipients at the same stage of development will enter the germline, the hematopoietic tissues, and the melanocyte lineage in the adult animal (Petitte et al., 1990). Recently, Naito et al. (1991) demonstrated that blastodermal cells from unincubated quail embryos could be introduced into stage X (E.G&K) chicken embryos to form interspecific chimeras. The extent of incorporation of the quail cells in these chimeras is unknown, although the expression of quail-like feather pigmentation indicates that cells of the melanocyte lineage were included in those that colonized the recipient embryo.

Although the results of Petitte et al. (1990) indicated that it was technically possible to produce a chicken that was chimeric in both somatic tissues and the germline using blastodermal cell transfer, several questions arose regarding the production of germline chimeras and their usefulness as a means of producing transgenic poultry. Among these were concerns about the need for specific combinations of donor and recipient genotypes, the properties of cells from specific regions within the donor embryo, the ability of cells from preoviposition embryos to form chimeras, and the ability of transfected blastodermal cells to form a chimeric embryo.

Table 1
Sex, Appearance, and Breeding Success of
Chimeras Made by Injecting Silkie Donor Cells
into White Leghorn Recipients[a]

Sex	Somatic or putative	Number of individuals	Number of offspring/individual
Male	Somatic	5	1092–1175
Female	Somatic	2	232, 240
Male	Putative	1	1150
Female	Putative	2	198, 221

[a] Donor cells were prepared by dissociating entire stage X (E.G.&K) blastoderms according to the method described by Petitte et al. (1990). None of the offspring possessed the marker traits from the Silkie donor cells.

Specificity of the Donor/Recipient Combination

In mice, it has been shown that particular lines are more suitable as recipients for the formation of germline chimeras than other lines. For example, when using CCE embryonic stem cells as the donor cell, blastocysts from C57BL/6 mice give rise to germline chimeras more consistently than other lines tested (Schwartzberg et al., 1989). To examine the influence of genotype on the formation of avian chimeras, embryos from Silkie and White Leghorn chickens were used reciprocally as donors and recipients. The line of Silkies used is homozygous for the dominant mutations polydactyly ($PoPo$), muffs and beards ($MbMb$), and crest ($CrCr$), segregates for the mutant and wild-type alleles at the two loci involved in walnut comb (R^-P^-), is homozygous recessive for dominant white (ii) and silkie feathering (hh), and is homozygous for the co-dominant mutations for feathered shanks and black skin. The melanocytes that colonize the dermis of the Silkie also colonize the connective tissue to pigment the visceral organs and musculature. Although both black skin and feathered shanks are controlled by multiple loci, they are present in varying degrees in a Silkie White Leghorn cross and, therefore, are useful markers of cells of Silkie origin in chimeras.

Several male and female somatic and putative chimeras were produced following the transfer of dispersed blastodermal cells from Silkie stage X (E.G&K) embryos into the subgerminal cavity of White Leghorn embryos (Table 1). At hatch, the somatic chimeras were distinguished by patches of rusty-brown and ash-gray down (Plate 1A*) and, in many cases, patches of black skin (Plate 1B*) whereas putative chimeras could not be distinguished from normal White Leghorn chicks. In all cases, the pigmented feathers disappeared from the Silkie/White Leghorn chimeras during the first posthatch

* Plates 1A and B appear following page 98.

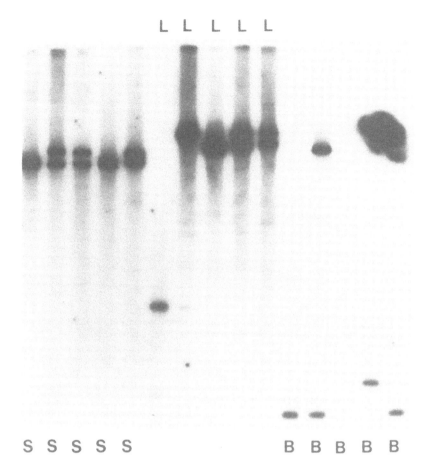

FIGURE 1. Characteristic fingerprints of blood DNA from Silkies (S), White Leghorns (L), and Barred Plymouth Rocks (B), prepared using M13mp18 as a probe with the conditions as described by Petitte et al. (1990) but with high-stringency washing at 65°C.

molt. One of the Silkie/Leghorn chimeras developed deposits of melanin in his legs during the second and third weeks posthatch (Plate 2*), and this pigmentation persisted throughout the remainder of his life. None of the other phenotypic markers characterizing the Silkie breed were observed in any of the chicks. These observations indicated that blastodermal cells from stage X (E.G&K) embryos contributed to the melanocyte population of the chimeras.

The contribution of the donor cells to hematopoietic cells and the testes was assessed by DNA fingerprinting. DNA from chimeras, Silkies, and White Leghorns was prepared from blood and semen, subjected to digestion, electrophoresis, and blotting, and probed with M13mp18 as described by Petitte et al. (1990). Under conditions of high stringency, breed specific bands were identified in DNA from Silkies and White Leghorns (Figure 1). In general,

* Plate 2 appears following page 98.

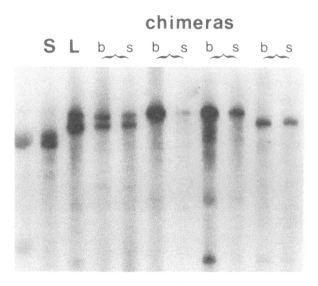

FIGURE 2.   Fingerprints of DNA from blood (b) and semen (s) from three Silkie/White Leghorn somatic chimeras (lanes 4–9) and one Silkie/White Leghorn putative chimera (lanes 10–11), demonstrating the absence of DNA from the donor Silkie cell lineage in three of these birds (lanes 4–7 and 10–11), and the presence and absence of DNA from the donor cells in blood and semen, respectively, of the chimera whose DNA is shown in lanes 8 and 9. The presence of Silkie DNA in lane 8 is indicated by the paired bands at an equivalent location to the prominent breed-specific bands in the Silkie control sample in lane 2. The chimera that developed dermal pigmentation in the legs is shown in lanes 6 and 7. Pooled DNA samples from 20 Silkies (S) and 20 White Leghorns (L) are shown in lanes 2 and 3, respectively.

DNA from Silkies has diagnostic bands that are smaller than those of White Leghorn DNA, although a small percentage of White Leghorns contain bands that are much smaller in size than those of Silkies (Figure 1, lane 6). Fingerprints of blood and semen DNA from three Silkie/Leghorn male somatic chimeras and one putative chimera are illustrated in Figure 2. One of the somatic chimeras exhibited Silkie-specific bands in blood, indicating that blastodermal cells from stage X (E.G&K) embryos contributed to adult hematopoietic tissues in this bird. By contrast, this analysis suggested that the donor cells had not contributed to the population of spermatogonia that subsequently formed spermatozoa in any of these birds, because Silkie-specific bands could not be distinguished in semen from any of these chimeras (Figure 2).

In order to confirm that the donor cells had not entered the germline and given rise to functional gametes, one putative and five somatic male chimeras were mated to White Leghorn hens and two putative and two somatic female chimeras were mated to White Leghorn cocks. The offspring from these

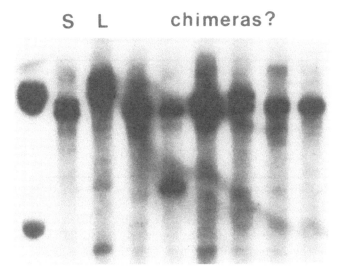

FIGURE 3. Fingerprints of blood DNA from White Leghorn/Silkie chimeras taken between the fourteenth and sixteenth day of incubation (lanes 4–9). Pooled DNA samples from 20 Silkies (S) and 20 White Leghorns (L) are shown in lanes 2 and 3, respectively. The lower band in lane 5 represents a band seen occasionally in White Leghorn samples (see the faint band in lane 3, which represents a mixture of 20 samples), but never in Silkie samples (lane 2).

matings were examined for the presence of the dominant and co-dominant traits that distinguished the Silkie donor cells. Between 1092 and 1175 offspring were obtained from the somatic and putative male chimeras, and between 198 and 240 offspring were obtained from the somatic and putative female chimeras (Table 1). None of the offspring possessed any of the marker traits, confirming the conclusion derived from the DNA fingerprints that the Silkie donor cells had not contributed to the germline of the chimera.

The introduction of dispersed blastodermal cells from White Leghorn embryos into Silkie recipients failed to produce any live chicks from 168 injections, whereas about 10–12% of the embryos survive to hatch when dispersed cells from Barred Rock embryos are injected into White Leghorn recipients. Thirty-seven White Leghorn/Silkie embryos survived to 14 days or more of incubation, but no live chicks were hatched. Blood DNA from six of the White Leghorn/Silkie embryos was subjected to fingerprint analysis (Figure 3). Both Silkie and White Leghorn-specific bands were apparent for two or three of these embryos, and only Silkie-specific bands were evident in blood from the remaining three putative chimeras. From these data it is evident that somatic chimerism occurred in all of the combinations of donor genotypes examined to date. The failure of White Leghorn/Silkie chimeras to hatch may reflect a genetic incompatibility between these two cell lineages,

Table 2

Effect of Opening the Shell of Eggs from Six Breeds of Chicken on
Development of Noninjected Embryos[a]

Breed	Number of eggs	Percent hatchability of nonwindowed eggs	Percent hatchability of windowed eggs
White Leghorn	30	87	37
Dwarf White Leghorn	16	75	17
Barred Plymouth Rock	22	60	0
Rhode Island Red	30	87	30
Light Sussex	28	82	28
Silkie	10	90	0

[a] The shells were opened and closed as described by Petitte et al. (1990).

although it is equally plausible that there are breed differences in the ability
of the embryo to develop after the eggshell is opened. This possibility has
been tested by comparing the survival of noninjected embryos in eggs that
have been opened using the same procedure that is employed during injection
of blastodermal cells (Table 2). These data confirm that development of Silkie
embryos is severely compromised when the shell is punctured to gain access
to the embryo. Furthermore, these data indicate that there are large breed
differences in the ability of embryos to complete development after the shell
has been opened, and that the choice of a White Leghorn recipient was
fortuitously a good one. The physiological consequences of opening an egg
that subsequently lead to embryonic mortality are, however, unknown.

Regional Specificity of the Donor Embryo

Detailed investigations by Eyal-Giladi and colleagues on axis formation during
early embryogenesis have revealed that between stages X and XIII (E.G&K),
the avian embryo can be divided into three regions, each with a specific
function. As described above, the area opaca serves mainly to facilitate blastoderm expansion and is the interface between the embryo and the yolk. The
area pellucida can be divided into the central disk, from which all tissues of
the embryo are derived, and the marginal zone, which corresponds to the
junction between the central disk and the area opaca. Cells in the marginal
zone are involved in axis formation and the induction of the primitive streak
(Azar and Eyal-Giladi, 1979; Khaner and Eyal-Giladi, 1986, 1989; Stern,
1990).

   Because of the differences in function among the three regions of the
early embryo, it was of interest to examine whether these differences were
also reflected in the ability of the cells from each area to form chimeras.
Therefore, cells from the central disk, marginal zone, area opaca, and entire
blastoderm of Barred Plymouth Rock embryos were injected into the

Table 3

Frequency of Chimerism at Hatch Following Introduction of Barred Plymouth Rock Blastodermal Cells Taken from the Area Opaca, Marginal Zone, Central Disk, or Intact Stage X (E.G&K) Embryo and Injected into White Leghorn Recipients

Source of cells	Number of somatic chimeras	Number of putative chimeras	Percent of somatic chimeras
Whole embryo	4	35	10
Area opaca	2	35	5
Marginal zone	1	24	4
Central disk	17	31	35

subgerminal cavity of White Leghorn recipients. Somaac chimeras were produced using cells from each of the regions, as indicated by the presence of black pigmentation in the down and the barred pigmentation pattern in adult feathers. The frequency of chimerism, however, was dependent on the source of cells from the donor embryo (Table 3). When cells from the central disk were used, the frequency of chimerism in embryos surviving to 14 days of age was significantly higher than that observed using cells from the whole embryo. The frequency of chimerism decreased from 35% to 4 – 5% when donor cells were isolated from the marginal zone or area opaca rather than the central disk. These observations suggest that somatic chimerism is generally the result of incorporation of donor cells from the central disk into the recipient. The number of donor cells from the central disk that were injected, however, had no effect on the frequency of somatic chimerism (Figure 4).

Since primordial germ cells are believed to arise exclusively from the central disk (Ginsburg and Eyal-Giladi, 1987), the frequency of germline transmission in the chimeras made from donor cells harvested from within and without this area was assessed. Incorporation of the donor cells was evaluated by mating both somatic and putative chimeras to Barred Plymouth Rocks and recording the number of black chicks, since they could arise only if Barred Plymouth Rock donor cells had entered the germline. Approximately 1000 offspring were obtained from each somatic and putative male chimera, and 100–300 offspring were obtained from each somatic and putative female chimera (Table 4). No black chicks were produced from any of these matings, indicating that, due to one or more of the reasons below, the donor cell lineage failed to form functional spermatozoa and oocytes. An insufficient number of donor cells destined to become functional gametes may have been transferred to the host embryo, although there was no difference in the frequency of somatic chimerism in recipients that received between 100 and 500 donor cells (Figure 4). Precursors of the primordial germ cells may not have been able to gain entrance into the germline of the recipient, but this suggestion seems unlikely because transferred primordial germ cells have given rise to

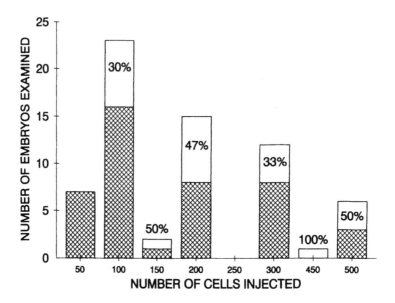

FIGURE 4. The number and proportion of somatic chimeras resulting from the injection of 100–500 cells from the central disk of Barred Plymouth Rock embryos into recipient White Leghorn embryos. The height of the bar represents the total number of embryos examined, the hatched portion represents the number of putative chimeras, and the open portion represents the number of somatic chimeras. The percentage of somatic chimeras is indicated for each group.

Table 4

Sex, Source of Donor Cells, Appearance, and Breeding Success of
Chimeras Made by Injecting Barred Plymouth Rock (BPR) Donor
Cells into White Leghorn (WL) Recipients[a]

Sex	Source of donor cells	Somatic or putative	Number of individuals	Number of offspring/individual
♂	Whole embryo	Somatic	1	865
♀	Whole embryo	Somatic	1	224
♂	Whole embryo	Putative	1	548
♀	Whole embryo	Putative	3	188–242
♂	Central disk	Somatic	5	1061–1118
♀	Central disk	Somatic	2	168, 183
♂	Central disk	Putative	3	1041–1113
♀	Central disk	Putative	2	186, 210
♂	Area opaca	Putative	5	661–1139
♀	Area opaca	Putative	8	150–283
♂	Marginal zone	Putative	1	Not tested
♀	Marginal zone	Putative	4	143–204

[a] All of these chicks were phenotypically hybrid BPR × WL offspring, indicating that the gametes were from the recipient germline in the chimera.

functional ova (Reynaud, 1976), and blastodermal cells from stage X (E.G&K) embryos have given rise to functional ova (Plate 3*) and spermatozoa when transferred to a recipient (Petitte et al., 1990). Alternatively, unknown physiological blocks may prevent processing of exogenous primordial germ cells as they differentiate from oogonia or spermatogonia into functional gametes. This may be a particular problem when the donor cells and recipient embryo form a male/female chimera, since it is known that female erythrocytes are absent from mixed-sex chimeras (Shaw et al., 1992). The absence of female cells in male chimeras may be attributed to their elimination during embryonic development or male/female chimeras may die during the second half of incubation (Shaw et al., 1992). Regardless of the precise mechanisms, experience has led us to believe that the major impediment to the production of germline chimeras is the low ratio of donor:recipient-derived cells that contribute to the chimera. Experiments are underway that are successfully addressing these problems.

In Vitro Manipulation of Donor Cells

The fact that blastodermal cells can be isolated *in vitro* before they are injected into a recipient embryo could facilitate modifications to the genome using a chimera as an intermediate step in the production of transgenic poultry. The procedures that are required between isolation of blastodermal cells and the establishment of a line of transgenic chickens and the potential advantages of this approach to alteration of the genome have been described by Etches et al. (1990) and are discussed in greater detail in Chapter 19. Briefly, genetic modifications would be introduced into isolated blastodermal cells, these genetically modified cells would be introduced into stage X (E.G&K) recipients, and chickens that carried the modification in their germline would be used as the foundation stock to establish lines of chickens expressing the modification. This approach has the theoretical advantages of utilizing homologous recombination to induce specific changes in the genome, and of allowing selection of appropriately modified cells prior to their introduction into the recipient embryo to reduce the expense of identifying and selecting transgenic chickens from large populations of live animals.

A strategy to produce transgenic, chimeric chickens formed by the transfer of selected donor cells of precisely modified genotype requires procedures for the long-term culture of blastodermal cells to allow time for selection of appropriately modified cells. It is essential that the cultured cells retain their ability to form chimeras during this process. Initial experiments to demonstrate this property were conducted with blastodermal cells from entire Barred Plymouth Rock embryos prepared according to the method described by Petitte et al. (1990). The cells were held at 37°C in a 5% CO_2 atmosphere overnight

---

* Plate 3 appears following page 98.

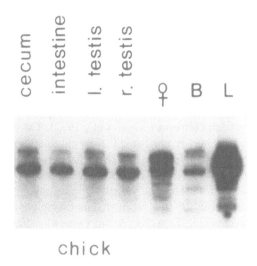

FIGURE 5. Fingerprint analysis of DNA from intestinal and gonadal tissues of the Barred Plymouth Rock offspring (lanes 1–4) from the female germline chimera shown in Figure 4. DNA from the chimera (♀), and pools of DNA from 20 Barred Plymouth Rocks (B) and 20 White Leghorns (L) are shown in lanes 5, 6, and 7, respectively. DNA from the different tissues of the offspring give identical fingerprints to each other, as expected, and to the pooled Barred Plymouth Rock sample (B), but not to the pooled White Leghorn sample (L), which is similar to that of the White Leghorn-like chimeric hen (♀).

before injection into recipient embryos, and one female and three male putative chimeras subsequently hatched. The putative chimeras were raised to sexual maturity and mated to Barred Plymouth Rocks to assess the incorporation of donor cells into the germline. The male putative chimeras produced between 462 and 683 offspring, all of which were typical Barred Plymouth Rock × White Leghorn hybrids, indicating that the donor cells had not been incorporated into the germline. The female putative chimera, however, produced 413 offspring, two of which were phenotypically Barred Plymouth Rock embryos (Plate 3), although both of these embryos died during the last 2–3 days of incubation. Fingerprint analysis of one of the dead, phenotypically Barred Plymouth Rock embryos revealed that DNA from several tissues contained bands that were normally present only for birds in the flock of Barred Plymouth Rocks that supplies the donor cells and that bands which are usually present for White Leghorns were absent (Figure 5). Taken together, these preliminary data supported the conclusion that dispersed blastodermal cells incubated for approximately 18 h prior to injection can retain the ability to enter into the germline. Experiments are in progress which indicate that donor cells can be cultured for longer periods than 18 h and still contribute to a chimera (R. J. Etches, A. Toner, and A. M. Verrinder Gibbins, unpublished results).

Table 5
Sex and Breeding Success of Putative
Chimeras Made by Injecting Donor
Cells from Barred Plymouth Rocks
(BPR) into White Leghorn (WL)
Recipients[a]

Sex	Number of individuals	Number of offspring/individual
Male	6	238–934
Female	13	71–175

[a] Donor cells were prepared by dissociating
the central disk of stage X (E.G&K)
blastoderms according to the method
described by Petitte et al. (1990), freezing
these cells in the presence of dimethyl
sulfoxide, thawing the cells, and injecting
them as described by Petitte et al. (1990).
All of these chicks were phenotypically
hybrid BPR × WL offspring, indicating
that the gametes were from the recipient
germline in the chimera.

In addition to providing access to the germline to facilitate the production
of transgenic chickens, blastodermal cells may provide an appropriate vehicle
for the conservation of genetic resources if they retain their ability to form
germline chimeras after storage in liquid nitrogen. If chimeras could be produced from previously frozen donor cells and these chimeras transmitted the
donor genotype at a reasonable frequency, *inter se* matings between them
would produce offspring whose genotype would be a faithful representation
of the population from which the cryopreserved blastodermal cells were obtained. To determine if this approach is feasible, blastodermal cells from
Barred Plymouth Rock chickens were suspended in a mixture of 80% (v/v)
fetal bovine serum (FBS), 10% (v/v) Dulbecco's modified essential medium
(DMEM), and 10% dimethyl sulfoxide, slowly frozen in a −70°C freezer,
and stored in liquid nitrogen for 1–15 days. The cells were thawed at 37°C,
resuspended in 80% FBS and 20% DMEM for 5 min, and transferred into a
mixture of 80% DMEM and 20% FBS prior to injection into recipients according to the protocol described by Petitte et al. (1990). Three somatic
chimeras (Plate 4*) and 19 putative chimeras were produced when cells from
the central disk of Barred Plymouth Rock donor embryos were frozen, thawed,
and injected into White Leghorn recipients. All of the somatic chimeras died
during the last week of development, and none of the putative chimeras
produced black chicks when mated to Barred Plymouth Rocks (Table 5).

---

*   Plate 4 appears following page 98.

Table 6

Frequency of Somatic Chimerism Following Injection of
Dispersed Blastodermal Cells from Stage V–VII (E.G&K)
Barred Plymouth Rock (BPR) Donor Embryos into Stage X
(E.G&K) White Leghorn (WL) Recipients[a]

Number of donor embryos	Number of chimeras	Number of embryos injected	Percent somatic chimeras
Stage V	1	27	4
Stage VI	4	46	9
Stage VII	6	41	15
Stage VIII	5	13	38

[a] All of these chicks were phenotypically hybrid BPR × WL offspring,
     indicating that the gametes were from the recipient germline in the
     chimera.

Assuming that the death of these embryos was due to perturbations caused
by injection and not to the prior treatment of the injected cells, these observations supported the proposal that frozen blastodermal cells might be used
to maintain genetic resources *in vitro*. Application of the concept awaits the
production of germline chimeras and the reconstitution of a gene pool by *inter
se* matings between them, but recent experiments (see conclusions) suggest
that this objective will soon be realized.

The Fate of Stage V–Stage VII Donor Cells in Stage X
Recipients

The formation of chimeras using blastodermal cells is predicated on the idea
that the early embryo contains a population of cells that are pluripotential,
and it is generally assumed that cells lose their pluripotentiality as development
proceeds. It seemed plausible, therefore, that cells from stage V (E.G&K)
embryos could colonize stage X (E.G&K) recipients more effectively than
cells from more advanced embryos. Experiments to examine this possibility
were conducted using embryos contained in eggs that were laid prematurely
following an intravenous injection of arginine vasotocin. Individual Barred
Plymouth Rock embryos from stages V–VIII (E.G&K) were isolated and
injected into the subgerminal cavity of stage X (E.G&K) White Leghorn
recipients. During stages V–VIII of embryonic development, the area pellucida sheds cells from its lower surface, although cellular morphology is
relatively constant during this period (Eyal-Giladi and Kochav, 1976; Watt
et al., 1992). Among those embryos that survived beyond 14 days of incubation, the frequency of phenotypic chimerism decreased as the stage of the
donor cells diverged from that of the recipient embryo (Table 6). These data
suggest that the developmental stage of the donor cells must be matched to

Table 7
Sex, Appearance, and Breeding Success of Chimeras
Made by Injecting Dispersed Blastodermal Cells
from Stage V–VII (E.G&K) Barred Plymouth Rock
Donor Cells into Stage X (E.G&K) White Leghorn
(WL) Recipients

Sex	Somatic or putative	Number of individuals	Number of offspring/individual
Female	Putative	10	44–179
Female	Somatic	2	161–180
Male	Putative	12	0–1125
Male	Somatic	2	54–995

some degree with the developmental stage of the recipient embryo to facilitate the formation of a chimera. The inability of the less mature donor cells to incorporate into these chimeras provides evidence that the interactions between the donor and recipient cells may be complex and may be dependent on expression of cell surface proteins that change between stages V and X (E.G&K), despite the apparently constant cell surface morphology that is revealed by ultrastructural analysis (Eyal-Giladi and Kochav, 1976; Watt et al., 1992). In addition, the assumption that blastodermal cells from less mature embryos retain greater pluripotentiality is inconsistent with these data, and consideration must be given to an alternative hypothesis that chimeras form because differentiated donor cells integrate into areas of the recipient that are committed to the same developmental destiny.

The possibility that blastodermal cells from early embryos could enter the germline of the recipient was examined by mating both putative and somatic chimeras that hatched and grew to sexual maturity to Barred Plymouth Rocks. None of the chimeras produced black chicks from these matings (Table 7), indicating that donor cells from stage V–VIII (E.G&K) embryos are not a rich source of cells destined for inclusion in the germline.

Chimeric Embryos Containing Transgenic Cells

Methods are required for the efficient introduction of exogenous genes into chicken blastodermal cells to produce transgenic, chimeric intermediates. A commercial cationic lipofection reagent has been shown to cause efficient transfection of chicken primary embryonic fibroblasts and blastodermal cells in culture or in suspension, perhaps because chick embryo cells survive in a lipid-rich environment in the yolk (Brazolot et al., 1991). Chicken blastodermal cells transfected with various *lacZ*-based constructs with *lacZ* expression under the control of a chicken β-actin/RSV promoter, a Zn^{2+}-inducible chicken metallothionein promoter, or a constitutive CMV promoter will express bacterial β-galactosidase activity in at least 1 out of every 25 cells subjected to transfection. Stable incorporation of the constructs into the genome was indicated for approximately 10% of transfected primary chicken fibroblasts that showed β-galactosidase activity.

In addition, the ability of transfected, noncultured blastodermal cells to contribute to embryonic chimeras was tested. Dispersed blastodermal cells were transfected with the various lacZ-based constructs and injected into recipient embryos. In the majority of cases, cells expressing bacterial β-galactosidase were localized in extraembryonic regions, particularly at the junction of the area pellucida vasculosa and the area opaca vasculosa. However, one embryo contained β-galactosidase-expressing cells in the prosencephalon, head ectoderm, and ventricle of the heart as well as in extraembryonic tissues (Brazolot et al., 1991). This clearly indicates that transfected blastodermal cells are capable of contributing to a chimeric embryo and represents a significant step toward a new method of developing transgenic chickens.

Conclusions

The knowledge that chimeras can be made using the population of cells present in stage X (E.G&K) chick embryos provides the foundation for a new approach to manipulation of the avian genome. These blastodermal cells should provide a practical means of gaining access to the genome, particularly when the cells can be maintained in vitro, and future work will be directed toward the development of an understanding of how the avian genome can be precisely modified to yield chickens with a desirable genotype. Genetic modifications to blastodermal cells can be introduced into somatic tissues in chimeras, and it seems likely that they can be introduced into the germline via chimeric intermediates. The major impediment to the production of effective chimeras has been the low proportion of donor cells in chimeras, particularly in the population of definitive gametes. Technical refinements that dramatically increase the contribution of donor cells to the recipient embryo have recently been developed (Etches et al., 1992) and these developments have increased the frequency of germline chimerism to a level that should support the kinds of genetic manipulations described in this manuscript. Further technical refinements of these discoveries will increase the frequency of incorporation of genetically modified blastodermal cells into the recipient embryo to yield a technology that is suitable for routine genetic selection. As these areas of research are explored, the production of transgenic birds using chimeric intermediates will become available to biologists and the poultry industry.

Acknowledgments

The authors wish to thank Shaver Poultry Breeding Farms, Ltd., for the generous donation of Barred Plymouth Rock and White Leghorn chickens,

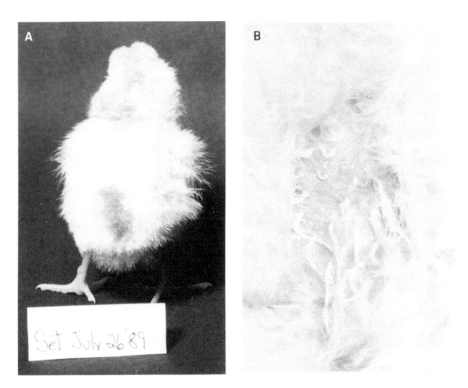

PLATE 1. Silkie/White Leghorn chimeras were distinguished at hatch by (A) patches of rusty brown or ash grey pigmentation in the down and (B) patches of pigmented skin.

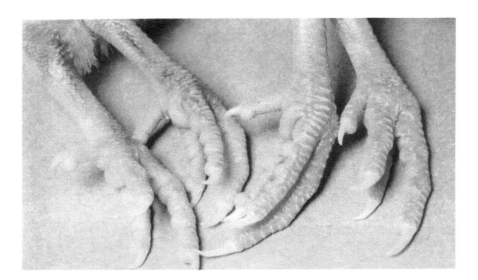

PLATE 2. The Silkie/White Leghorn chimera shown on the left developed pigmented areas in the dermis, providing evidence that precursors of the melanocytes, which migrate into the dermis in Silkie embryos, were transferred to the White Leghorn embryo. Legs from a typical White Leghorn are shown on the right for comparison.

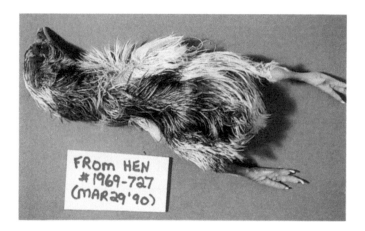

PLATE 3. The Barred Plymouth Rock progeny of the female germline chimera. This embryo was contained in a double yolked egg and consequently died in the final stages of incubation.

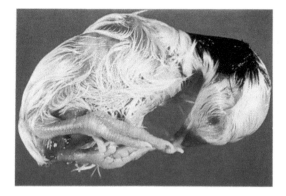

PLATE 4. Three somatic chimeras produced by injecting dispersed cells from stage X (E. G & K) Barred Plymouth Rock embryos which had been frozen and thawed into White Leghorn recipients according to the method described by Petitte. (From Petitte, J. N., et al., *Development*, 108, 185, 1990.)

and Gordon Ridler for the foundation pair of Silkies. This work was supported by grants from the Natural Sciences and Engineering Research Council (NSERC) and the Ontario Ministry of Agriculture & Food to RJE and from NSERC, the Ontario Egg Producer's Marketing Board, Agriculture Canada, and the University Research Incentive Fund to AMVG.

References

Azar, Y., and Eyal-Giladi, H. (1979). Marginal zone cells: The primitive streak-inducing component of the primary hypoblast in the chick. *J. Embryol. Exp. Morphol.*, 52, 79–88.

Bradley, A., Evans, M. J., Kaufman, M. H., and Robertson, E. (1984). Formation of germ-line chimeras from embryo-derived tertocarcinoma cell lines. *Nature* (London), 309, 255–256.

Brazolot, C. L., Petitte, J. N., Etches, R. J., and Verrinder-Gibbins, A. M. (1991). Efficient transfection of chicken cells by lipofection and introduction of transfected blastodermal cells into the embryo. *Mol. Reprod. Deve.*, 30, 304–312.

Didier, E. (1977). Chimérisme somatique et germinal après la greffe intracoelomique du mésoderme splanique de caille chez l'embryon de poulet. *C. R. Acad. Sci. Paris*, 284, 671–674.

Didier, E., and Fargeix, N. (1976). Germinal population of gonads in some chimerical embryos obtained by connecting pieces of Japanese quail and domestic chick blastoderms. *Experientia*, 32, 1333–1334.

Dieterlen-Lievre, F., Hajji, K., and Martin, C. (1985). Is the time of meiosis initiation programmed in the germ cells or the ovarian stroma? A study in avian chimeras. *Arch. Anat. Microsc. Morphol. Exp.*, 74, 20–22.

Etches, R. J., Petitte, J. N., Verrinder-Gibbins, A. M., Brazolot, C. L., and Liu, G. (1990). Production of chimeric chicks by embryonic cell transfer and the prospects for gene manipulation. In: *Avian Incubation and Embryology*, Tullett S., Ed. Butterworths, London, pp. 305–309.

Etches, R. J., Carscience, R. S., Shaw, D. L., and Verrinder Gibbins, A. M. (1992). Improved efficiency of incorporation of donor blastodermal cells in chimeric chickens and their detection using a female specific DNA probe. In: *Advances in Gene Technology: Feeding the World in the 21st Century*, Whelan W. J. et al., Eds. IRL Press, p. 51.

Evans, M. J., and Kaufman, M. H. (1981). Establishment in culture of pluripotential cells from mouse embryos. *Nature (London)*, 292, 154–156.

Eyal-Giladi, H., and Kochav, S. (1976). A complementary normal table and a new look at the first stages of the development of the chick. 1. General morphology. *Deve. Biol.*, 49, 321–337.

Eyal-Giladi, H., Ginsburg, M., and Farbarov, M. (1981). Avian primordial cells are of epiblastic origin. *J. Embryol. Exp. Morphol.*, 65, 139–147.

Ginsburg, M., and Eyal-Giladi, H. (1986). Temporal and spatial aspects of the gradual migration of primordial germ cells from the epiblast into the germinal crescent in the avian embryo. *J. Embryol. Exp. Morphol.*, 95, 53–71.

Ginsburg, M., and Eyal-Giladi, H. (1987). Primordial germ cells of the young chick blastoderm originate from the area pellucida irrespective of the embryo-forming process. *Development*, 101, 209–219.

Hajji, K., Martin, C., Perramon, A., and Dieterlen-Lievre, F. (1988). Sexual phenotype of avian chimeric gonads with germinal and stromal cells of the opposite genetic sexes. *Biol. Structure Morphol.*, 1, 107–116.

Khaner, O., and Eyal-Giladi, H. (1986). The embryo-forming potency of the posterior marginal zone in stage X through XII of the chick. *Dev. Biol.*, 115, 275–281.

Khaner, O., and Eyal-Giladi, H. (1989). The chick's marginal zone and primitive streak formation. I. Coordinative effect of induction and inhibition. *Dev. Biol.*, 134: 206–214.

Martin, G. (1981). Isolation of a pluripotent cell line from early mouse embryos cultured in medium conditioned by teratocarcinoma stem cells. *Proc. Natl. Acad. Sci. USA*, 78, 7634–7638.

Marzullo, G. (1970). Production of chick chimeras. *Nature*, 225, 72–73.

Mitrani, E., and Eyal-Giladi, H. (1982). Cells from the early chick embryo in culture. *Differentiation*, 21, 56–61.

Naito, M., Watanabe, M., Kinutani, K., Nirasawa, K., and Oishi, T. (1991). Production of quail-chick chimeras by blastoderm cell transfer. *Br. Poultry Sci.*, 32, 79–86.

Petitte, J. N., Clark, M. E., Liu, G., Verrinder-Gibbins, A. M., and Etches, R. J. (1990). Production of somatic and germline chimeras in the chicken by transfer of early blastodermal cells. *Development*, 108, 185–189.

Petitte, J. N., Clark, M. E., and Etches, R. J. (1991). Assessment of functional gametes in chickens after transfer of primordial germ cells. *J. Reprod. Fert.*, 92, 225–229.

Reynaud, G. (1969). Transfert de cellules germinales primordiales de dindon à l'embryon de poulet par injection intravasculaire. *J. Embryol. Exp. Morphol.*, 21, 485–507.

Reynaud, G. (1976). Capacités reproductrices et descendance de poulets ayant subi un transfert de cellules germinales primordiales durant la vie embryonnaire. *Wilhelm Roux's Arch.*, 179, 85–110.

Rogulska, T., Ozdzenski, W., and Komar, A. (1971). Behaviour of mouse primordial germ cells in the chick embryo. *J. Embryol. Exp. Morphol.*, 25, 155–164.

Sanders, E. J. and Dickau, E. A. (1981). Morphological differentiation of an embryonic epithelium in culture. *Cell Tissue Res.*, 220, 539–548.

Schwartzberg, P. L., Goff, S. P., and Robertson, E. J. (1989). Germ-line transmission of a *c-abl* mutation produced by targeted gene disruption in ES cells. *Science*, 246, 799–803.

Shaw, D. L., Carscience, R. S., Etches, R. J., and Verrinder Gibbins, A. M. (1992). The fate of donor blastodermal cells in male chimeric chickens. *Biochem. Cell. Biol.*, in press.

Shuman, R. M. (1981). Primordial germ cell transfer in the chicken, Gallus domesticus. M.S. thesis, University of Minnesota, St. Paul, MN.

Simkiss, K., Rowlett, K., Bumstead, N., and Freeman, B. M. (1989). Transfer of primordial germ cell DNA between embryos. *Protoplasma*, 151, 164–166.

Singh, R. P., and Meyer, D. B. (1967). Primordial germ cells in blood smears from chick embryos. *Science*, 156, 1503–1504.

Stern, C. D. (1990). The marginal zone and its contribution to the hypoblast and primitive streak of the chick embryo. *Development*, 109, 667–682.

Sutasurya, L. A., Yasugi, S., and Mizuno, T. (1983). Appearance of primordial germ cells in young chick blastoderms cultured in vitro. *Dev. Growth Differ.*, 25, 517–521.

Swartz, W. J., and Domm, L. V. (1972). A study on the division of primordial germ cells in the early chick embryo. *Am. J. Anat.*, 135, 51–70.

Swift, C. H. (1914). Origin and early history of the primordial germ-cells in the chick. *Am. J. Anat.*, 15, 483–516.

Tachinante, F. (1974). Sur les échanges interspécifiques de cellules germinales entre le poulet et la caille, en culture organotypique et en greffes coelomiques. *C.R. Acad. Sci. Paris*, 278, 1895–1898.

Thomas, K. R., and Capecchi, M. R. (1987). Site-directed mutagenesis by gene targeting in mouse embryo-derived stem cells. *Cell*, 51, 503–512.

Urven, L. E., Erickson, C. A., Abbott, U. K., and McCarrey, J. R. (1988). Analysis of germ line development in the chick embryo using an antimouse EC cell antibody. *Development*, 103, 299–304.

Watt, J. M., Petitte, J. N., and Etches, R. J. (1992). Early development of the chick embryo. in press.

Wentworth, B. C., Tsai, H., Hallett, J. H., Gonzales, D. S., and Rajcic-Spasojevic, G. (1989). Manipulation of avian primordial germ cells and gonadal differentiation. *Poultry Sci.*, 68, 999–1010.

Chapter 7

The Use of Avian Chimeras in Developmental Biology

Francoise Dieterlen-Lièvre and Nicole Le Douarin

Summary. Vertebrate development is characterized by extensive and complex cell migrations, during which cell interactions occur that are responsible for changes in cell behavior, fate, and terminal differentiation. In order to unravel these events, cells must be tagged with an easily detectable, permanent marker. The combination of quail and chick cells provides such a marker (Le Douarin, 1969). The marker is based on the presence in quail cell nuclei of a nucleolus-associated mass of heterochromatin, which is absent from chick nuclei. Chimeras are constructed surgically *in ovo* by introducing or substituting a selected rudiment or territory from one species into the other. The chimeras, constructed during the second to fifth day of incubation depending on the pattern, are allowed to develop for various periods, and eventually to hatching. Selected topics in which such chimeras have been instrumental will be described here: development of the peripheral nervous system studied through exchanges of neural tube and neural crest segments between the two species; central nervous system mapping analyzed with similar transplants performed at the level of brain vesicles resulting in the transfer of behavioral or genetic traits; analysis of the development of the immune and hemopoietic system relying on grafts of organ rudiments, the stroma of which is colonized by extrinsic stem cells; origin and traffic of these stem cells, development of immune tolerance, as well as formation of the vascular tree.

Introduction

The use of chimeras was initiated by Spemann in amphibian embryos (see Spemann, 1938). This artifice was later applied in birds with the aim of tracing the migrations of neural crest cells (Weston, 1963). Construction patterns became diversified only when the quail/chick marker made avian chimeras popular; numerous chimeric schemes were then designed to suit specific aims. The marker system is based on the presence of a prominent mass of heterochromatin in the nuclei of quail cells (Le Douarin, 1969, 1973). The structure of the heterochromatic component varies with cell types, being either restricted to one mass or fragmented into two or three elements associated with the nucleolus. In the chick, in contrast, heterochromatin is distributed into small dots. This marker, often referred to as the quail nucleolar marker, offers many assets: It does not need any specific preparative manipulation, it is indefinitely stable, and it is easily detected (Figure 1). The quail/chick label inspired similar interspecies chimeric systems for amphibians

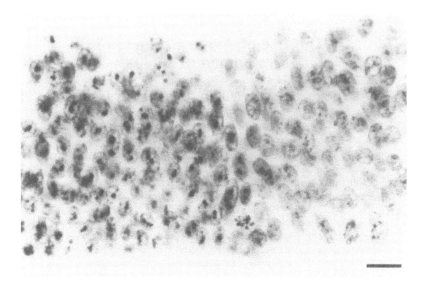

FIGURE 1. The quail/chick marker. The left half of this midbrain section is chick and the right half is quail. DNA has been stained by means of the Feulgen-Rossenbeck reaction. Note that fusion between quail and chick tissue is perfect. Bar = 10 μm.

FIGURE 2. Two brain chimeras and a control chick after hatching. The chimeras have been engrafted during the second day of incubation with the dorsal prosencephalon from a quail embryo. Quail melanoblasts have colonized the feathers, marking the level of the graft.

(Thiébaud, 1983) and for mice (Rossant and Frels, 1980). Neither system has been exploited to the extent of the quail/chick chimeras, however.

With the advent of quail/chick chimeras, the avian model became promoted to the rank of a paradigm, being, among upper vertebrates, the most amenable to sophisticated surgery. This is due to the accessibility of the avian embryo, its development in a plane, and its large size. Transplantations are performed according to a number of techniques: Ectopic grafting of rudiments in the body wall, the coelom, or on the chorioallantoic membrane makes it possible to demonstrate and study colonization of organ rudiments by extrinsic cells; orthotopic substitution of a rudiment or a blastodisc territory allows the analysis of cell origin, cell migration, and fates. The outcome may be evaluated during embryonic life or later (Figure 2). In the latter case, features that develop only after birth may be studied, for instance, behavioral traits or immune tolerance.

In the present overview, which is not intended to be exhaustive, a few chapters of this story will be illustrated, to highlight the achievements and to demonstrate the usefulness of these chimeras in the age of gene study and transfer.

Mapping the Derivatives of the Peripheral Nervous System

Historically, avian chimeras were first used in conjunction with tritiated thymidine labeling of DNA to elucidate the ontogeny of cells from the neural crest (Weston, 1963). A segment of the neural tube can be taken out from embryos on the first to second day of development and replaced by an equivalent segment bearing a marker (Figure 3). The operation is carried out at the most posterior level of the closing neural tube. At that level, neural crest cells are located at the apex of the neural tube and have not initiated their migration. The transitory structure will very rapidly disintegrate, releasing cells that migrate either as a sheet (mesencephalic and rhombencephalic level) or as individuals (spinal cord level) among mesodermal cells. These cells multiply as they migrate, reach precise sites of arrest, give rise to a number of definite derivatives inserted among cells of other origins, and establish with them structural and functional associations. Many of these derivatives were known prior to the advent of the quail/chick marker. However, some of them were not recognized, some were in dispute, and some were wrongly attributed a neural crest origin. Furthermore, the migration pathways were unknown.

The neural crest derivatives were mapped at two body levels: the cephalic (Le Lièvre and Le Douarin, 1975) and truncal (Le Douarin and Teillet, 1973). The cephalic neural crest was shown to give rise to neural and glial cells (classical derivatives of the crest), but also to most supporting tissues of the head, neck, and upper trunk. The latter contribution could be traced to a

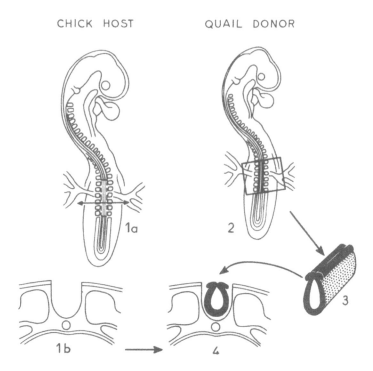

CHICK HOST   QUAIL DONOR

FIGURE 3.   Grafting scheme to produce neural chimeras. Here orthotopic, homochronic grafting (see text) is depicted at the ''adrenomedullary'' level, i.e., the level of the neural crest from
which cells colonize the adrenal medulla. The neural tube bearing the neural crest at its apex is
treated with trypsin prior to engrafting, so that any contaminating mesodermal cells are eliminated.

special subpopulation of neural crest cells, the so-called mesectoderm, which
plays the part implemented in the rest of the body by the mesoderm. The
only non-neural crest contribution is from the paraxial mesoderm, which gives
rise to a very minor proportion of the head supporting tissues (Couly et al.,
1991). Thus the mesectoderm is at the origin of cartilage, bone, muscle, and
mural cells of the blood vessels. Endothelial cells, however, never originate
from the neural crest (Le Lièvre and Le Douarin, 1975).

The mapping experiments made it possible to recognize several domains
in the truncal neural crest (Figure 4): The ''vagal'' level (level of somites 1
to 7) provides cholinergic neuronal cells and glial derivatives to the whole
gut. The ''medullary'' level (somites 18 to 24) provides the medullary cells
of the adrenal gland. The truncal level (somites 7 to the most caudal ones)
provides sensory neurones and glial cells of the dorsal root ganglia. The
posterior truncal level (after somite 28) provides part of the neural cells of
the postumbilical gut, which thus receives a dual contribution (Le Douarin
and Teillet, 1973).

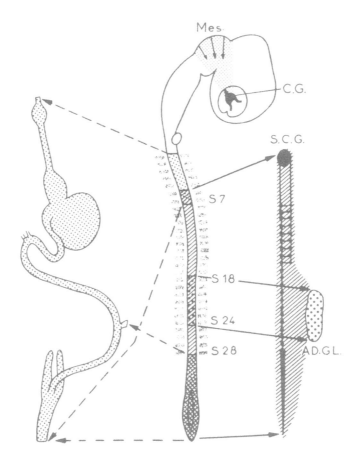

FIGURE 4. Neural crest fate map, indicating the level of origin on the neural axis of adrenal medulla and autonomic ganglia. ADGL, adrenal gland; CG, ciliary ganglion; Mes, mesencephalon; S, somite; SCG, superior cervical ganglion.

These fate maps (i.e., the structures to which each level of the neural crest gives rise) were established by substituting chick neural tube segments with quail donor segments derived from precisely the same level (orthotopic exchanges). These exchanges were performed between embryos from the same stage (homochronic exchanges). Other experiments were performed to determine whether various potentialities were present at all levels of the neural crest. For instance, a neural tube segment located at the level of somites 1–7 (vagal level) was transplanted at level of somites 18–24 (adrenomedullar level). As the neural crest is a transitory structure, it has to be retrieved as soon as the neural tube forms at the particular level. In the example above, the donor has 7 pairs of somites, while the host has 24 pairs of somites. These exchanges are defined as heterotopic and heterochronic. They showed that

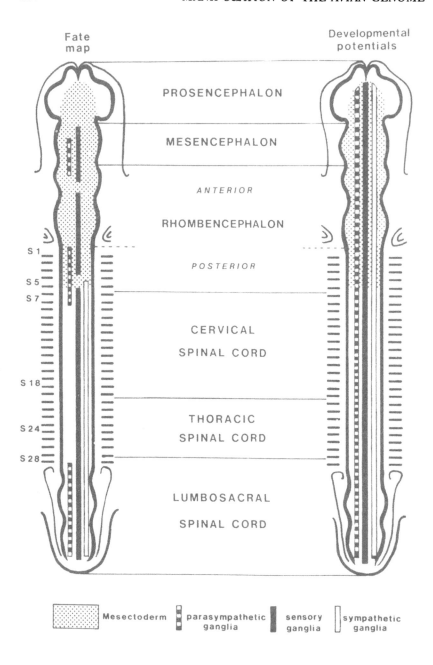

FIGURE 5.   Comparison between actual axial origin of neural crest derivatives and potentialities of each level. Note that only mesectodermal potentialities are restricted to a defined level.

neural crest from various levels was capable, when transplanted to a heterotopic level, of giving rise to new derivatives, specified by their experimental site of origin. Thus developmental potentialities are more extensive than potentialities inferred from the fate map (Figure 5). In other words, most potentialities are present at all levels of the crest, but will not materialize into corresponding differentiations unless the cells depart from the crest at the appropriate level. Migration pathways will then lead them to the sites of arrest that permit these differentiations. The only exception concerns the cephalic neural crest, where the unique mesectodermal subpopulation is located. From a very early stage in development, this subpopulation seems already segregated from other neural crest elements (Le Douarin and Teillet, 1974; Le Lièvre, 1974, 1978). The potentialities thus identified in the chimeras were later investigated *in vitro* by cloning the cells in the mesencephalic neural crest (Baroffio et al., 1988; Dupin et al., 1990) or by dye-marking a single neural crest cell *in situ* (Broner-Fraser and Fraser, 1988). The results demonstrate that, though the crest appears multipotential as a population, individual cells diverge very early. Some cells are capable of yielding large clones comprising up to 2×10^4 cells, which differentiate into all neuronal (catecholaminergic or cholinergic), glial, pigment, and mesectodermal phenotypes. Others give rise to smaller or very small clones (down to one cell) containing only a restricted assortment of phenotypes or one phenotype.

Mapping the Central Nervous System

Cell migration and lineage segregation in the brain are complex and poorly understood. Though experimental investigations using quail/chick chimeras began only recently, they have already led to intriguing conclusions. In these experiments, cephalic neural folds were exchanged between embryos of the two species at very early stages (Couly and Le Douarin, 1988, 1990). This scheme, which could involve either one or both sides of the brain, revealed that elements that lie adjacent to each other in functional units of the adult animal have a common embryological origin. It had always been thought that these elements converge during development when about to establish functional links. For instance, these experiments have provided an alternative interpretation to previous ideas about the development of the hypophysis and hypothalamus. Rathke's pouch, a rudiment emitted by the roof of the mouth, was supposed to unite secondarily with the floor of the diencephalon. The chimera experiments have clearly demonstrated that the presumptive territories of the hypothalamus and the antehypophysis are neighboring areas respectively located in the neural fold and neural plate (Figure 6) at the stage of one pair of somites. As the head undergoes the morphogenetic movements that bend it ventrally and elongate it caudally, the region where Rathke's pouch is going to emerge is carried away. The same process is at work in the development

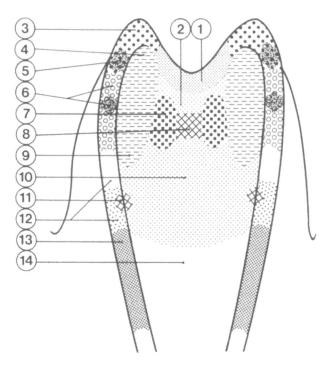

FIGURE 6.   Fate map of the anterior neural primordium of the three- to four-somite-stage avian embryo. (1) adenohypophysis, (2) hypothalamus, (3) ectoderm of nasal cavity, (4) floor of telencephalon, (5) olfactive placode, (6) ectoderm of upper beak and egg tooth, (7) optic vesicles, (8) neurohypophysis, (9) roof of telencephalon, (10) diencephalon, (11) hemiepiphysis, (12) ectoderm of calvaria and caudal prosencephalic neural crest, (13) rostral mesencephalic neural crest, (14) mesencephalon. Note that the presumptive adenohypophysis (1) and hypothalamus (2) are contiguous.

of the olfactory placode and the olfactory center (Figure 6). Thus functional units actually arise as developmental units (Le Douarin, 1986). Mapping the origins of cell types in the cerebellum and the hindbrain has been initiated recently with the chimera technique (Hallonet et al., 1990; Tan and Le Douarin, 1991).

Study of Behavioral Traits

Transplantation of segments of the CNS has been performed extensively from the quail into the chick and from the chick into the quail. Despite the different sizes of the adult animal in the two species, excellent integration of the graft's and the host's structures was usually achieved. All these studies have been conducted during embryonic life; behavioral studies were then undertaken on hatched neural chimeras. To date, two models have been selected for study: the song of juvenile males, and an epileptic mutation.

These experiments are fraught with two major difficulties; only 10% of operated embryos hatch, and the foreign tissues are usually rejected in these xenospecific grafts, as will be seen below. The first difficulty has been overcome by constructing many chimeras, a few of which have hatched. Furthermore, the humidity of the incubator should be high, and the window in the egg shell should be small; the eggs should be incubated with their long axis horizontal, and the operated eggs should be moved manually, twice a day. Deaths will occur, however, during the later part of incubation and at the time of hatching. The problem of foreign tissue rejection can be alleviated by injecting testosterone into the hatchlings. Within a couple of days, animals of both sexes develop the juvenile male crow. Young male chickens and quails crow in very distinctive fashions. The quail emits a segmented song, while the chick emits a single squeak. The song patterns, analyzed by means of sonograms, also have different amplitudes. Various levels of the quail brain were substituted for corresponding segments of the chick embryos. The conclusion could be drawn that the diencephalon and mesencephalon are necessary to determine the quail song pattern (Balaban et al., 1988). Such chimeras produce the segmented quail crow using the chick phonation system. Interestingly, the typical quail posture is determined by a different brain level, i.e., the rhombencephalon (E. Balaban and M. A. Teillet, personal communication).

A second behavioral trait, an epileptic mutation of the chicken, is being studied currently (Teillet et al., 1991). This mutation has many common characters with generalized human epilepsy. The crises are induced by intermittent light stimulation. Brain regions from the epileptic strain were transplanted into other chick embryos. The parts of the brain involved in the seizures are the prosencephalon and mesencephalon, the transplantation of which provokes full seizures. Transplantation of prosencephalon alone makes the chimera hypersensitive to intermittent light but initiates only the first phases of the seizure.

Development of the Immune and Hemopoietic Systems

An important feature of these systems is that the stromal cells and blood stem cells belong to distinct embryonic lineages. The stem cells colonize the hemopoietic organ rudiments early on: In the bone marrow they settle as a self-renewable pool; in the thymus, the colonization process continues as long as lymphoid cells are produced. This specific feature of the hemopoietic system was first uncovered using chorioallantoic grafts between chick embryos, in which the sex chromosomes served as a marker (Moore and Owen, 1965). This labeling system had several drawbacks; in particular, only mitotic cells could be diagnosed. The quail/chick marker made it possible to trace the origin of all cells, demonstrating a rhythmic pattern of colonization in the thymus (Le Douarin and Jotereau, 1973; Jotereau and Le Douarin, 1978;

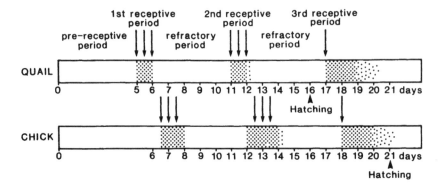

FIGURE 7. Colonization patterns of thymus. In the two species, the recurrence of stem cell entry occurs according to a regular rhythm.

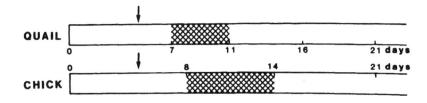

FIGURE 8. Colonization patterns of the bursa of Fabricius. This organ receives only one wave of stem cells during embryonic life.

Coltey et al., 1987). This pattern is characterized by short waves of attractivity that are followed by longer periods when cells are not recruited into the thymus (Figure 7). The precision and regularity of the rhythm is astonishing. It appears probable that cyclic colonization is also the rule in mammals, though the different phases overlap, making the pattern less discernible (Jotereau et al., 1987). Birds have another primary lymphoid organ, the bursa of Fabricius, devoted to the differentiation of B lymphocytes. The colonization pattern of the bursa has been found to comprise a unique phase that lasts several days in both species (Figure 8) and ends before hatching. At the same time, as the colonization ceases, stem cells capable of colonizing the bursa disappear from the animal. The bursa functions for several weeks, producing a large contingent of pre-B cells that settle into the bone marrow, where they undergo their final maturation. The bursa is composed of tightly packed follicles arranged around a central lumen, which is a diverticulum of the cloaca. It could be demonstrated that each follicle is populated by very few stem cells (one to three), which were traced through two alleles of a B-cell surface antigen, Bu-1, as a marker. The chimeric patterns were either embryonic parabionts or embryos whose contingent of B precursors were depleted with the drug cyclophosphamide before restoration with a mixed suspension of Bu-1a and Bu-1b precursors (Pink et al., 1985).

Tolerance

When neural chimeras hatched and were raised for a few months, they displayed signs of "autoimmune" disease. The neural tube had been grafted at the wing level, and the first sign was drooping of the wings; a few days later, paralysis extended to the legs, making the animal unable to stand and death ensued when the chimera became unable to feed. The spinal cord in these animals was heavily infiltrated with inflammatory cells at the level of the graft but also in the neighboring chick segments (Kinutani et al., 1989). In order to study the rejection process in circumstances that would not kill the animal, the wing was selected as a test organ. It was grafted *in situ* in 5-day embryos. The hosts hatched. Inflammatory signs appeared early, between 2 and 3 weeks depending on the animals, and within a few days the grafted wing became autoamputated. The rejection in this xenogeneic system is indeed extremely acute. The delay in response when neural tissue is involved is attributed to the blood-brain barrier, which eventually breaks down. Rejection of the wing could be alleviated or altogether bypassed by engrafting the chick host embryo with a quail thymic rudiment 2 days before grafting the quail wing bud (Ohki et al., 1989). The thymus was also grafted *in situ*, after destroying the host rudiments. This is a difficult operation, as birds have two thymic rudiments on each side of the neck; often some host thymus is retained and the foreign tissue engrafts partially. It was found that one-third of the total thymic tissue was sufficient to ensure tolerance. At the time of grafting, the thymus rudiments are present as an epithelial cord, and do not contain cells of the blood lineage. The cellular origins of the differentiated thymus were studied at the time of sacrifice, by means of monoclonal antibodies recognizing MHC class II antigens of the chick or the quail, proving that the quail tissue only gave rise to thymic epithelium. This established the capacity of that thymic component to educate lymphocytes for self-recognition. This capacity was previously attributed exclusively to accessory cells of the thymus (dendritic cells and macrophages), which belong to the blood lineage. A similar demonstration was later obtained in mice, in experiments inspired by the quail/chick model (Salaün et al., 1990).

Origin of Blood Stem Cells

Since the hemopoietic organs all receive their contingent of stem cells from an extrinsic source, the origin of these cells in the embryo had to be determined. Moore and Owen (1967) were the first to detect that stem cells arise extrinsically to the hemopoietic organs, with one exception, the yolk sac.

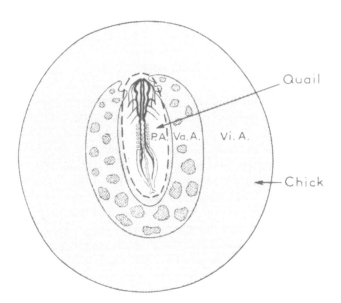

FIGURE 9. Grafting scheme for construction of "yolk sac chimeras" according to Martin (1972). The stippled line indicates the suture between the two components. PA, pellucid area; VaA, vascular area; ViA, vitelline area.

These authors proposed that all stem cells arise in the embryo's primordial hemopoietic organ, the yolk sac. This hypothesis was tested by means of "yolk sac chimeras", in which the embryonic area of a quail was grafted onto the extraembryonic area of a chick blastodisc (Figure 9). The extraembryonic area gave rise to the yolk sac in the next 2 days, and the hemopoietic organs of these chimeras were entirely populated by quail cells (Dieterlen-Lièvre, 1975; Martin et al., 1978). The blood of these chimeras was studied using polyclonal antibodies directed against quail or chick erythrocytes in a hemolysis assay. These chimeras had 95–100% chick cells until the fifth day of embryonic development, then quail erythrocytes were released in the blood in increasing proportions (Beaupain et al., 1979). Some chimeras had up to 80% quail red cells on the thirteenth day of embryonic development; however, the extent of replacement of chick cells by quail was highly variable. This experiment was repeated using yolk sac chick/chick chimeras, in which immunoglobulin allotypes and major histocompatibility complex antigens served as markers. Yolk sac stem cells never gave rise to B lymphocytes (Lassila et al., 1978). Furthermore, in that type of chimera, erythrocyte replacement was very regular and very swift: On the tenth day of embryonic development, only 20% of the red cells were progeny from yolk sac stem cells, and these had disappeared altogether by hatching (Lassila et al., 1982).

Thus it is clear that, in birds, yolk sac stem cells give rise to the early erythrocytes of the primitive series; they do contribute to the first series of so-called fetal red cells, and thereafter they become extinct without giving rise to stem cells with life-long renewal capacity. This demonstration opened the search for the site of emergence of intraembryonic stem cells. A new grafting pattern was used, with the dorsal mesentery as the site of implantation. In this region, prominent foci of hemopoiesis develop between the fifth and eight days of embryonic development in both species. It was found that the aorta taken from quail embryos between the third and fourth days of embryonic development and grafted in the chick dorsal mesentery gave rise to grafted hemopoietic cell foci (Dieterlen-Lièvre, 1984). Moreover when cells from this region were dissociated and seeded in a semisolid medium, they gave rise to colonies. Macrophage colonies, granulocyte colonies, or mixed colonies were obtained, depending on the growth activities present in the medium. In contrast, when cells of the rest of the embryo deprived of the aortic region were seeded, no colonies were obtained (Cormier et al., 1986; Cormier and Dieterlen-Lièvre, 1988). Hemopoietic progenitors in the paraaortic region were three to four times more numerous than in the newborn bone marrow.

Origin of Endothelia

Angiogenesis, the process through which new vessels arise from preexisting ones, has been intensely analyzed in adults both *in vivo* and *in vitro*, particularly during tumor formation. On the other hand, endothelial cells have to appear *de novo* in the embryo, a process now conventionally designated as vasculogenesis. Tracing of endothelial cells has been achieved in quail/chick chimeras by means of two monoclonal antibodies (Mabs) that recognize the quail hemangioblastic (hemopoietic and endothelial) lineage, aMB1 and QH1 (Péault et al., 1983, 1988; Dieterlen-Lièvre, 1984). These Mabs are highly sensitive tools and have no affinity for cells of the chick. Both vasculogenesis and angiogenesis have been found to occur in avian chimeras analyzed by means of the QH1 antibody (Pardanaud et al., 1987, 1989). Vasculogenesis occurred during the differentiation of organs that are made up of mesoderm and endoderm, while angiogenesis was the rule for organs that are constituted of mesoderm plus ectoderm. The two processes were demonstrated by two-way grafting of the rudiments: Any QH1 + cell in a quail rudiment developing in a chick host was donor-derived, while any QH1 + cell in a chick rudiment grafted into a quail host was host-derived, i.e., extrinsic to the rudiment. In lung, pancreas, intestine, or spleen, endothelial cells developed from intrinsic progenitors present in the mesoderm of the rudiments; in contrast, in the limb bud, endothelial cells from the host colonized the bone marrow, muscles, and connective tissue surrounding the bones. The extrinsic origin of bone marrow endothelia had previously been evinced in the same type of chimeric explants (Jotereau and Le Douarin, 1978) with the nucleolar marker. However, only a fraction of the endothelial cell nuclei bear the distinctive heterochromatin mass. The quail hemangioblastic lineage-specific Mab makes it clear that in grafted quail limb buds, no positive cells are present, while in chick grafted buds, all endothelial cells are positive. Interestingly the hemopoietic cells followed a different pattern: In both internal and external organ rudiments, they had an extrinsic origin, i.e., they were always provided by the host and colonized the rudiments. Thus they clearly have a different origin than the endothelial cells in the case of internal organs. This disjunction is particularly striking in the case of the spleen, which is pervaded by a dense vessel network and is an active hemopoietic organ during embryonic life.

These experimental data have led us to extend the hypothesis put forward by Wilt (1965), who proposed that hemopoiesis is enhanced in the extraembryonic islands and in the mesoderm by the tightly associated endoderm. The effect of endoderm, according to Wilt, might be either trophic or inductive. It is clear now that a similar effect operates on the precursors of endothelium.

A last comment is timely. Reciprocal grafting discloses distinct origins of endothelial cells and hemopoietic cells during organogenesis, whereas descriptive studies suggest a common ontogenetic origin of these cell lineages. In the extraembryonic area during the second day of embryonic development, blood islands arise as apparently homogeneous aggregates of cells, the peripheral ones then becoming endothelial and the central ones hemopoietic. One day later, aggregates of hemopoietic cells are clustered on the ventrolateral, luminal aspects of the aortic endothelium. No experimental data prove or disprove yet whether such physical associations result from the existence of a common precursor.

Conclusion

The quail/chick chimera technology has contributed to the comprehensive understanding of the ontogeny of practically all embryonic cell lineages. The information acquired in the avian model has been extended often to the mammalian embryo, with strategies devised along the guidelines already established for the bird.

For researchers using the mammalian embryo, transgenesis has provided a rewarding tool. Successful means of creating transgenic chickens are not yet routine but may be close at hand. However, the lengthy generation time of the chicken (6 months) will hinder the establishment of transgenic families. In that context, we surmise that surgical chimeras will be a tool for genetic studies. They might be constructed by introducing defined groups of cells with genetic material modified by *in vitro* transfection, for instance, to analyze the interactions between cells producing and cells receiving a signal.

References

Balaban, E., Teillet, M. A., and Le Douarin, N. M. (1988). Application of the quail-chick chimera system to the study of brain development and behavior. *Science*, 241, 1339–1342.

Baroffio, A., Dupin, E., and Le Douarin, N. M. (1988). Clone-forming ability and differentiation potential of migratory neural crest cells. *Proc. Natl. Acad. Sci. USA*, 85, 5325–5329.

Beaupain, D., Martin, C., and Dieterlen-Lièvre, F. (1979). Are developmental hemoglobin changes related to the origin of stem cells and site of erythropoiesis? *Blood*, 53, 212–225.

Broner-Fraser, M., and Fraser, S. E. (1988). Cell lineage analysis reveals multipotency of some avian neural crest cells. *Nature*, 335, 161–167.

Coltey, M., Jotereau, F. V., and Le Douarin, N. M. (1987). Evidence for a cyclic renewal of lymphocyte precursor cells in the embryonic chick thymus. *Cell Diff.*, 22, 71–82.

Cormier, F., and Dieterlen-Lièvre, F. (1988). The wall of the chick embryo aorta harbours M-CFC, G-CFC, GM-CFC and BFU-E. *Development*, 102, 279–285.

Cormier, F., De Paz, P., and Dieterlen-Lièvre, F. (1986). In vitro detection of cells with monocytic potentiality in the wall of the chick embryo aorta. *Dev. Biol.*, 118, 167–175.

Couly, G., and Le Douarin, N. M. (1988). The fate-map of the cephalic neural primordium at the presomitic to the 3-somite stage in the avian embryo. *Development*, 103, suppl. 101–113.

Couly, G., and Le Douarin, N. M. (1990). Head morphogenesis in embryonic avian chimeras: Evidence for a segmental pattern in the ectoderm corresponding to the neuromeres. *Development*, 108, 543–558.

Couly, G. F., Coltey, P. M., and Le Douarin, N. M. (1992). The developmental fate of the cephalic mesoderm in quail-chick chimeras. *Development*, 14, 1–15.

Dieterlen-Lièvre, F. (1975). On the origin of haemopoietic stem cells in the avian embryo: An experimental approach. *J. Embryol. Exp. Morphol.*, 33, 607–619.

Dieterlen-Lièvre, F. (1984). Emergence of intraembryonic blood stem cells studied in avian chimeras by means of monoclonal antibodies. *Comp. Dev. Comp. Immunol.*, suppl. 3, 75–80.

Dupin, E., Baroffio, A., Dulac, C., Cameron-Curry, P., and Le Douarin, N. M. (1990). Schwann cell differentiation in clonal culture of the neural crest as evidenced by the anti-SMP monoclonal antibody. *Proc. Natl. Acad. Sci. USA*, 87, 1119–1123.

Hallonet, M., Teillet, M. A., and Le Douarin, N. M. (1990). A new approach to the development of the cerebellum provided by the quail-chick marker system. *Development*, 108, 19–31.

Jotereau, F., and Le Douarin, N. M. (1978). The developmental relationship between osteocytes and osteoclasts: A study using the quail-chick nuclear marker in endochondral ossification. *Dev. Biol.*, 63, 253–265.

Jotereau, F., Heuze, F., Salomon-Vie, V., and Gascan, H. (1987). Cell kinetics in the fetal mouse thymus: Precursor cell input, proliferation and emigration. *J. Immunol.*, 138, 1026–1030.

Kinutani, M., Tan, K., Coltey, M., Kitaoka, K., Nagano, Y., Takashima, Y., and Le Douarin, N. M. (1989). Avian spinal cord chimeras. Further studies on the neurological syndrome affecting the chimeras after birth. *Cell Differ. Dev.*, 26, 145–162.

Lassila, O., Eskola, J., Toivanen, P., Martin, C., and Dieterlen-Lièvre, F. (1978). The origin of lymphoid stem cells studied in chick yolk sac-embryo chimaeras. *Nature*, 272, 353–354.

Lassila, O., Martin, C., Toivanen, P., and Dieterlen-Lièvre, F. (1982). Erythropoiesis and lymphopoiesis in the chick yolk-sac-embryo chimeras: Contribution of yolk sac and intraembryonic stem cells. *Blood*, 49, 377–381.

Le Douarin, N. M. (1969). Particularités du noyau interphasique chez la caille japonaise (Coturnix coturnix japonica). Utilisation de ces particularités comme "marquage biologique" dans les recherches. *Bull. Biol. Fr. Belg.*, 103, 435–452.

Le Douarin, N. M. (1973). A biological cell labeling technique and its use in experimental embryology. *Dev. Biol.*, 30, 217–222.

Le Douarin, N. M. (1986). Cell line segregation during peripheral nervous system ontogeny. *Science*, 231, 1515–1522.

Le Douarin, N., and Jotereau, F. (1973). Origin and renewal of lymphocytes in avian embryo thymuses. *Nature New Biol.*, 246, 25–27.

Le Douarin, N. M., and Teillet, M. A. (1973). The migration of neural crest cells to the wall of the digestive tract in avian embryo. *J. Embryol. Exp. Morphol.*, 30, 31–48.

Le Douarin, N. M., and Teillet, M. A. (1974). Experimental analysis of the migration and differentiation of neuroblasts of the autonomic nervous system and of neurectodermal mesenchymal derivatives, using a biological cell marking technique. *Dev. Biol.*, 41, 162–184.

Le Lièvre, C. (1974). Rôle des cellules mésectodermiques issues des crêtes neurales céphaliques dans la formation des arcs branchiaux et du squelette viscéral. *J. Embryol. Exp. Morphol.*, 31, 453–477.

Le Lièvre, C., (1978). Participation of neural crest-derived cells in the genesis of the skull in birds. *J. Embryol. Exp. Morphol.*, 47, 17–37.

Le Lièvre, C., and Le Douarin, N. (1975). Mesenchymal derivatives of the neural crest: Analysis of chimaeric quail and chick embryos. *J. Embryol. Exp. Morphol.*, 34, 125–154.

Martin, C. (1972). Technique d'explantation *in ovo* de blastodermes d'embryons d'oiseaux. *C. R. Soc. Biol.*, 116, 283.

Martin, C., Beaupain, D., and Dieterlen-Lièvre, F. (1978). Developmental relationships between vitelline and intra-embryonic haemopoiesis studied in avian "yolk sac chimeras." *Cell Diff.*, 7, 115–130.

Moore, M. A. S., and Owen, J. J. T. (1965). Chromosome marker studies on the development of the haemopoietic system in the chick embryo. *Nature*, 208, 958–989.

Moore, M. A. S., and Owen, J. J. T. (1967). Stem cell migration in developing myeloid and lymphoid systems. *Lancet*, 2, 658–659.

Ohki, H., Martin, C., Corbel, C., Coltey, M., and Le Douarin, N. M. (1989). Effects of early embryonic grafting of foreign tissues on the immune response of the host. In: *Recent Advances in Avian Immunology Research*, edited by B. S. Boghal. Alan R. Liss, New York, pp. 3–17.

Pardanaud, L., Altmann, C., Kitos, P., Dieterlen-Lièvre, F., and Buck, C. (1987). Vasculogenesis in the early quail blastodisc as studied with a monoclonal antibody recognizing endothelial cells. *Development*, 100, 339–349.

Pardanaud L., Yassine F., and Dieterlen-Lièvre F. (1989). Relationship between vasculogenesis, angiogenesis and hemopoiesis during avian ontogeny. *Development*, 105, 473–485.

Péault, B, Coltey, M., and Le Douarin, N. M. (1988). Ontogenic emergence of a quail leukocyte/endothelium surface antigen. *Cell Differ.*, 23, 165–174.

Péault, B., Thiery, J. P., and Le Douarin, N. M. (1983). A surface marker for the hemopoietic and endothelial cell lineages in the quail species defined by a monoclonal antibody. *Proc. Natl. Acad. Sci. USA*, 80, 2976–2980.

Pink, J. R., Vainio, O., and Rijnbeek, A. M. (1985). Clones of B lymphocytes in individual follicles of the bursa of Fabricius. *Eur. J. Immunol.*, 15, 83–87.

Rossant, J., and Frels, W. I., (1980). Interspecific chimeras in mammals: Successful production of live chimeras between *Mus musculus* and *Mus caroli*. *Science*, 208, 419–420.

Salaün, J., Bandeira, A., Khazaal, I., Calman, F., Coltey, M., Coutinho, A., and Le Douarin, N. M. (1990). Thymic epithelium tolerizes for histocompatibility antigens. *Science*, 247, 1471–1474.

Spemann, H. (1938). *Embryonic Development and Induction.* New Haven, CT: Yale University Press.

Tan, K., and Le Douarin, N. M. (1991). Development of the nuclei and cell migration in the medulla oblongata. *Anat. Embryol.,* 183, 321–343.

Teillet, M. A., Naquet, R., Le Gal La Salle, G., Merat, P., Schulerr, B., and Le Douarin, N. M. (1991). Transfer of genetic epilepsy by embryonic brain grafts in the chicken. *Proc. Natl. Acad. Sci. USA,* 88, 6966–6970.

Thiébaud, Ch. H. (1983). A reliable new cell marker in *Xenopus. Dev. Biol.,* 98, 245–249.

Weston, J. A. (1963). A radioautographic analysis of the migration and localization of trunk neural crest cells in the chick. *Dev. Biol.,* 6, 279–310.

Wilt, F. H. (1965). Erythropoiesis in the chick embryo: The role of endoderm. *Science,* 147, 1588–1590.

Chapter 8

Transfection of Chick Embryos Maintained Under *In Vitro* Conditions

Helen Sang, Clare Gribbin, Christine Mather, David Morrice, and Margaret Perry

Summary. We have developed a method for culture of the chick fertilized ovum to hatch. This method enables access to the earliest stages of embryo development, which normally occur within the oviduct of the hen. The procedure has three stages, to allow for the differing requirements of the embryo. The fate of plasmid DNA injected into fertilized ova prior to the first cleavage division was followed by culturing injected embryos for up to 7 days. Southern transfer analysis of total DNA extracted from injected embryos suggested that the introduced DNA replicated during the first 24 h of development and was subsequently gradually lost. The results of injecting the construct pHFBGCM, which contains the *lacZ* gene under control of the CMV immediate-early promoter, have been followed by staining for β-galactosidase activity. Expression of the reporter gene was first seen after 12 h of development. After 24 h, at the onset of blastulation, more than 90% of embryos contained stained cells within the area pellucida. The number of cultures with stained cells within the embryo, rather than in the extraembryonic blastoderm, decreased substantially after the primitive streak stage. After 7 days in culture, only 7% of surviving embryos had stained cells within embryonic tissues. These results suggest that, if plasmid DNA is integrated into the host chromosomes, integration is a relatively rare event. Possible future directions for development of the system described are discussed.

Introduction

The earliest stages of chick development are relatively inaccessible, as they take place within the oviduct of the hen (see Chapters 2 and 3). The first cell division takes place at the isthmus of the oviduct, 4 to 5 h after fertilization of the ovum. Subsequent cell divisions, which occur in the uterus while the shell membranes are laid down and calcification of the shell take place, are very rapid. By the time the egg is laid, the embryo consists of at least 60,000 cells (Foulkes, 1990). In order to explore the possibilities for genetic manipulation of poultry by introducing cloned DNA into the single-celled embryo, we have developed a method for culture of the fertilized ovum to hatch. In this chapter we outline the culture method and describe the results we have obtained from injection of DNA into fertilized ova. Finally, we discuss the possible future applications of the methods described.

In Vitro Culture of Chick Embryos

Culture of chick embryos for limited periods has been possible for many years. Waddington (1932) showed that embryos from 1-day incubated eggs explanted onto solid medium continued to develop for about 2 days. The method of New (1955) for culturing early embryos in aqueous medium using the vitelline membrane as support has been used in many developmental investigations. Embryos cultured with the yolk and albumen intact can survive for longer periods. Embryos from 3-day incubated eggs survived up to 15 days in petri dishes (Auerbach et al., 1974) and have been successfully maintained up to the period prior to hatch in plastic sacs (Dunn and Boone, 1976). Embryos cultured under these conditions are hypocalcemic and do not hatch (Rowlett, 1991). The egg shell has been shown to be required to obtain hatched chicks from quail (Ono and Wakasugi, 1984) and the domestic hen (Rowlett and Simkiss, 1987).

    Perry (1988) developed a method that enables the culture of chick fertilized ova through to hatch. It involves the use of three culture systems, which are designed to fulfill the requirements of the developing embryo: system I for the oviductal stages of development from fertilization to the blastoderm at time of lay; system II for embryogenesis from day 1 to day 4; system III for the period of embryonic growth from day 4 to day 22. The developing embryo is transferred from one system to the next at the appropriate times.

The Three-System Culture Method

Fertilized ova are obtained from laying hens that are sacrificed 2 to 3 h after the preceding egg was laid. The ovum is removed from the magnum and transferred to a glass container. Culture medium consisting of thin albumen and a solution of salts and glucose, to allow for the differences in composition of albumen of uterine and laid eggs, is added, and the culture vessel is sealed with saran wrap (Figure 1). During this phase of development it is important that the embryo is not submerged in culture medium. The culture is incubated at 41 – 42°C for 24 h. We refer to the cultures during the first 24 h of development as day 0 embryos, to avoid confusion with staging of embryo development from the time of lay (day 1). Embryos are then transferred to a recipient shell prepared by drilling off the sharp end of a shell from a freshly laid egg. The shell is filled with medium, sealed with cling film secured by nylon rings, and incubated at 38°C with intermittent rocking through an angle of 90° in hourly cycles (Figure 2). After 3 days the embryos are again transferred, based on the procedures described by Rowlett and Simkiss (1987), to a large recipient shell enabling an air space to be left above the embryo. The shell is sealed with cling film and incubation continued at 38°C until hatch.

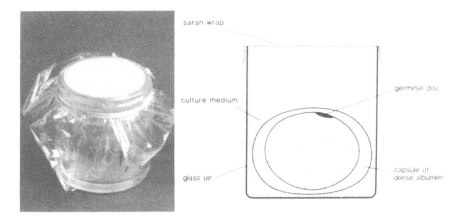

FIGURE 1. System I for culture of the fertilized ovum to blastoderm formation (24 h after removal from the magnum, day 0 to day 1) in a glass container. Diagram from Perry (1988).

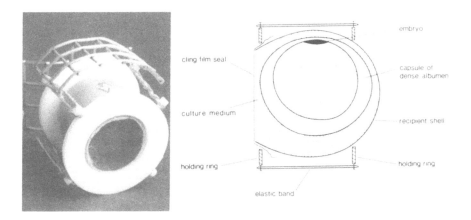

FIGURE 2. System II for culture of the chick embryo, from the laid egg stage, during embryogenesis (day 1 to day 4) and early growth phase (day 4 to day 8). Diagram from Perry (1988).

The embryos are turned through an angle of 30° for 5 days and then maintained stationary for the remainder of the incubation period (Figure 3). The glass jar used in system I may be replaced with a recipient egg shell, covered with a petri dish lid, and incubated in 5% CO_2 (Perry and Mather, 1991). No chicks have yet been hatched from embryos cultured from fertilized ova in a single shell, with no transfer steps.

Table 1 shows the survival rate of embryos cultured from fertilized ova through to hatch using the three culture systems in sequence. Approximately

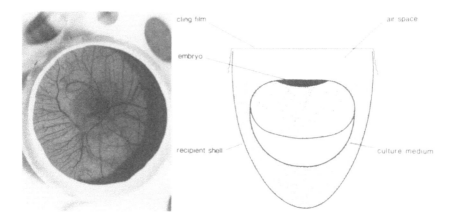

FIGURE 3. System III for culture of the chick embryo from day 4 to hatch. Diagram from Perry (1988).

Table 1
Survival of 112 Chick Embryos Grown in
Culture Following Removal from the
Magnum

Days of incubation	Number of embryos	Live embryos/cultured ova (%)
0	112	
4	77	69
9	44	39
19	34	30
22	17	15

Posthatch survival of cultured chick embryos = 10%
(11/112)

Data from Perry and Mather (1991).

70% of embryos survived to day 4 of incubation, but then some losses occurred as a result of transfer to the third culture system. There was a peak of embryo loss in the few days prior to hatch. The overall posthatch survival rate was about 10%. The embryos that died late in development had a series of defects that contributed to their failure to hatch. These included poor position in the shell, incomplete development of the chorioallantoic membrane, residual albumen, small size, and incomplete retraction of the yolk sac. These features are often found if normal eggs are not turned adequately during the first 8 days of incubation.

Turning of eggs during this period is critical for formation of the subembryonic fluid. The turning regime used may be less effective for cultured

Table 2
Mean Weight (g) of Subembryonic Fluid
Removed from 7-Day Embryos (1) from
Intact Eggs, (2) Cultured in System II from
Day 1, and (3) Cultured in System I then
Transferred to System II, Subjected to
Different Angles of Turning During
Incubation[a]

Angle	0°	30°	90°
1. Intact eggs	15.2	16.7	16.8
2. Day 1 to day 7	13.4	14.5	ND
3. Day 0 to day 7	11.8	14.2	ND

[a] REML analysis demonstrated that the differences
in weights of subembryonic fluid were significantly
affected by both the angle of turning and the culture
system ($P < 0.001$).

embryos than for intact eggs, because the internal structure of the egg is disrupted by transfer to recipient shells. We have compared the formation of subembryonic fluid between intact incubated eggs and embryos cultured from fertilized ova, using systems I and II, and from laid eggs, using system II (Table 2). The mean weight of subembryonic fluid was greater with a greater angle of turning in all cases and was significantly less in cultured embryos than in the intact eggs. We are currently testing modifications of the turning procedure to try to increase the hatch rate of healthy chicks cultured from fertilized ova. Naito et al. (1990) modified the culture method and obtained a hatch rate of over 30% by removing the dense albumen capsule from around the yolk during the transfer from system I to II. We have not been successful in obtaining an improvement in hatch rate using this modification. Naito and Perry (1989) have shown that the three-stage culture system can also be used for culture of embryos from the early cleavage stage. Embryos in early cleavage can be obtained as soft-shelled eggs from the uterus by induced oviposition, which does not involve sacrifice of the hens.

The three culture systems can be used separately or in pairs if complete culture from fertilized ova to hatched chicks is not required. Table 3 shows the success we have achieved in obtaining embryos or chicks using different stages of the culture method. Embryos taken from 3-day incubated eggs were cultured in system III with a hatch rate of 40%. Over 20% of embryos from laid eggs cultured using system II and transfer to system III hatched successfully. Embryos from laid eggs cultured in system II only had a high survival rate until day 7 but could not be maintained longer. Approximately 70% of embryos from fertilized ova cultured in system I then transferred to

Table 3
Survival Rate of Chick Embryos in Culture

Culture system	Phase of development[a]	Number of cultures	Number of embryos (%)	Number of chicks (%)[b]
III	Day 4 to hatch	64	—	26 (41)
II	Day 1 to day 7	47	39 (83)	—
II to III	Day 1 to hatch	49	—	11 (22)
I to II	Day 0 to day 7	43	29 (67)	—

[a]   The embryos were staged in days from fertilization (day 0).
[b]   Approximately 50% of the hatchlings were healthy.

Data from Perry (1988).

system II survived to 7 days. We have used this method of culture of fertilized ova for 7 days, using systems I and II, to analyze the fate of DNA injected at the single-cell stage.

Injection of DNA

DNA is injected into the cytoplasm of the germinal disk of ova removed from the magnum, prior to culture in system I (Sang and Perry, 1989, Figure 4). Injection into zygote pronuclei is not possible yet because they are masked by yolk spheres in the cytoplasm and they are difficult to distinguish from the supernumerary nuclei resulting from polyspermic fertilization. The injections are into the center of the germinal disk, where the zygote pronuclei are usually located (Perry, 1987). An air-filled microinjection system with needles of tip diameter 10 μm is used. One to 5 nL of DNA (25–500 pg) dissolved in water is injected at a depth of between 50 and 200 μm below the vitelline membrane. Injection of precleavage embryos reduces their survival rate at 7 days to approximately 50%. We have carried out two series of experiments with different plasmid constructs to determine the fate of injected DNA. In

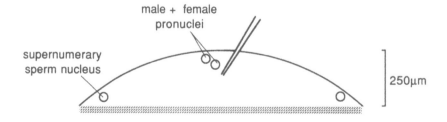

FIGURE 4.   Diagram of a vertical section through the germinal disk, showing the pronuclei, supernumerary nuclei, and approximate site of injection.

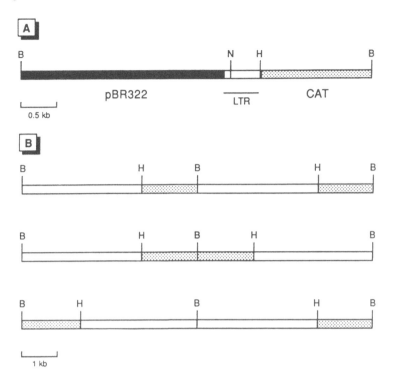

FIGURE 5. (A) Map of pRSVcat, linearized at the single *Bam*HI site. B. *Bam*HI; H, *Hin*dIII; LTR, RSV promoter/enhancer; CAT, chloramphenicol acetyl transferase gene. (B) Possible orientations of pRSVcat molecules after random ligation of *Bam*HI-cut molecules. Sizes of restriction fragments generated by *Hin*dIII digestion are indicated: Head-to-tail ligation will result in 5-kb *Hin*dIII fragments, head-to-head ligation in 3.4-kb *Hin*dIII fragments, and tail-to tail ligation in 6.8-kb *Hin*dIII fragments.

the first series of experiments the fate of the plasmid was mainly followed by DNA analysis (Sang and Perry, 1989) and in the second series a construct expressing β-galactosidase was analyzed histochemically (Perry et al., 1991).

Injection of pRSVcat

The construct pRSVcat (Figure 5A), which contains the bacterial chloramphenicol acetyl transferase (CAT) gene under the control of the Rous sarcoma virus (RSV) promoter, was injected in linear form (*Bam*HI-digested) into fertilized ova and the embryos cultured for different periods of up to 7 days. Within 24 h of injection, the plasmid DNA was detected as high-molecular-weight molecules (Figure 6A). If this DNA was subsequently digested with *Hin*dIII, which cuts once within pRSVcat, three restriction fragments derived from the plasmid were detected (Figure 6B). These restriction fragments can

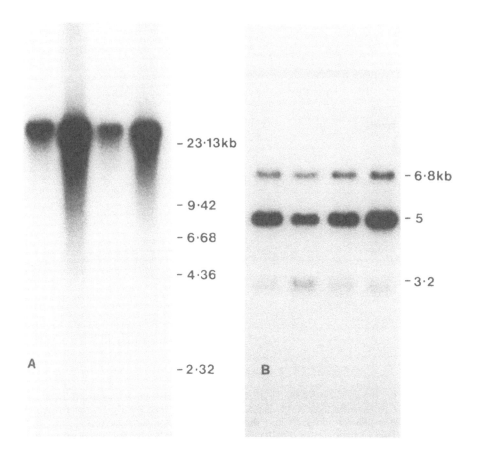

FIGURE 6. (A) DNA extracted from four individual embryos that were cultured for 24 h after injection into fertilized ova of pRSVcat cut with *Bam*HI. Undigested DNA was separated in a 0.7% agarose gel, Southern blotted, and probed with plasmid vector (pUC9). Evidence of hybridization is seen at the limiting mobility of the gel. (B) DNA extracted from four embryos cultured for 24 h after injection of pRSVcat was digested with *Hin*dIII and the fragments were separated on a 1% agarose gel, Southern blotted, and probed with plasmid vector (pUC9). The sizes of the three characteristic *Hin*dIII fragments generated from ligation of *Bam*HI-linearized monomers of pRSVcat are indicated (see Figure 5). From Sang and Perry (1989).

be generated by ligation of the linear plasmid molecules in tandem arrays in direct or inverse orientations (Figure 5B). These observations indicated that the linear DNA molecules ligated in random orientations to produce large concatemers of the injected plasmid. High-molecular-weight DNA was also detected after 24 h of incubation when pRSVcat was injected in monomeric circular form. These molecules contained direct repeats of the plasmid, suggesting that recombination had occurred between the monomers. The amount of pRSVcat present in embryos at different times after injection of linear

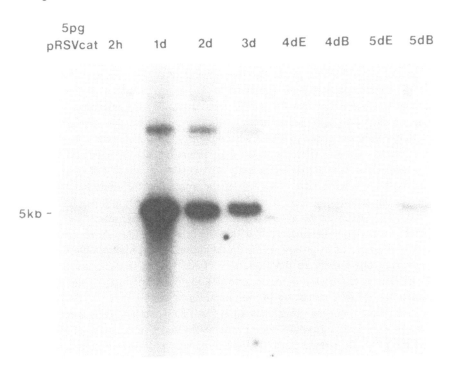

FIGURE 7. Embryos were injected with approximately 50 pg of *Bam*HI-linearized pRSVcat and cultured for 2 h or 1, 2, 3, 4, or 5 days. The day 4 and day 5 samples were separated into embryo (E) and extraembryonic membranes (B). Equivalent amounts of DNA extracted from each sample were digested with *Bam*HI, separated in a 1% gel, Southern blotted, and probed with plasmid vector (pUC9). The pRSVcat DNA apparently replicated between 2 and 24 h and then the amount present in the embryos gradually decreased with prolonged culture.

molecules was estimated. The injected DNA apparently replicated, possibly by as much as 50-fold, during the first 24 h of development, but was then gradually diluted out and lost during the subsequent growth of the embryos (Figure 7). After 7 days of incubation, plasmid DNA was detectable in 30% of the cultures, but only 7% of embryos contained detectable plasmid sequences when embryos and extraembryonic membranes were analyzed separately. Restriction digests of DNA from 61 six- and seven-day cultures were analyzed by Southern transfer to look for restriction fragments that could be generated by chromosomal integration of the plasmid. No junction fragments were detected, suggesting that the plasmid DNA persisted episomally. If copies of the plasmid did integrate, such events were rare and the embryos were mosaic. Expression of the CAT gene was detected in 2-day and 7-day embryos by assaying for CAT enzyme activity.

Expression of *lacZ* Constructs

A *lacZ* expression construct was used to examine the spatial and temporal distribution of DNA injected into the germinal disk. Histochemical detection of β-galactosidase activity was used to derive a more detailed picture than could be obtained by analyzing DNA, and in particular to establish if the site of injection corresponded to areas of the germinal disk that give rise to the embryo. The construct pHFBGCM, containing *lacZ* under the control of the cytomegalovirus (CMV) immediate early promoter, was injected and embryos examined for β-galactosidase activity after periods of 3 h to 7 days in culture. Stained cells were first observed in the center of the embryos 9 h after injection, when they were in mid-cleavage (Eyal-Giladi and Kochav, 1976, stage IV). After 24 h, at the beginning of blastulation, more than 90% of embryos contained stained cells within the area pellucida. Stained cells were present singly or in groups in the epithelial and subepithelial layers in areas covering one-quarter to three-quarters of the blastoderm (Figure 8A). The variation in expression between embryos may be explained by the different amounts of DNA persisting in individual embryos.

In some embryos, after incubation for 2 days, stained cells were seen in the prospective embryonic region around Hensen's node at the anterior end of the primitive streak. Large stained cells were prominent in the area opaca in most embryos at this stage (Figure 8B). The expression of *lacZ* declined sharply after 3 days in culture, at the stage corresponding with the end of gastrulation. Stained cells were often seen in the extraembryonic membranes, but rarely in the tissues of the embryo itself. By 7 days in culture, *lacZ* expression was detected in only 1 of 12 surviving embryos. Clones of cells in this embryo may represent a late integration event.

These results confirm the conclusions from the Southern transfer analysis of embryos injected with pRSVcat. Plasmid DNA injected into fertilized ova is gradually diluted out and lost during embryonic growth and development. The persistance of exogenous DNA in the extraembryonic membranes to later stages of development than in the embryo itself may be related to the lower rates of cell division in these tissues than in the developing embryo. There is little evidence for chromosomal integration of injected DNA. Naito et al. (1991a) introduced a *lacZ* expression construct, containing the RSV enhancer and chicken β-actin promoter, into fertilized ova in similar experiments. They also observed a mosaic pattern of expression in most embryos. One embryo, out of a total of 55, stained for β-galactosidase activity in many of the cells of the embryo, possibly as the result of an early integration event. We have not taken embryos injected at the single-cell stage through to hatch because the evidence is that integration of injected DNA is very unlikely and therefore little benefit could be gained from analyzing hatched chicks.

A

B

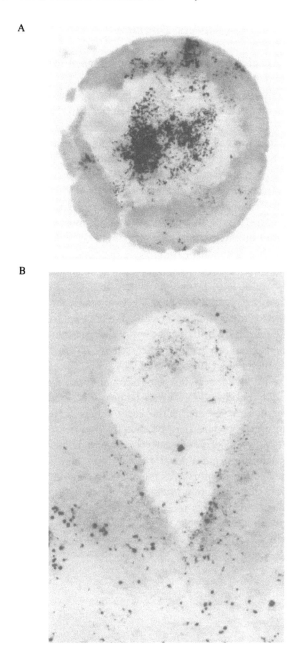

FIGURE 8. (A) Chick embryo at early blastoderm (Eyal-Giladi and Kochav, 1976, stage X), 24 h after injection of pHFBGCM, stained for β-galactosidase activity. (B) Chick embryo cultured for 2 days after injection of pHFBGCM. Stained cells can be seen in the area opaca and in the area pellucida around Hensen's node and in the germinal crescent, anterior to the primitive streak.

Future Directions

The chick embryo culture method enables experimental manipulation of the oviductal stages of the very early chick embryo. The hatch rate of healthy chicks using a combination of the three culture is systems is rather low. Modifications of the method, such as that described by Naito et al. (1990) and optimization of incubation conditions, currently being defined in our laboratory, may be expected to raise the hatch rate to an estimated 30%. A consistent hatch rate higher than this estimate is unlikely to be achieved, as losses are inevitable during transfer steps and variation in the condition of ova on collection from the oviduct influences embryo survival. Simplifications to the method, for example, reducing the number of transfers of embryos between culture vessels, will also make the method more practicable.

DNA injection into the cytoplasm of the single-cell chick embryo rarely, if at all, leads to chromosomal integration of the foreign DNA. The system in its present state of development is not a viable method for producing transgenic birds. Methods to promote efficient integration of injected DNA or to exploit its episomal persistence must be developed if the present approach to gene transfer into birds is to be successful. We are exploring the possibility of promoting integration by utilizing the virus-specific integration system of retroviruses, which has been precisely defined by *in vitro* studies (Katz et al., 1990). Other methods of introducing DNA, e.g., lipofection (Felgner et al., 1987), may result in higher frequencies of integration. The use of episomal vectors to propagate introduced DNA may be possible, although such vectors for use in higher eukaryotes are in the early stages of development. The injection and culture procedure may be directly useful as a transient assay method for the expression of genes early in development. We are currently exploring this possibility using a developmentally regulated promoter. Naito et al. (1991b) used cultured embryos as recipients in cell transfer experiments, based on the successful production of chimeric chicks by Petitte et al. (1990). The use of cultured embryos as recipients for injection of cells could also allow the use of embryos between stage V and stage X (Eyal-Giladi and Kochav, 1976), to investigate the influence of age of recipient on production of chimeric birds.

References

Auerbach, R., Kubai, R., Knighton, D., and Folkman, J. (1974). A simple procedure for the long-term cultivation of chick embryos. *Dev. Biol.*, 41, 391–394.

Dunn, B. E., and Boone, M. A. (1976). Growth of the chick embryo *in vitro. Poultry Sci.*, 55, 1067–1071.

Eyal-Giladi, H., and Kochav, S. (1976). From a cleavage to primitive streak formation: A complementary normal table and a new look at the first stages of the development of the chick. 1. General morphology. *Dev. Biol.*, 49, 321–337.

Felgner, P. L., Gadek, T. R., Holm, M., Roman, R., Chan, H. W., Wenz, M., Northrop, J. P., Ringold, G. M., and Danielson, M. (1987). Lipofection: A highly efficient, lipid-mediated DNA-transfection procedure. *Proc. Natl. Acad. Sci. USA*, 84, 7413–7417.

Foulkes, A. G. (1990). The unincubated avian blastoderm — Its characterisation and an investigation of developmental quiescence. Ph. D. thesis, University of Southampton, England.

Katz, R. A., Merkel, G., Kulkosky, J., Leis, J., and Skalka, A. M. (1990). The avian retroviral In protein is both necessary and sufficient for integrative recombination in vitro. *Cell*, 63, 87–95.

Naito, M., and Perry, M. M. (1989) Development in culture of the chick embryo from cleavage to hatch. *Br. Poultry Sci.*, 30, 251–256.

Naito, M., Agata, K., Otsuka, K., Kino, K., Ohta, M., Hirose, K., Perry, M. M., and Eguchi, G. (1991a). Embryonic expression of β-actin-lacZ hybrid gene injected into the fertilised ovum of the domestic fowl. *Int. J. Dev. Biol.*, 35, 69–75.

Naito, M., Nirasawa, K., and Oishi, T. (1990). Development in culture of the chick embryo from fertilised ovum to hatching. *J. Exp. Zool.*, 25, 322–326.

Naito, M., Watanabe, M., Kinutani, M., Nirasawa, K., and Oishi, T. (1991b). Production of quail-chick chimeras by blastoderm cell transfer. *Br. Poultry Sci.*, 32, 79–86.

New, D. A. T. (1955). A new technique for the cultivation of the chick embryo *in vitro. J. Embryol. Exp. Morphol.*, 3, 320–331.

Ono, T., and Wakasugi, N. (1984). Mineral content of quail embryos cultured in mineral-rich and mineral-free conditions. *Poultry Sci.*, 63, 159–166.

Perry, M. M. (1987) Nuclear events from fertilisation to the early cleavage stages in the domestic fowl (*Gallus domesticus*). *J. Anat.*, 150, 99–109.

Perry, M. M. (1988) A complete culture system for the chick embryo. *Nature*, 331, 70–72.

Perry, M. M., and Mather, C. M. (1991). Satisfying the needs of the chick embryo in culture, with emphasis on the first week of development. In: *Avian Incubation*, edited by S. G. Tullett. Poultry Science Symposium, 22, 91–105, Butterworth-Heinemann, London.

Perry, M. M., Morrice, D., Hettle, S., and Sang, H. (1991). Expression of exogenous DNA during the early development of the chick embryo. *Roux's Arch. Dev. Biol.*, 200, 312–319.

Petitte, J. N., Clark, M. E., Liu, G., Verrinder Gibbins, A. M., and Etches, R. J. (1990). Production of somatic and germline chimeras by transfer of early blastoderm cells. *Development*, 108, 185–189.

Rowlett, K. (1991). Embryo growth and development in culture. In: *Avian Incubation*, edited by S. G. Tullet. Poultry Science Symposium, 22, 107–124, Butterworth-Heinemann, London.

Rowlett, K., and Simkiss, K. (1987). Explanted embryo culture: in vitro and in ovo techniques for the domestic fowl. *Br. Poultry Sci.*, 28, 91–101.

Sang, H., and Perry, M. M. (1989). Episomal replication of cloned DNA injected into the fertilised ovum of the hen, *Gallus domesticus. Mol. Reprod. Dev.*, 1, 98–106.

Waddington, C. H. (1932). Experiments on the development of chick and duck embryos cultivated *in vitro. Phil. Trans. Roy. Soc.*, 221B, 179–230.

Chapter 9

Avian Leukosis Retroviruses and Gene Transfer into the Avian Genome

Donald W. Salter, William S. Payne, Lyman B. Crittenden, Mark J. Federspiel, Christos J. Petropoulos, James A. Bradac, and Stephen Hughes

Summary. To date, avian retroviruses remain the only method for transferring foreign genes into the avian germline. In this report, our results using avian leukosis retroviruses as vectors to transfer and express foreign gene constructs in chicken somatic and germ cells are summarized. In addition, the characteristics of two interesting transgenic chicken lines with desirable disease-resistance properties are discussed. Our attempt to develop Japanese quail as a transgenic animal model for poultry is also summarized.

Introduction

Many avian geneticists have dreamed of bypassing conventional selection procedures by inserting desirable genes into the avian germline. The report of the insertion of a growth hormone gene into the mouse germline that resulted in an increased growth rate and final size was the first step in the fulfillment of this dream (Palmiter et al., 1982). However, the transfer of foreign DNA into the avian germline has lagged far behind the successes in mammalian species. This lack of progress is due primarily to the inaccessibility of the pronuclei in the avian embryo at the time DNA is injected in standard transgenic techniques. Germline insertion for poultry has been achieved by retroviral infection of young embryos (Salter et al., 1987; Bosselman et al., 1989a; Lee and Shuman, 1990) and congenital infection of very early embryos by viremic hens (Chen et al., 1990; Crittenden and Salter, 1990b). Nevertheless, substantial progress has been made in alternative methods such as primordial germ cell (PGC) isolation, retroviral infection, and transfer (Wentworth et al., 1989; Simkiss et al., 1989, 1990; Nakamura et al., 1991), blastodermal cell isolation, transfection, and transfer (Gibbins et al., 1990; Petitte et al., 1990; Naito et al., 1991a; see also Chapter 6), sperm transfection (see Chapter 10), and transfection of very young embryos (Sang and Perry, 1989; Perry and Sang, 1990; Naito et al., 1991b; and Chapter 8).

Retroviruses as Tools for the Insertion of Foreign Genes into the Avian Germline

We and others proposed the use of replication-competent and -defective avian retroviral vectors to transfer foreign DNA into the avian germline (Crittenden and Salter, 1985, 1986; Shuman and Shoffner, 1986; Hughes et al., 1986;

Freeman and Bumstead, 1987). We believed that this approach would succeed for the following reasons.

1. A retrovirus replicates by inserting a double-stranded DNA copy of its genome into the host genome. Figures 1A and 1B show the avian leukosis virus (ALV) DNA proviral genome structure and ALV virion structure, respectively. Briefly, the avian retroviral life cycle begins (Figure 1C) with the binding of the virion via the receptor-binding glycoprotein (SU) to specific receptors in the avian cell membrane. After internalization and release of the nucleoprotein (NC)-capsid (CA)-matrix (MA)-protease (PR)-RNA genome complex into the cytoplasm, the RNA genome is converted to a double-stranded proviral DNA by reverse transcriptase (RT) (Figure 1D). Viral DNA migrates to the cell nucleus (Figure 1E), where it integrates randomly via integrase (IN) into the host cell DNA (Figure 1F) as a provirus. At Figure 1G, the proviral DNA directs its own synthesis by transcribing progeny RNA genomes and viral messenger RNAs for translation into viral precursor proteins, which are further processed into mature viral proteins. The viral proteins (MA, CA, NC, PR, RT, IN) and progeny RNA genomes assemble at the cell membrane with the viral glycoproteins (SU and TM) and bud from the cell membrane (Figures 1H and 1I) to begin again. The reader is referred to reviews by Varmus (1988), Sanes (1989), and Crittenden (1991) for further details on the retrovirus life cycle. Retroviruses can be genetically engineered to serve as vectors that transmit natural or foreign genes into the host genome [reviewed by Varmus (1988) and Sanes (1989)]. Jaenisch (1976) first demonstrated that retroviruses could be used as vectors to transfer DNA into somatic and germ cells of early mouse embryos.

2. Germ cells of most or all animals contain many different proviruses that are transmitted in a Mendelian manner. The avian species are no exception, having perhaps hundreds of different germline-inserted proviruses. These proviruses are clonal within the chicken DNA, and can be silent or expressed to yield virus gene products or infectious retroviruses [reviewed by Smith (1986) and Crittenden (1991)]. Since the avian germline is accessible to natural infection, we believed it should be accessible to artificial infection by avian retroviral vectors.

Germline Approach for Chickens

The lack of experimental information about the infection of unfertilized and newly fertilized ova persuaded us to try a simple approach to infect germ cells with avian retroviruses. Some facts were known about the generation of PGCs during the incubation of the fertilized egg. PGCs appear in the germinal crescent after about 18 h of incubation and can be found in the blood at about 3 days incubation (Wentworth et al., 1989; Simkiss et al., 1989, 1990). We reasoned that if high-titer, replication-competent avian leukosis virus (ALV) was placed near the blastoderm just before incubation began,

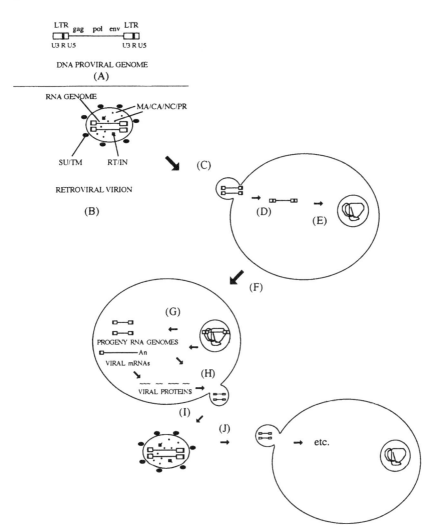

FIGURE 1. Structure of proviral genome and virion and retroviral life cycle. A. Structure of the avian leukosis virus (ALV) proviral DNA genome after it integrates into the chicken cell genome (see part G). The proviral genome contains three coding regions, *gag*, *poly*, and *env*, which code for group-specific antigen, polymerase, and envelope proteins. The U3, R, and U5 regions of the long terminal repeats (LTR) at each end of the proviral genome provide functions necessary for the promotion, initiation, and polyadenylation of messenger transcripts and contain sequences necessary for integration into the host cell genome. B. Structure of the retroviral virion. The retrovirus is made up of two copies of a single-stranded RNA genome enclosed in viral proteins and covered with an envelope [derived from the cell (not shown)] containing viral glycoproteins. The ALV *gag* gene codes for the matrix (MA), capsid (CA), nucleoprotein (NC), and protease (PR) proteins; the *pol* gene codes for the reverse transcriptase (RT) and integrase (IN) proteins; the *env* gene codes for the receptor-binding (SU) and transmembrane (TM) proteins. C–J. Retroviral life cycle. See text for description. Figure and nomenclature adapted from Sanes (1989) and Crittenden (1991).

the embryo would be infected and the virus would spread and infect some of the PGCs. Simply testing for genetic transmission of proviral DNA to progeny by using recombinant DNA methodology would provide evidence of germline insertion.

Germline Method for Chickens

The reader is referred to the original publications and reviews by us and other authors for details of the methods used to place replication-competent and -defective avian retroviruses in or near the developing avian embryo in newly laid eggs (Salter et al., 1986, 1987; Salter and Crittenden, 1988, 1989, 1991; Crittenden and Salter, 1985, 1986, 1988, 1990a, 1990b; Crittenden et al., 1989; Bosselman et al., 1989a, 1989b, 1990a, 1990b; Briskin et al., 1991; Lee and Shuman, 1990; Shuman, 1991), unfertilized ova (Shuman and Shoffner, 1986; Shuman, 1991), or PGCs (Simkiss et al., 1990), or congenital infection of very early embryos by viremic hens (Chen et al., 1990, Kopchick et al., 1991; Crittenden and Salter, 1990b). We routinely inject 50 to 100 µL of high-titer, virus-containing solutions (10^4 to 10^5 infectious units) or virus-producing chicken embryo fibroblasts (1×10^6 cells) blindly into the yolk near the developing blastoderm of unincubated fertile eggs from line 0 chickens. Line 0 chickens lack endogenous proviruses that are closely related to the ALV vectors (Astrin et al., 1979), thereby enabling us to use a simple dot-blot procedure to detect low-frequency transmission of proviral DNA in large numbers of progeny (Salter et al., 1986). Infected male chicks are raised to maturity and mated to specific-pathogen-free (SPF) line 0 females. The progeny are then analyzed for proviral DNA. The availability of subgroup A susceptible SPF line 0 chickens and relatively benign subgroup A recombinant ALV were important to our success. Benign recombinant ALV can be produced from pathogenic exogenous ALV by exchanging the LTR (see Figure 1A), which is the portion of the provirus genome known to be responsible for most of the pathogenicity, with the smaller LTR from the low-pathogenic subgroup E endogenous ALV (Hughes et al., 1986, 1987, 1990; Greenhouse et al., 1988; Petropoulos and Hughes, 1991; Wright and Bennett, 1986).

Successful Insertion of Retroviral Genes into the Chicken Germline

We reported evidence for successful germline insertion at a symposium of the 1985 Poultry Science Annual Meeting (Salter et al., 1986). Line 0 viremic males were generated by *in ovo* injection of fertile, unincubated eggs with replication-competent subgroup A wild-type and recombinant ALV. Approximately one-fourth of these ALV viremic males transmitted proviral DNA to their progeny at frequencies ranging from 1% to 11%. Restriction enzyme analysis of genomic DNA from these positive progeny revealed clonal proviruses similar to germline-inserted endogenous proviral sequences described previously. This pattern of proviral integration was quite different from nonclonal proviruses in the DNA of the viremic male parents or congenitally infected progeny from viremic females (Salter et al., 1986, 1987). Mendelian segregation of each ALV provirus was demonstrated by backcrossing each of the positive progeny to SPF line 0 chickens (Salter et al., 1987; Crittenden et al., 1989). A detailed report of the 23 different transgenic lines (labelled *alv*1 through *alv*23) produced in the original study was published (Crittenden et al., 1989). Several interesting lines of transgenic chickens were created.

1. As expected, the pattern of germline-inserted retrovirus expression in the transgenic lines reflected that of the endogenous ALV retroviruses. All but two of the transgenic lines, *alv*6 and *alv*11, produced infectious, replication-competent retroviruses. Molecular and biological properties of these two transgenic lines carrying replication-defective proviruses and their uses in our disease-resistance studies will be discussed more fully below.

2. At least one transgenic line, *alv*4, contained a proviral insertion on the Z sex chromosome.

3. Two of the first-generation progeny contained two (*alv*14 and *alv*15) and three (*alv*9, *alv*10, and *alv*11) germline-inserted proviral genomes, respectively, which then segregated in the next generation. Two of the germline-inserted proviruses, *alv*9 and *alv*11, appeared to be linked with a recombination frequency of less than 0.2 (Crittenden and Salter, 1990b).

4. In two transgenic lines expressing infectious replication-competent retroviruses, *alv*3 and *alv*19, additional copies of proviral DNA were spontaneously inserted into the germline at different sites in the next generation's genome. These lines can be used to generate different transgenic lines of chickens containing new proviral inserts for genome mapping or insertional mutagenesis studies (Crittenden et al., 1989; Crittenden and Salter, 1990b).

*alv*6 and *alv*11: Model Systems for the Study of Pathogen-Derived Resistance Genes

Sanford and Johnson (1985) proposed that information about the replicative cycle of a pathogen and its nucleotide sequence could be used to construct genes that could inhibit the replication of that pathogen. For example, transcripts that block critical functions could inhibit the replication of that pathogen. There are at least five different ALV envelope subgroups (A–E) based on interference of infection between viruses of the same subgroup (Hanafusa, 1965; Steck and Rubin, 1966; Vogt and Ishizaki, 1966) and neutralizing antibody. It is generally believed that interference is the result of the viral envelope glycoprotein produced in the infected cell blocking the receptor and preventing infection by virus of the same subgroup. Based on the interference phenomenon, we (Crittenden and Salter, 1985, 1986) and others (Freeman and Bumstead, 1987) proposed that a viral gene coding for the subgroup A envelope glycoprotein would be a good candidate disease resistance gene for insertion into the chicken germline. This pathogen-derived disease-resistance gene (Sanford and Johnson, 1985) has application in the poultry industry, since pathogenic subgroup A ALV has a detrimental effect on chicken productivity (Gavora et al., 1980). We were able to test this proposal using the transgenic line, *alv*6, which did not produce infectious ALV *in vitro* or *in vivo* but still expressed the subgroup A envelope glycoprotein. We produced a second transgenic line, *alv*11, by breeding one of the first-generation progeny that originally carried three different proviral inserts, *alv*9, *alv*10, and *alv*11 (Crittenden et al., 1989). The properties of these two transgenic lines are as follows.

1.   The *alv*6 insert appears to be a full-length provirus with no obvious major deletion in its genome (Crittenden et al., 1989). We have not used methods that are sensitive enough to detect small deletions or base changes. Full-length genomic RNA is synthesized in some tissues (Federspiel et al., 1991). Little or no viral capsid protein coded for by the *gag* gene is detected in tissues of birds carrying this provirus (see Figures 1A and 1B). The status of the reverse transcriptase and integrase proteins coded for by the *pol* gene has not been examined. Chicken embryo fibroblasts (CEF) carrying this insert do not appear to produce intact virions (Crittenden et al., 1989). As described below, the subgroup A *env* gene is highly expressed.

2.   The *alv*11 insert contains a 0.5-kb deletion within the *pol* gene (Crittenden et al., 1989; see Figure 1A). Genomic RNA containing this deletion is synthesized in some tissues (Federspiel et al., 1991). Capsid protein coded for by the *gag* gene is detected in the tissues of birds carrying this insert. The status of reverse transcriptase and integrase proteins coded for by the *pol* gene has not been examined but, based on the deletion in the *pol* gene, active reverse transcriptases probably not made. CEFs carrying this insert produce intact virions (Crittenden et al., 1989). As described below, the subgroup A *env* gene is moderately expressed.

In Vitro Studies of *alv*6 and *alv*11

In retroviruses, messenger RNA coding for the envelope glycoprotein group is a spliced message that differs from the polycistronic RNA encoding group-specific antigen and polymerase proteins. Thus, we wanted to measure the presence of envelope receptor-binding glycoprotein (see Figure 1B) directly, because of the potential usefulness of these transgenic chickens in our disease

Table 1

Relative Resistance of *alv*6 and *alv*11 CEF to Subgroup A
RSV Infection

Chicken embryo fibroblasts	RSV titer[a]	Relative resistance
*alv*6	8.8×10^1	4430
*alv*11	3.6×10^4	11
Line 0	3.9×10^5	1

[a] Average of 4 to 6 individual embryos.

resistance studies that were discussed previously. CEFs carrying either *alv*6, *alv*11, or no proviral insert (line 0) were challenged with varying dilutions of subgroup A Rous sarcoma virus (RSV). Transformed foci were counted after 5 days. As can be seen in Table 1, fibroblasts carrying either *alv*6 or *alv*11 inserts were approximately 4000-fold and 10-fold more resistant to subgroup A ALV than line 0 fibroblasts, respectively. This degree of resistance is directly proportional to the level of *env* message in *alv*6 and *alv*11 9-day embryos (Federspiel et al., 1991). In a separate study (data not shown but see Crittenden et al., 1989; Salter and Crittenden, 1989), line 0, *alv*11, and *alv*6 CEFs were equally susceptible to subgroup B RSV. Thus, significant amounts of subgroup A receptor-binding glycoprotein are made in cells carrying the *alv*6 and *alv*11 replication-defective proviruses.

In Vivo Studies of *alv*6 and *alv*11

We next extended our research to *in vivo* disease-resistance studies with these transgenic birds. Subgroup A exogenous ALV induces lymphoid leukosis (LL), which is characterized by bursal lymphomas with metastasis to other organs. Males homozygous for *alv*6 and hemizygous for *alv*11 were mated to SPF line 0 females to yield hemizygous *alv*6, *alv*11, and line 0 chicks. All chicks received 10^3 to 10^4 iu of the prototype subgroup A ALV, RAV-1, intraabdominally at hatch, and were housed intermingled in cages within a pen. ALV viremia and neutralizing antibody to ALV were determined at intervals until the end of the experiment (21 weeks). The results of this experiment are shown in abbreviated form in Table 2. Based on the lack of infectious ALV and antibody to ALV, chickens carrying the *alv*6 never became infected with subgroup A ALV. Almost all of the line 0 control chickens became infected. Most of the chickens carrying the *alv*11 insert appeared to be infected; however, there was a significant difference in the number of antibody-positive *alv*11 and line 0 birds at 21 weeks (63% versus 95%, respectively). More important, both *alv*6 and *alv*11 chickens had significantly less LL than line 0 chickens (2%, 2%, and 17%, respectively). These data are consistent with previous data for line 0 and hemizygous *alv*6 chickens challenged with a field strain of subgroup A ALV at hatch and with

Table 2
Interference of RAV-1 Infection and Oncogenicity in *alv*6, *alv*11, and
Line 0 Chickens

Line	Viremia at 4 weeks	Antibody at 21 weeks	Lymphoid leukosis[a]
Line 0	2/59	38/40 (95%)	10/59 (17%)
*alv*11	17/44	27/43 (63%)	1/44 (2%)
*alv*6	0/62	0/61 (0%)	1/62 (2%)

[a]   Total through 21 weeks.

horizontal exposure from infected hatchmates for 40 weeks (100% and 0% infected with 56% and 0% LL, respectively; Salter and Crittenden, 1989). Since the *alv*6 chickens lacked any evidence of ALV infection, this residual level of pathogenesis in the *alv*6 chickens and, perhaps, *alv*11 chickens may reflect the low background of LL previously seen in line 0 chickens (Crittenden et al., 1979) or some unexpected problem with constitutive envelope glycoprotein expression. More information concerning this spontaneous induction of LL in *alv*6 and *alv*11 chickens will be given below. It was surprising, considering the low resistance of *alv*11 to *in vitro* challenge by subgroup A RSV, that a significant proportion of *alv*11 chickens were not infected or had ALV-induced pathogenicity. The resistance to pathogenicity but not infection may reflect the relative level of subgroup A envelope expression in the different tissues. We have found that bursal tissue RNA from 1-day-old and thymus and spleen tissue from 28-day-old *alv*11 chicks contain high levels of envelope message (data not shown, but see Federspiel et al., 1991).

Pathogenic subgroup A ALV is maintained in field chicken populations by congenital transmission of ALV from viremic hens to progeny through shedding in the egg with subsequent horizontal spread to susceptible hatchmates (Okazaki et al., 1979). We therefore studied whether the expression of subgroup A envelope glycoprotein from the *alv*6 insert during the transit of the fertilized egg through the oviduct of infected hens could influence the rate of congenital infection. Viremic hens tolerant to a field strain of subgroup A ALV (RPL-41) were mated to hemizygous males carrying the *alv*6 proviral insert. Progeny were hatched and tested for the presence or absence of the *alv*6 insert. Viremia, antibody, and LL were monitored at appropriate intervals throughout the 36-week experiment. Table 3 shows an abbreviated summary of the data (Crittenden and Salter, 1992). At 16 weeks, over 90% of the chicks that lacked the *alv*6 insert were virus positive, but only about 30% of the *alv*6-positive chicks were positive. Only 18% of the *alv*6-positive chickens were diagnosed with LL, as compared to 51% of the *alv*6-negative chickens at the end of the experiment. Although resistance to subgroup A infection by congenital transmission is not as absolute as infection of day-old chicks, significant effects can be obtained by expressing the envelope glycoprotein gene in the developing chick embryo.

Table 3
Interference of RPL-41 Infection and Oncogenicity in
Congenitally Infected *alv*6 + and *alv*6 − Chickens

Progeny	Viremia (16 weeks)	Lymphoid leukosis (36 weeks)
*alv*6 +	19/66 (29%)	12/66 (18%)
*alv*6 −	45/49 (92%)	25/49 (51%)

Other Studies on the Defective Proviral Inserts

We understood from the beginning of these studies that transgenic chickens carrying infectious retroviruses would be unacceptable to the poultry industry (Crittenden and Salter, 1985, 1986). Thus, generating transgenic chickens containing replication-defective proviral inserts were a significant improvement over transgenic chickens expressing fully infectious ALV. Nevertheless, even these defective inserts can cause problems. We knew that infectious subgroup A ALV could be generated by recombination with endogenous retroviral genes. We have shown recombination of active endogenous virus (EV) genes with the *alv*6 proviral insert to produce infectious subgroup A ALV by mating *alv*6-carrying males to EV21-carrying females (L. B. Crittenden, unpublished results). However, the resultant recombinant virus will have endogenous LTRs and should be minimally pathogenic.

An unexpected problem has become apparent in breeder populations of transgenic chickens carrying *alv*6 and *alv*11 replication-defective proviruses. A significant incidence of LL has been detected in older chickens carrying these proviruses. Table 4 summarizes some preliminary evidence comparing the incidence of LL in the same generation of *alv*6 birds under two different housing conditions and with and without vaccination with bivalent vaccine against Marek's Diseas Virus (MDV). It is quite apparent that significant numbers of chickens died of LL when housed in pens and vaccinated against MDV with bivalent [HVT (serotype 3) and SB-1 (serotype 2)] vaccine compared with unvaccinated chickens housed in plastic isolators. It appears significant in the former case to have approximately 20% of the birds dying of LL with an average death date of around 270 days, and in the later case no birds dying within the 280 days and only one with a bursal tumor detected at necropsy at the termination of the experiment. Similarly, older *alv*11 birds housed in pens and vaccinated with bivalent MDV vaccine had a significant incidence of LL (data not shown). No evidence of subgroup A ALV activation or infection with other avian retroviruses have been found. A phenomenon termed "disease augmentation" has recently been described for enhancement of LL production and mortality in certain lines of chickens that were vaccinated with bivalent MDV vaccine containing SB-1 (serotype 2) and HVT (serotype 3) and exposed to pathogenic ALV (Bacon et al., 1989). The enhancement has been shown to caused by serotype 2 MDV, and its mechanism is under

Table 4

Incidence of Lymphoid Leukosis (LL) in Transgenic Chickens Carrying the
alv6 Transgene

Virus	Housing	Vaccine	Transgene	LL incidence	Mean age of LL mortality (days)
None	Pen	Bivalent MDV	*alv6*	7/35 (20%)	270[a]
None	Isolator	None	*alv6*	1/26 (4%)	280[b]
None	Isolator	None	Line 0	0/21 (0%)	NA
RPL-42	Isolator	None	*alv6*	0/36 (0%)	NA
RPL-42	Isolator	None	Line 0	22/39 (56%)	148[a]

[a] All died with LL.
[b] Bursal tumor at termination of experiment. None died with LL.

intense investigation. Although there is only circumstantial evidence that
serotype 2 MDV is playing a role in the increased incidence of LL in *alv6*
and *alv*11 breeders, this year's *alv6* and *alv*11 breeders were vaccinated with
monovalent MDV vaccine (HVT), and the incidence of LL is quite low (less
than 2%). Long-term studies are in progress to supplement this preliminary
evidence and to provide tumor material for investigating the mechanism(s)
of this phenomenon.

Japanese Quail: An Avian Model System for Germline Studies with ALV Vectors?

The mouse is a convenient model system for transgenic studies in large farm
animals (Pursel et al., 1989). This model system simplifies preliminary studies
on promoter/gene functions that would be difficult and expensive in large
mammals. It would be equally advantageous to have a model system for
poultry. The Japanese quail is a relatively small bird, requiring less housing
and less food with production of less waste. It develops rapidly to sexual
maturity (approximately 6 weeks compared to 24 weeks or more for chickens)
and has a rather pleasant song to some avian researchers (as compared to a
pen full of roosters and hens!). These characteristics have made it the avian
model for genetic selection. In addition, Japanese quail lack endogenous ALV
genes that are closely related to ALV (D. W. Salter, unpublished results). We
wanted to determine whether the ALV vectors could be used to make transgenic Japanese quail. The ability to make transgenic quail would allow us to
assay new replication-competent and -defective ALV vectors more easily, less
expensively, and in less time than in chickens. Despite an intense effort, we
have concluded that Japanese quail will not serve as a model system for avian
germline studies with ALV vectors. The following statements can be made
to summarize the efforts that were made in two different laboratories.

1. Even though Japanese quail are susceptible to subgroup A ALV infection (Hanafusa, 1975; D. W. Salter, unpublished results), we found that it was impossible to infect fertile, unincubated quail eggs with recombinant ALV retroviruses carrying the RAV-0 type LTR and the subgroup A envelope gene using either high-titer viral supernatants or infected CEF or quail embryo fibroblasts.

2. It was difficult to infect quail embryos with high-titer supernatant containing exogenous wild-type subgroup A ALV or recombinant ALV carrying the RAV-1 type LTR and subgroup A envelope gene (5 positive of almost 500 chicks assayed). We were only able to obtain significant numbers of infected quail chicks by injecting 10^6 exogenous ALV-infected CEFs or ALV-infected QT-6 quail cells per fertile, unincubated egg (79 positive of almost 600 assayed).

3. None of the viremic males produced in (2) transmitted proviral DNA characteristic of germline insertion to their progeny (over 2500 progeny from over 30 viremic males were tested).

4. Viremic females produced in (2) did infect a significant number of their progeny through congenital transmission (40 positive progeny of over 800 progeny tested from over 20 viremic females).

5. We have previously shown with chickens that males congenitally infected from viremic females transmit genetically at a higher frequency than males infected at day of set (Crittenden and Salter, 1990b; D. W. Salter and L. B. Crittenden, unpublished results). In addition, Chen et al. (1990) and Kopchick et al. (1991) used this technique to produce transgenic chickens. None of the viremic male progeny from viremic quail females in (4) transmitted proviral DNA characteristic of germline insertion to their progeny (over 2100 progeny from over 20 viremic males were tested).

Thus, Japanese quail will not be a useful model system for transgenic poultry using methods and ALV vectors that have been employed successfully in chickens. However, there has been one report describing a line of transgenic Japanese quail that was produced by injecting a defective reticuloendotheliosis (REV) vector and helper virus into fertile eggs (Lee and Shuman, 1990). In addition, methods based on transfection or retroviral infection of PGCs, embryonic stem cells and fertilized ova may be successful for quail.

ALV Retroviral Vectors for Studying Promoter Function

The analysis of promoter function, e.g., tissue specificity or developmental regulation, often requires that the putative promoter sequences with reporter genes are tested in transgenic animals. Traditionally, this has been done by microinjection of appropriate promoter/gene constructs into the male pronucleus of early mouse embryos. However, as described previously, similar studies cannot be done in birds. A series of replication-competent avian retrovirus vectors have been constructed that can be used to introduce DNA sequences into the chromosomes of avian cells. These vectors can efficiently transcribe genes from foreign promoters. The reader is referred to previous publications that describe in detail the construction, nomenclature, and *in vitro* testing of these vectors (Hughes et al., 1986, 1987, 1990; Greenhouse et al., 1988; Petropoulos and Hughes, 1991). Briefly, the v-*src* gene of the Schmidt-Ruppin A strain of RSV has been replaced with a unique *Cla*I restriction enzyme site at the end of the *env* gene (see Figure 1A). Adapter plasmids have been constructed to allow the conversion of any promoter/gene combination into a *Cla*I fragment for insertion into the ClaI restriction enzyme site in these retroviral vectors (Hughes et al., 1987). Recently, a more active *pol* gene from the Bryant high-titer RSV was inserted into the vector (Petropoulos and Hughes, 1991). The resultant proviral DNA has been termed RCASBP, which is the abbreviation for (R)eplication-(C)competent, (A)LV LTR with (S)plice acceptor and (B)ryant (P)olymerase. The corresponding vector with the endogenous Rous-associated virus type (0) LTR is designated RCOSBP. Genes inserted into the unique *Cla*I site of the RCASBP and RCOSBP vectors are expressed as spliced, subgenomic transcripts initiating from the promoter within the proviral LTR. By removing the splice acceptor sequences that are present immediately upstream from the *Cla*I site in these vectors, new vectors, termed RCANBP and RCONBP, (N)o splice acceptor, are generated that allow the transcription of inserted genes from an internal promoter. These vectors allow one to choose whether to transcribe from the LTR or an internal promoter. They have been shown to transcribe reporter genes efficiently *in vitro* (Hughes et al., 1987; Greenhouse et al., 1988; Petropoulos and Hughes, 1991).

Previously, Petropoulos et al. (1989) demonstrated that the chicken α-actin promoter fused to a chloramphenicol acetyltransferase (CAT) gene was transcribed and expressed in a tissue-specific manner in the heart and skeletal muscle of transgenic mice. For *in vivo* studies, the RCANBP vectors have been constructed with the CAT gene linked to the the α-actin promotor. The α-actin promoter/CAT gene cassettes were inserted in either the forward or backward orientation relative to the orientation of the viral genes. The RCASBP/CAT vectors, which express CAT from the LTR promoter via a spliced message, were used as controls. RCASBP/CAT and RCANBP/α-actin CAT virus vectors were injected into fertile, unincubated chicken eggs, and various tissues from chicks positive for ALV capsid antigen were assayed for CAT activity. A number of the RCANBP/α-actin CAT-positive birds demonstrated tissue-specific expression of CAT in skeletal muscle. CAT expression was found in many tissues from some of the RCASBP/CAT-positive birds (data not shown, but see Petropoulos et al., 1992).

Future Prospects for Avian Retroviral Vectors

To date, replication-competent and -defective retroviral vectors represent the only method for transferring genes into the germ cells of poultry. However, the amount of foreign genetic information that can be carried by these vectors is limited to about 2 kb for the replication-competent vectors and about 6 kb for the defective vectors. To insert genes with associated regulatory elements whose size exceeds these limits, it is necessary to develop methods of germline insertion that do not involve retroviral vectors. In addition, nonretroviral methods may be needed because the public may not accept products from transgenic poultry made with viral vectors. This symposium addressed some of these methods. Still, the development of nonretroviral methods may be difficult. Retroviral vectors can certainly be used to pursue the goals of basic research, and to define the capabilities and limitations of inserting genes into the avian germline.

Acknowledgments

The expert technical assistance of Kent Helmer, Cecyl Fischer, Leonard Provencher, Marilyn Newton, Carrie Cantwell, and Jim Pulaski is gratefully acknowledged. Research sponsored in part by the NCI, DHHS, under contract no. NO1-CO-74101 with ABL, USDA-ARS, USDA grants 85-CRCR-1-1725 (to LBC) and 87-CRCR-1-2445 (to DWS), and grant US-811-84 from BARD, The US-Israel Binational Agricultural Research and Development Fund. Donald W. Salter, William S. Payne, Lyman B. Crittenden, and Mark J. Federspiel are former employees of the USDA-ARS Avian Disease and Oncology Laboratory, East Lansing, Michigan where a portion of the *in vivo* research was performed. The contents of this publication do not necessarily reflect the views or policies of the DHHS, nor does mention of trade names, commercial products, or organizations imply endorsement by the U.S. government.

References

Astrin, S. M., Buss, E. G., and Hayward, W. S. (1979). Endogenous viral genes are non-essential in the chicken. *Nature (London)*, 282, 339–341.

Bacon, L. D., Witter, R. L., and Fadly, A. M. (1989). Augmentation of retrovirus-induced lymphoid leukosis by Marek's disease herpesviruses in white Leghorn chickens. *J. Virol.*, 63, 504–512.

Bosselman, R. A., Hsu, R.-Y., Boggs, T., Hu, S., Bruszewski, J., Ou, S., Kozar, L., Martin, F., Green, C., Jacobsen, F., Nicolson, M., Schultz, J. A., Semon, K. M., Rishell, W., and Stewart, R. G. (1989a). Germline transmission of exogenous genes in the chicken. *Science*, 243, 533–535.

Bosselman, R. A., Hsu, R.-Y., Boggs, T., Hu, S., Bruszewski, J., Ou, S., Souza, L., Kozar, L., Martin, F., Nicolson, M., Schultz, J. A., Semon, K. M., Rishell, W., and Stewart, R. G. (1989b). Replication-defective vectors of reticuloendotheliosis virus transduce exogenous genes into somatic stem cells of the unincubated chicken embryo. *J. Virol.*, 63, 2680–2689.

Bosselman, R. A., Hsu, R.-Y., Boggs, T., Hu, S., Nicolson, M., Briskin, M. J., Schultz, J. A., Rishell, W., and Stewart, R. G. (1990a). Insertion and expression of model genes in the chicken germline using a replication-defective REV vector. In: *Proceedings of the 4th World Congress on Genetics Applied to Livestock Production*, Edinburgh, edited by W. G. Hill, R. Thompson, and J. A. Woolliams, pp. 94–96.

Bosselman, R. A., Hsu, R.-Y., Briskin, M. J., Boggs, T., Hu, S., Nicolson, M., Souza, L. M., Schultz, J. A., Rishell, W., and Stewart, R. G. (1990b). Transmission of exogenous genes into the chicken. *J. Reprod. Fert., Suppl.* 41, 183–195.

Briskin, M. J., Hsu, R.-Y., Boggs, T., Schultz, J. A., Rishell, W., and Bosselman, R. A. (1991). Heritable retroviral transgenes are highly expressed in chickens. *Proc. Natl. Acad. Sci. USA*, 88, 1736–1740.

Chen, H. Y., Garber, E. A., Mills, E., Smith, J., Kopchick, J. J., DiLella, A. G. & Smith, R. G. (1990). Vectors, promoters, and expression of genes in chick embryos. *J. Reprod. Fert., Suppl.* 41, 173–182.

Crittenden, L. B. and Salter, D. W. (1985). Genetic engineering to improve resistance to viral diseases of poultry: A model for application to livestock improvement. *Can. J. Anim. Sci.* 65, 553–562.

Crittenden, L. B. and Salter, D. W. (1986). Gene insertion: current progress and long-term goals. *Avian Dis.* 30, 43–46.

Crittenden, L. B. and Salter, D. W. (1988). Insertion of retroviral vectors into the avian germ line. In *Proc. 2nd Int. Conf. Quant. Genet.* B. S. Weir, E. J. Eisen, M. M. Goodman & G. Namkoong, eds. Sinauer Associates Sunderland, ME, p. 207–214.

Crittenden, L. B. and Salter, D. W. (1990a). Expression of retroviral genes in transgenic chickens. *J. Reprod. Fert., Suppl.* 41, 163–171.

Crittenden, L. B. and Salter, D. W. (1990b). Expression and mobility of retroviral inserts in the chicken germ line. In *Proc. UCLA Symposia Transgenic Models Medicine and Agriculture*. Wiley-Liss, Boston, pp. 73–87.

Crittenden, L. B. (1991). Retroviral elements in the genome of the chicken: implications for poultry genetics and breeding. *Crit. Rev. Poult. Biol.*, 3, 73–109.

Crittenden, L. B., and Salter, D. W. (1992). A transgene, *alv*6, that expresses the envelope of subgroup A avian leukosis virus reduces the rate of congenital transmission of a field strain of avian leukosis virus. *Poultry Sci.*, 71, 799–806.

Crittenden, L. B., Salter, D. W., and Federspiel, M. J. (1989). Segregation, viral phenotype, and proviral structure of 23 avian leukosis virus inserts in the germ line of chickens. *Theor. Appl. Genet.*, 77, 505–515.

Crittenden, L. B., Witter, R. L., Okazaki, W., and Neiman, P. E. (1979). Lymphoid neoplasms in chicken flocks free of infection with exogenous avian tumor viruses. *J. Natl. Cancer Inst.*, 63, 191–200.

Federspiel, M. J., Provencher, L., Crittenden, L. B., and Hughes, S. H. (1991). Experimentally introduced defective endogenous proviruses are highly expressed in chickens. *J. Virol.*, 65, 313–319.

Freeman, B. M., and Bumstead, N. (1987). Transgenic poultry: Theory and practice. *World Poultry Sci. Assoc. J.*, 43, 180–190.

Gavora, J. S., Spencer, J. L., Gowe, K. S., and Harris, D. I. (1980). Lymphoid leukosis virus infection: Effects on production and mortality and consequences in selection for high egg production. *Poultry Sci.*, 59, 2165–2178.

Gibbins, A. M. V., Brazolot, C. L., Petitte, J. N., Liu, G., and Etches, R. J. (1990). Efficient transfection of chicken blastodermal cells and their incorporation into recipient embryos to produce chimeric chicks. In: *Proceedings of the 4th World Congress on Genetics Applied to Livestock Production*, Edinburgh, edited by W. G. Hill, R. Thompson, and J. A. Woolliams, pp. 119–122.

Greenhouse, J. J., Petropoulos, C. J., Crittenden, L. B., and Hughes., S. H. (1988). Helper-independent retrovirus vectors with Rous-associated virus type O long terminal repeats. *J. Virol.*, 62, 4809–4812.

Hanafusa, H. (1965). Analysis of the defectiveness of Rous sarcoma virus. III. Determining influence of a new helper virus on the host range and susceptibility to interference of RSV. *Virology*, 25, 248–255.

Hanafusa, H. (1975). Avian RNA tumor viruses. *Cancer*, 2, 49–90.

Hughes, S. H., Greenhouse, J. J., Petropoulos, C., and Sutrave, P. (1987). Adapter plasmids simplify the insertion of foreign DNA into helper-independent retroviral vectors. *J. Virol.*, 61, 3004–3012.

Hughes, S. H., Kosik, E., Fadly, A. M., Salter, D. W., and Crittenden, L. B. (1986). Design of retroviral vectors for the insertion of foreign DNA into the avian germ line. *Poultry Sci.*, 65, 1459–1462.

Hughes, S. H., Petropoulos, C. J., Federspiel, M. J., Sutrave, P., Forry-Schaudies, S., and Bradac, J. A. (1990). Vectors and genes for improvement of animal strains. *J. Reprod. Fert., Suppl.* 41, 39–49.

Jaenisch, R. (1976). Germline integration and Mendelian transmission of the exogenous Moloney leukemia virus. *Proc. Natl. Acad. Sci. USA*, 73, 1260–1264.

Kopchick, J. J., Mills, E., Rosenblum, C., Taylor, J., Macken, F., Leung, F., Smith, J., and Chen, H. (1991). Methods for the introduction of recombinant DNA into chicken embryos. In: *Transgenic Animals*, edited by N. L. First and F. P. Haseltine. Butterworth-Heinemann, Boston, pp. 275–295.

Lee, M.-R., and Shuman, R. M. (1990). Transgenic quail produced by retrovirus vector infection transmit and express a foreign marker gene. In: *Proceedings of the 4th World Congress on Genetics Applied to Livestock Production*, Edinburgh, edited by W. G. Hill, R. Thompson, and J. A. Woolliams, pp. 97–110.

Naito, M., Agata, K., Otsuka, K., Kino, K., Ohta, M., Hirose, K. Perry, M. M., and Eguchi, G. (1991a). Embryonic expression of beta-actin-lacZ hybrid gene injected into the fertilized ovum of the domestic fowl. *Int. J. Dev. Biol.*, 35, 69–75.

Naito, M. Watanabe, M., Kinutani, M., Nirasawa, K., and Oishi, T. (1991b). Production of quail-chick chimaeras by blastoderm cell transfer. *Br. Poultry Sci.*, 32, 79–86.

Nakamura, M., Maeda, H., and Fujimoto, T. (1991). Behavior of chick primordial germ cells injected into the blood stream of quail embryos. *Okajimas Folia Anat. Jpn.*, 67, 473–478.

Okazaki, W., Burmester, B. R., Fadly, A., and Chase, W. (1979). An evaluation of methods for eradication of avian leukosis virus from a commercial breeder flock. *Avian Dis.*, 23, 688–697.

Palmiter, R. D., Brinster, R. L., Hammer, R. E., Trumbauer, M. E., Rosenfeld, M. G., Birnberg, N. C., and Evans, R. M. (1982). Dramatic growth of mice that develop from eggs microinjected with metallothionein-growth hormone fusion genes. *Nature (London)*, 300, 611–615.

Perry, M. M., and Sang, H. M. (1990). *In vitro* culture and approaches for DNA transfer in the chick embryo. In: *Proceedings of the 4th World Congress on Genetics Applied to Livestock Production*, Edinburgh, edited by W. G. Hill, R. Thompson, and J. A. Woolliams, pp. 115–118.

Petitte, J. N., Clark, M. E., Liu, G., Gibbins, A. M. V., and Etches, R. J. (1990). Production of somatic and germline chimeras in the chicken by transfer of early blastodermal cells. *Development*, 108, 185–189.

Petropoulos, C. J., and Hughes, S. H. (1991). Replication-competent retrovirus vectors for the transfer and expression of gene cassettes in avian cells. *J. Virol.*, 65, 3728–3737.

Petropoulos, C. J., Payne, W., Salter, D. W., and Hughes, S. H. (1992). Appropriate *in vivo* expression of a muscle-specific promoter using avian retroviral vectors for gene transfer. *J. Virol.*, 66, 3391–3397.

Petropoulos, C. J., Rosenberg, M. P., Jenkins, N. A., Copeland, N. G., and Hughes, S. H. (1989). The chicken skeletal muscle a-actin promoter is tissue specific in transgenic mice. *Mol. Cell. Biol.*, 9, 3785–3792.

Pursel, V. G., Pinkert, C. A., Miller, K. F., Bolt, D. J., Campbell, R. G., Palmiter, R. D., Brinster, R. L., and Hammer, R. E. (1989). Genetic engineering of livestock. *Science*, 244, 1281–1287.

Salter, D. W., and Crittenden, L. B. (1988). Gene insertion into the avian germ line. *Occ. Publ. Br. Soc. Anim. Prod. No. 12*, pp. 32–57.

Salter, D. W., and Crittenden, L. B. (1989). Artificial insertion of a dominant gene for resistance to avian leukosis virus into the germ line of the chicken. *Theor. Appl. Genet.*, 77, 457–461.

Salter, D. W., and Crittenden, L. B. (1991). Insertion of a disease resistance gene into the chicken germline. In: *Transgenic Animals*, edited by N. L. First and F. P. Haseltine. Butterworth-Heinemann, Boston, pp. 125–131.

Salter, D. W., Smith, E. J., Hughes, S. H., Wright, S. E., and Crittenden, L. B. (1987). Transgenic chickens: Insertion of retroviral genes into the chicken germ line. *Virology*, 157, 236–240.

Salter, D. W., Smith, E. J., Hughes, S. H., Wright, S. E., Fadly, A. M., Witter, R. L., and Crittenden, L. B. (1986). Gene insertion into the chicken germ line by retroviruses. *Poultry Sci.*, 65, 1445–1458.

Sanes, J. R. (1989). Analyzing cell lineage with a recombinant retrovirus. *Trends in Neuro-Sciences*, 12, 21–28.

Sanford, J. C., and Johnson, S. A. (1985). The concept of parasite-derived resistance-deriving resistance genes from the parasites own genome. *J. Theor. Biol.*, 113, 395–405.

Sang, H., and Perry, M. M. (1989). Episomal replication of cloned DNA injected into the fertilized ovum of the hen, *Gallus domesticus*. *Mol. Reprod. Devel.*, 1, 98–106.

Shuman, R. M. (1991). Production of transgenic birds. *Experientia*, 47, 897–905.

Shuman, R. M., and Shoffner, R. N. (1986). Gene transfer by avian retorviruses. *Poultry Sci.*, 65, 1437–1444.

Simkiss, K., Rowlett, K., Bumstead, N., and Freeman, B. (1989). Transfer of primordial germ cell DNA between embryos. *Protoplasma*, 151, 164–166.

Simkiss, K., Vick, L., Luke, G., Page, N., and Savva, D. (1990). Infection of primordial germ cells with defective retrovirus and their transfer to the developing embryo. In: *Proceedings of the 4th World Congress on Genetics Applied to Livestock Production*, Edinburgh, edited by W. G. Hill, R. Thompson, and J. A. Woolliams, pp. 111–114.

Smith, E. J. (1986). Endogenous avian leukemia virus. In: *Avian Leukosis*, edited by G. F. de Boer. Nijhof, Boston, pp. 101–120.

Steck, F. T., and Rubin, H. (1966). The mechanism of interference between an avian eosis virus and Rous sarcoma virus 1. Establishment of interference. *Virology*, 29, 628–641.

Varmus, H. (1988). Retroviruses. *Science*, 240, 1427–1435.

Vogt, P. K., and Ishizaki, R. (1966). Patterns of viral interference in the avian leukosis and sarcoma complex. *Virology*, 30, 368–374.

Wentworth, B. C., Tsai, H., Hallet, J. H., Gonzalez, D. S., and Raici-Spassojevic, G. (1989). Manipulation of avian primordial germ cells and gonadal differentiation. *Poultry Sci.*, 68, 999–1010.

Wright, S. E., and Bennett, D. D. (1986). Region coding for subgroup specificity of envelope of avian retroviruses does not determine lymphomagenicity. *Virus Res.*, 6, 173–180.

Chapter 10

CHox-cad Characterization and Rooster Sperm Preservation as a First Step in the Generation of Transgenic Chickens with Modified Homeobox Genes

Yosef Gruenbaum, Ayala Frumkin, Zehava Rangini, Elie Revel, Sinai Yarus, Yael Margalit, Hasan Khatib, Ariel Darvsi, and Abraham Fainsod

Summary.The characterization of a chicken homeobox gene that contains a homeobox sequence belonging to the *cad* family of homeodomains (the homeobox-translated region) is described. *CHox-cad* genomic and cDNA clones were isolated and analyzed. Northern and *in situ* hybridization analyses revealed that the *CHox-cad* gene probably encodes a single transcript of 2.6 kb that first appears in epiblastic cells when gastrulation begins, in the primitive streak, and in definitive endoderm. At later stages of development, *CHox-cad* is expressed in the endodermal lining of the developing gut. In addition, one target site of *CHox-cad* was identified. The idea of a possible role of homeobox genes in regulating quantitative traits (QTs) was examined in layer-cross and broiler populations, using *CHox-cad* as a probe. The results of this study revealed significant marker associations with several QTs. In order to further characterize *CHox-cad* regulation and the effects of its ectopic expression in the chicken, we have started to produce transgenic chickens using sperm cells as vectors. Some properties of the buffered semen diluent that is used in these experiments are described.

Introduction

Homeobox sequences are 183-bp exonic sequences present in multiple copies in *Drosophila* and vertebrate genomes (Scott et al., 1989). In *Drosophila*, homeobox sequences have been localized mainly in known developmental genes [see Gehring (1987a, 1987b) for reviews]. In those cases in *Drosophila* homeobox genes where mutations are not available, expression patterns suggest that the homeobox genes are involved in embryonic development or differentiation. Based on the high levels of homology between *Drosophila* and vertebrate homeobox sequences, it was proposed that these genes may also play a role in embryonic development or cellular differentiation of vertebrates. Comparison of the genomic organization of the *Drosophila* and vertebrate homeobox genes has shown that many of them are organized in clusters (Acampora et al., 1989; Duboule and Dolle, 1989; Fritz et al., 1989; Graham et al., 1989; Wedden et al., 1989; Njolstad et al., 1990). The vertebrate clusters are homologous to the fly clusters (Duboule and Dolle, 1989; Graham et al., 1989), the main difference being that in vertebrates the ancestral cluster apparently underwent several duplications (Hart et al., 1987; Kappen et al., 1989; Schughart et al., 1989). In addition to the homeobox gene clusters, in *Drosophila* there are a number of isolated homeobox genes scattered through the genome. Examples are *caudal* (Mlodzik et al., 1985), *H 2.0* (Barad et al., 1988), and *msh* (Robert et al., 1989). For some of the *Drosophila* nonclustered homeobox genes, vertebrate homologs have been isolated such as *Cdx 1* and *CHox-cad* for *caudal* (Duprey et al., 1988; Frumkin et al., 1991), *CHox E* for *H 2.0* (Rangini et al., 1991), and *Hox 7.1* for *msh* (Hill et al., 1989; Robert et al., 1989).

The Chicken *CHox-cad* Homeobox Gene

Isolation and Initial Characterization of *CHox-cad* Genomic Clone

A chicken genomic library in λEMBL4 was screened under low-stringency hybridization conditions with a number of homeobox probes that included the *Drosophila antp* and *sex combs reduced* homeobox sequences and the murine *Hox 1.5* and *Hox 3.1* genes as described in Rangini et al. (1989). More than 15 different recombinant phages were isolated in this manner; one phage, λGG4, which preferentially hybridized to the *sex combs reduced* probe, was selected for further characterization. A 701-bp *Sac*I genomic fragment of clone λGG4 was sequenced, revealing that indeed this phage contains a homeobox (Figure 1A; Frumkin et al., 1991) belonging to the *caudal* family of homeoboxes (Figure 1C). The *caudal* family of homeobox genes includes the *Drosophila caudal* gene itself (Mlodzik et al., 1985), as well as the murine genes *Cdx 1* (Duprey et al., 1988) and *Cdx 2* (James and Kazenwadel, 1991) and the *Caenorhabditis elegans* gene *ceh-3* (Burglin et al., 1989). The level of homology between the homeodomains in this family is very high (Figure 1C). An interesting feature also gleaned from the genomic sequence is that the *CHox-cad* homeobox is interrupted by an intron. The intron is 128 bp in length and is localized 118 bp from the beginning of the homeobox, thereby breaking the homeodomain between amino acids 44 and 45 (Figure 1C). The sequences flanking the intron are in good agreement with the consensus splice donor and acceptor sequences (Shapiro and Senapathy, 1987). In *Drosophila*, relatively few homeobox genes are known whose homeobox sequences are interrupted by introns. The introns in *Drosophila* are localized in two positions along the homeodomain. One location is between amino acids 17 and 18, as in the case of the *engrailed* and *invected* genes (Poole et al., 1985). The second location is between amino acids 44 and 45, as in *labial* (Mlodzik et al., 1988), *Abdominal-B* (DeLorenzi et al., 1988), *Distal-less* (Cohen et al., 1989), *S59* (Dohrmann et al., 1990), and *H 2.0* (W. McGinnis and M. Barad, personal communication). In vertebrates, as in *Drosophila*, introns have been localized to two positions in the known homeobox sequences. The first position is identical to one of the known positions in *Drosophila*, interrupting the homeodomain between amino acids 44 and 45. Only one other vertebrate homeobox gene in addition to *CHox-cad* is known with an intron in this position, the chicken *CHox E* gene (Rangini et al., 1991). The other four instances of introns in the homeobox in vertebrates localize between amino acids 46 and 47, as in the murine genes *Evx 1*, *Evx 2* (Bastian and Gruss, 1990) and *S8* (Opstelten et al., 1991) and the *Xenopus* gene *Xhox 3* (Ruiz i Altaba and Melton, 1988).

CHox-cad cDNA Cloning

A cDNA library was prepared in λgt10 from poly(A)$^+$ RNA of H&H stage 12–13 chicken embryos (Hamburger and Hamilton, 1951). This cDNA library was screened with the *Sac*I genomic fragment containing the homeobox sequence. Two of the isolated cDNA clones of 2.6 kb (pC33) and 1.6 kb (pC27) were selected for sequencing. Sequence analysis revealed that C27 was a partial cDNA clone of the same transcript as C33; Figure 1A shows the relative position of C33 and C27. C33 is 2486 bp long, out of which 250 bp are 5′ untranslated, and 1492 bp are 3′ untranslated (Figure 1B). This cDNA clone contains a 744-bp-long open reading frame capable of coding for a protein consisting of 248 amino acid residues including the *CHox-cad* homeodomain. The C27 clone is 1542 bp long and codes for the last 16 amino acids of the putative *CHox-cad* protein, but consists mainly of the 3′ untranslated region also present in C33 (Figure 1A). Comparison of the C33 and C27 sequences revealed 12 differences between the two clones (Figure 1B). Both clones contain an AATAAA polyadenylation signal 19 bp from the end of the cDNA (Birnstiel et al., 1985). Northern analysis using different fragments of the C33 cDNA clone as probes has suggested that the first 124 bp in C33 may not belong to the normal *CHox-cad* transcript, as they do not hybridize to embryonic RNA; they may have resulted from a cloning artefact (A. Frumkin, Y. Gruenbaum, and A. Fainsod, unpublished results).

CHox-cad Expression During Embryogenesis

In order to determine the temporal pattern of expression of the *CHox-cad* gene, embryonic RNA was analyzed by Northern blot hybridization. Northern hybridizations with probes upstream or downstream to the homeobox revealed a 2.6-kb transcript. The onset of transcript accumulation correlates with the onset of gastrulation. Maximal transcript levels are observed when the primitive streak is fully developed (H&H stage 5), and the levels decrease so that by H&H stage 26, the 2.6-kb transcript disappears (Frumkin et al., 1991).

The spatial pattern of *CHox-cad* expression was analyzed by *in situ* hybridization to embryo sections. The *Sac*I genomic sequence and the 3′ untranslated region from the C33 cDNA clone were used as probes and gave identical results in all experiments. In H&H stage 5 embryos, *CHox-cad* transcripts are localized to the epiblast, the primitive streak, and the definitive endodermal cells (Figure 2). Analysis of serial sections at this early

A

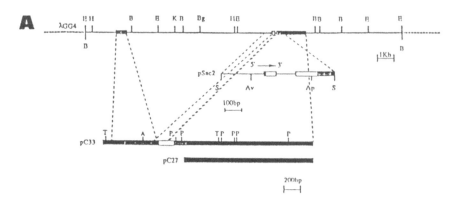

B

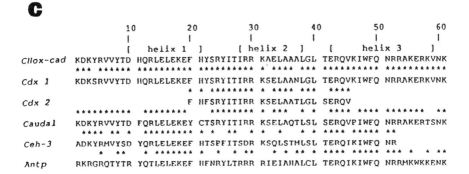

FIGURE 1. A. Restriction maps of genomic and cDNA clones of *CHox-cad*. The homeobox is marked as a white box. The rest of the protein coding sequences are marked as gray boxes, and the untranslated regions are black boxes. Av, *Ava*l; Ap, *Apa*l; B, *Bam*Hl; Bg, *Bgl*ll; E, *Eco*Rl; H, *Hind*ll, K, *Kpn*l; P, *Pst*l; T, *Taq*l. The location of the pSac1 fragment is shown. B. cDNA sequence of *CHox-cad* derived from pC33 and pC27 cDNAs. The putative CHox-cad protein sequence is shown underneath. Differences between pC33 and pC27 sequences are shown above the sequence. The homeobox sequence is double underlined. The first 124 bp in the pC33 clone, which may be due to a cloning artefact, are underlined. C. Comparison of *CHox-cad* homeodomain to *D. melanogaster caudal* and *antp*, *M. musculus Cdx 1* and *Cdx 2*, and *C. elegans Ceh-3* homeodomains. Conserved amino acids between CHox-cad homeodomains and the homeodomains presented are marked with stars.

developmental stage also revealed that the *CHox-cad* transcripts are present in the embryo in a gradient fashion: The maximal level of *CHox-cad* transcripts is observed at the posterior end of the embryo, and their level decreases anteriorly. Rostral to Hensen's node, *CHox-cad* transcripts cannot be detected by *in situ* hybridization (Figure 2). Analysis of H&H stage 18–19 embryo sections for *CHox-cad* transcripts showed localization of the mRNA in the lining of the digestive tract and the yolk sac (Figure 3). The epithelial lining of the gut and the yolk sac is of endodermal origin. Fate mapping of the chicken embryo during gastrulation stages has shown that epiblast cells that migrate into the primitive streak through Hensen's node are destined in their majority to become definitive endodermal cells (Nicolet, 1970; Fontaine and Le Douarin, 1977). Cells that migrate through the more posterior regions of the primitive streak will become mostly mesodermal cells. After undergoing gastrulation, the endodermal cells will migrate and line the embryonic gut. Comparison of the pathway of migration and localization of endodermal cells in the chick embryo and the *CHox-cad* spatial pattern of expression suggests that *CHox-cad* is expressed in the early endodermal lineage. Because *CHox-cad* transcripts were also revealed in the epiblast and the primitive streak, the gene may be expressed in precursor endodermal cells in the epiblast and may remain active in these cells as they migrate through the primitive streak. The possibility of labeling precursor endodermal and mesodermal cells in the epiblast has been shown by Stern and Canning (1990) utilizing the *HNK-1*

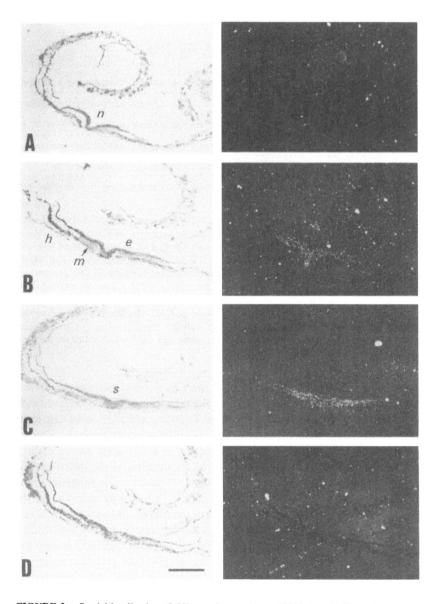

FIGURE 2.   Spatial localization of *CHox-cad* transcripts at H&H stage 5. The embryos were serially cross-sectioned and the sections were hybridized with ^{35}S-labeled single-strand anti-sense (A–C) or, as a control, sense (D) RNA probes. Sections A–C are arranged from the anterior to the posterior. Each section is presented as bright field, left; dark field, right: A, anterior to Hensen's node; B, midway along the primitive streak; C, caudal cross section. Abbreviations: e, epiblast; h, hypoblast; m, mesoderm; n, neural folds; s, primitive streak.

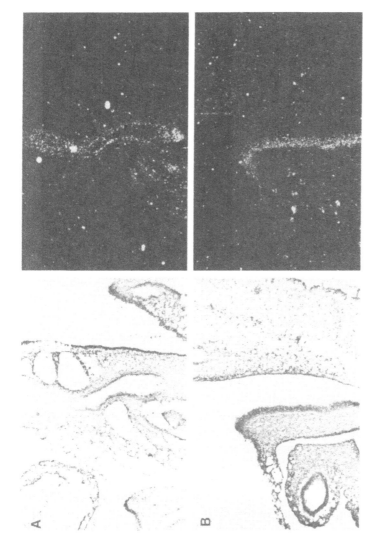

FIGURE 3. Spatial localization of *CHox-cad* transcripts at H&H stage 19. A. Parasaggital section at the level of the aortic arches. B. Section through the region where the yolk sac ends. Each section is presented as bright field, left, and dark field, right.

antibody. At present, possible expression of *CHox-cad* in the precursor mesodermal cells in the epiblast and as they migrate through the primitive streak
cannot be ruled out. If this were the case, *CHox-cad* would be turned off in
mesodermal cells as they migrate out of the primitive streak.

Transcripts of the *Drosophila caudal* gene are initially seen as a gradient
of RNA that is deposited by the female into the egg. The maximal point of
this gradient is in the posterior end of the embryo. This maternal transcript
gradient is replaced by zygotic transcripts localized to the posterior end of
the embryo. At later stages of development, the *caudal* transcripts are localized
in the posterior midgut and the Malphigian tubules. The posterior midgut is
of endodermal origin. In addition to *CHox-cad*, two other vertebrate homeobox genes of the *caudal* family have been described: *Cdx 1* and *Cdx 2*.
Both genes were cloned from murine tissues, and both are expressed in the
intestinal epithelium of the mouse. The main difference between the two
genes is the period in the life of the mouse during which they are expressed
mainly between days 14 and 17 in the endodermal epithelium: *Cdx 1* is
expressed during embryogenesis, while *Cdx 2* is expressed in the adult intestine. The chicken gene, *CHox-cad*, represents a novel member of this
family, as it is expressed only during early embryogenesis, and its expression
in the gut precedes in time the expression of *Cdx 1*. These comparisons of
the temporal patterns of the three vertebrate *caudal*-type genes together with
sequence comparisons suggest that the vertebrate genome contains multiple
members of the *caudal* family of homeobox genes.

CHox-cad DNA Binding

In order to initiate studies of the DNA-binding specificity of the *CHox-cad*
protein product, expression of the protein was obtained in bacteria. A fusion
protein including the whole *CHox-cad* protein fused to the carboxy terminus
of glutathione-*S*-transferase was expressed from constructs in the pGEX vector
(Smith and Johnson, 1988). Protein extracts from bacteria expressing the
fusion protein were utilized in electrophoretic mobility shift assays. The high
level of homology between the *caudal* and *CHox-cad* homeodomains raised
the possibility that both proteins would bind the same target sequences. In
order to test this hypothesis, we prepared an oligonucleotide the sequence of
which included the *caudal* binding site in the *fushi tarazu* promoter (Dearolf
et al., 1989). Mobility shift assays using this sequence as a target for the
CHox-cad fusion protein revealed that *CHox-cad* can bind the same sequence
as the *Drosophila caudal* protein (Figure 4). This result paves the way for
studies aimed at isolating target sequences of the *CHox-cad* protein in the
chicken genome.

CHox-cad as a Marker for Quantitative Traits

Studies on the possibility that homeobox genes might also be involved in the
regulation of quantitative traits were started. Restriction fragment length

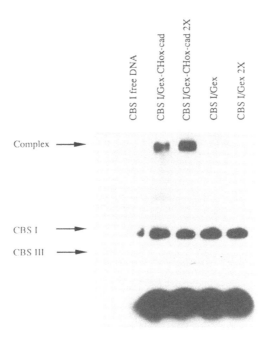

FIGURE 4. *CHox-cad* DNA binding. The CHox-cad glutathione-*S*-transferase fusion protein (Gex-CHox-cad) was tested for its capability to bind a known target sequence of the *Drosophila caudal* protein. An oligonucleotide containing the caudal binding site in the *fushi tarazu* promoter was utilized (CBS1,5'-TTTATGTCTTTATGA). As a negative control, a bacterial extract that produces only the glutathione-*S*-transferase was used (GEX).

polymorphism (RFLP) analysis of *Bam*Hl-digested chicken blood DNA samples, using the *CHox-cad* gene as a probe, revealed two *CHox-cad* alleles of 9.8 kb (designated S) and 9.0 kb (designated F) (Figure 5). The distribution of the two alleles was examined in the Anak 80 broiler line and in a layer-cross between White Leghorn and Silver H. The distribution of the the two alleles in the two populations differed significantly. In particular, in the broiler population, chickens that were homozygous for the F allele (FF) were relatively rare compared to the SS genotype, while the opposite was found in the layer-cross. *CHox-cad* RFLP-associated effects were found in both populations. The results of a one-tail *t*-test in the broiler population is shown in Table 1; in the Anak 80 broiler population, significant RFLP-associated effects were found for mature body weight and percentage of egg production.

Preservation and Cryopreservation of Rooster Semen

In order to use sperm cells as vectors for the production of transgenic chickens, we have developed a buffered solution, termed ERS, that can retain the viability of sperm cells during the period in which the sperm cells are incubated with DNA (60–90 min). When poultry semen was diluted in ERS, which

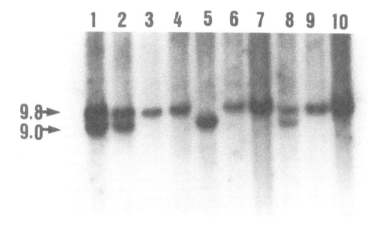

FIGURE 5. Restriction fragment length polymorphism (RFLP) analysis of *CHox-cad* alleles. DNA samples were extracted from the blood of 10 different chickens (lanes 1–10), digested with *Bam*HI, and subjected to Southern blot analysis using the genomic *Sac*I fragment as a probe.

Table 1
CHox-cad Gene as a Marker for Quantitative Traits[a]

Genotype	SS	FS	FF	*t*-test (FF-SS)
No.	63	53	9	
Weight	3369.4 ± 46.8	3309.2 ± 51.0	2997.8 ± 123.9	S
Age	149 ± 1.0	195.3 ± 1.1	186.9 ± 2.7	NS
Lay	56.6 ± 2.2	63.5 ± 2.3	73.5 ± 5.7	S

[a] Mean value (±S.E.) for each *CHox-cad* RFLP genotype of the traits body weight, age at first egg, and percentage of egg production. Since the prior tendency of genotypes is known, a one-tail *t*-test was carried out. Abbreviations: SS, homozygous for the 9.0-kb allele; FF, homozygous for the 9.8-kb allele; FS, heterozygous; NS, not significant; S, significant at a level of 0.05 or lower

contains NaCl, KCl, $CaCl_2$, $NaHCO_3$, and $MgSO_4$, as well as several other additives, the motility and morphology of the sperm cells was similar to that in commercially available extenders (Bakst, 1990). The sperm cells in semen diluted with ERS in the ratio 1:9 (semen:ERS) remained motile for up to 14 days of incubation at 20°C or 4°C (Figure 6). However, following the first 2 days of incubation, motility was slowed and the number of cells with abnormal morphology increased.

Preliminary results indicate that, in chickens, dilution of the semen 1:3 or even 1:8 (semen:ERS) resulted in levels of fertility that were quite similar to those obtained by insemination with undiluted semen. When each of several known cryopreservatives was added to the ERS solution, the sperm cells remained motile following cryopreservation and thawing. When the ERS

ERS buffer
Sperm Diluted 1:9 in ERS

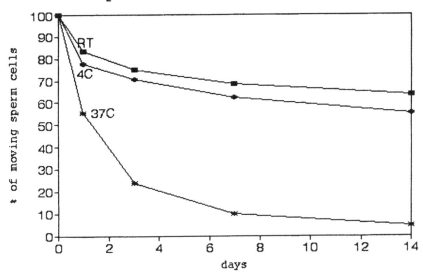

FIGURE 6. Relative motility of rooster sperm cells diluted 1:9 in ERS. Freshly collected rooster semen (200 μL) was diluted in ERS and kept for 14 days at the various temperatures. Each day, a sample of the incubated sperm cell suspension was analyzed microscopically and the percent of motile cells was compared to that in the initial, freshly collected semen.

solution contained 6–15% glycerol, motility was maintained at approximately 80–90% (Figure 7). Although the percentages of the different cryopreservants that were added to the ERS solution were somewhat different for different poultry species, similar observations were obtained for semen derived from turkey, goose, Muscovy duck, and ostrich. Preliminary results with chickens indicated that cryopreserved semen remained fertile. However, under the experimental conditions that were used, the levels of fertility were still significantly lower than those obtained by inseminations with fresh semen (E. Revel, S. Yarus, Y. Gruenbaum, and A. Fainsod, unpublished results).

Future Direction

In order to express *CHox-cad* ectopically in chickens cells, we have subcloned the 2.6-kb cDNA sequences under the regulation of the SV40 early promoter. The generation of transgenic chicken embryos that express *CHox-cad* ectopically will involve inseminating chickens with semen diluted 1:9 in ERS and incubated with the linearized form of the construct. The results of the transgene expression will be analyzed in embryos at different stages of development.

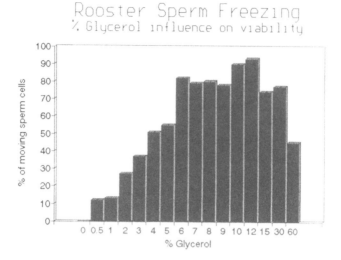

FIGURE 7.   Relative motility of rooster sperm cells following storage at − 196°C and thawing, using different concentrations of glycerol. Freshly collected rooster semen was diluted 1:9 in ERS. Glycerol was added to the diluted semen to the desired final concentration, and the diluted semen was left for 10–15 min at 10°C. The tubes containing the semen were placed in a styrofoam box in a freezer (− 80°C). Following 8–14 h of incubation at − 80°C, the tubes were placed in a liquid nitrogen tank. Thawing was performed by placing the tubes on ice. Similar results were obtained by placing the diluted semen above liquid nitrogen fumes. Addition of protein to the ERS buffer resulted in increased motility of the cells.

References

Acampora, D., D'Esposito, M., Faiella, A., Pannese, M., Migliaccio, E., Morelli, F., Stornaiuolo, A., Nigro, V., Simeone, A., and Bonceinelli, E. (1989). The human HOX family. *Nucleic Acids Res.*, 17, 10385–10402.

Barad, M., Jack, T., Chadwick, R., and McGinnis, W. (1988). A novel, tissue specific, *Drosophila* homeobox gene. *EMBO J.*, 7, 2151–2161.

Bakst, M. R. (1990). Preservation of avian cells. In: *Poultry Breeding and Genetics*, edited by R. D. Crawford. Elsevier, Amsterdam, pp. 91–108.

Bastian, H., and Gruss, P. (1990). A murine *even-skipped* homologue, *Evx 1*, is expressed during early embryogenesis and neurogenesis in a biphasic manner. *EMBO J.*, 9, 1839–1852.

Birnstiel, M. L., Busslinger, M., and Strub, K. (1985). Transcription termination and 3′ processing: The end is in site! *Cell*, 41, 349–359.

Burglin, T. R., Finney, M., Coulson, A., and Ruvkun, G. (1989). *Caenorhabditis elegans* has scores of homeobox-containing genes. *Nature*, 341, 239–243.

Cohen, S. M., Bronner, G., Kuttner, F., Jurgens, G., and Jackle, H. (1989). *Distal-less* encodes a homeodomain protein required for limb development in *Drosophila*. *Nature*, 338, 432–434.

Dearolf, C. R., Topol, J., and Parker, C. S. (1989). The *caudal* gene product is a direct activator of the *fushi tarazu* transcription during *Drosophila* embryogenesis. *Nature*, 341, 340–343.

DeLorenzi, M., Ali, N., Saari, G., Henry, C., Wilcox, M., and Bienz, M. (1988). Evidence that the Abdomial-B r element function is conferred by a trans-regulatory homeoprotein. *EMBO J.*, 7, 3223–3231.

Dohrmann, C., Azpiazu, N., and Frasch, M. (1990). A new *Drosophila* homeobox gene is expressed in mesodermal precursor cells of distinct muscles during embryogenesis. *Genes Dev.*, 4, 2098–2111.

Duboule, D., and Dolle, P. (1989). The structural and functional organization of the murine HOX gene family resembles that of *Drosophila* homeotic genes. *EMBO J.*, 8, 1497–1505.

Duprey, P., Chowdhury, K., Dressler, G. R., Balling, R., Simon L. D., Guenet, J., and Gruss, P. (1988). A mouse gene homologous to the *Drosophila* gene *caudal* is expressed in epithelial cells from the embryonic intestine. *Genes Dev.*, 2, 1647–1654.

Fontaine, J., and Le Douarin, N. M. (1977). Analysis of endoderm formation in the avian blastoderm by the use of quail-chick chimaeras. *J. Embryol. Exp. Morphol.*, 41, 209–222.

Fritz, A. F., Cho, K. W. Y., Wright, C. V. E., Jegalian, B. G., and DeRobertis, E. M. (1989). Duplicated homeobox genes in *Xenopus*. *Dev. Biol.*, 131, 584–588.

Frumkin, A., Rangini, Z., Ben-Yehuda, A., Gruenbaum, Y., and Fainsod, A. (1991). A chicken *caudal* homologue, *CHox-cad*, is expressed in the epiblast with posterior localization and in the early endodermal lineage. *Development*, 112, 207–219.

Gehring, W. J. (1987a). The homeobox: Structural and evolutionary aspects. In: *Molecular Approaches to Developmental Biology*, pp. 115–129.

Gehring, W. J. (1987b). Homeo boxes in the study of development. *Science*, 236, 1245–1252.

Graham, A., Papalopulu, N., and Krumlauf, R. (1989). The murine and *Drosophila* homeobox gene complexes have common features of organization and expression. *Cell*, 57, 367–378.

Hamburger, Y., and Hamilton, H. L. (1951). A series of normal stages in the development of the chick embryo. *J. Morphol.*, 88, 49–92.

Hart, C. P., Fainsod, A., and Ruddle, F. H. (1987). Sequence analysis of the murine *Hox-2.2*, *-2.3*, and *-2.4* homeoboxes: Evolutionary and structural comparisons. *Genomics*, 1, 182–195.

Hill, R. E., Jones, P. F., Rees, A. R., Sime, C. M., Justice, M. J., Copeland, N. G., Jenkins, N. A., Graham, E., and Davidson, D. R. (1989). A new family of mouse homeobox-containing genes: Molecular structure, chromosomal location, and developmental expression of *Hox-7.1*. *Genes Dev.*, 3, 26–37.

James, R., and Kazenwadel, A. (1991). Homeobox gene expression in the intestinal epithelium of adult mice. *J. Biol. Chem.*, 266, 3246–3251.

Kappen, C., Schughart, K., and Ruddle, F. H. (1989). Two steps in the evolution of Antennapedia-class vertebrate homeobox genes. *Proc. Natl. Acad. Sci. USA*, 86, 5459–5463.

Mlodzik, M., Fjose, A., and Gehring, W. J. (1985). Isolation of *caudal*, a *Drosophila* homeo box-containing gene with maternal expression, whose transcripts form a concentration gradient at the pre-blastoderm stage. *EMBO J.*, 4, 2961–2969.

Mlodzik, M., Fjose, A., and Gehring, W. J. (1988). Molecular structure and spatial expression of a homeobox gene from the *labial* region of the Antennapedia-complex. *EMBO J.*, 7, 2569–2578.

Nicolet, G. (1970). Analyse autoradiographique de la localisation des differentes ebauches presomtives dans la ligne primitive de l'embryon de Poulet. *J. Embryol. Exp. Morphol.*, 23, 79–108.

Njolstad, P. R., Molven, A., Apold, J., and Fjose, A. (1990). The zebrafish homeobox gene *hox-2.2*: Transcription unit, potential regulatory regions and in situ localization of transcripts. *EMBO J.*, 9, 515–524.

Opstelten, D.-J. E., Vogels, R., Robert, B., Kalkhoven, E., Zwatkruis, F., De Laaf, L., Destree, O. H., Deschamps, J., Lawson, K. A., and Meijlink, F. (1991). The mouse homeobox gene, *S8,* is expressed during embryogenesis predominantly in mesenchyme. *Mechanisms Dev., 34,* 29–42.

Poole, S. J., Kauvar, L. K., Drees, B., and Kornberg, T. (1985). The *engrailed* locus of *Drosophila:* Structural analysis of an embryonic transcript. *Cell,* 40, 37–43.

Rangini, Z., Ben-Yehuda, A., Shapira, E., Gruenbaum, Y., and Fainsod, A. (1991). *CHox E,* a chicken homeogene of the H2.0 type exhibits dorsoventral restriction in the proliferating region of the spinal cord. *Mechanisms Dev.,* 35, 13–24.

Rangini, Z., Frumkin, A., Shani, G., Guttmann, M., Eyal-Giladi, H., Gruenbaum, Y., and Fainsod, A. (1989). The chicken homeobox genes *CHox1* and *CHox3:* Cloning, sequencing and expression during embryogenesis. *Gene,* 76, 61–74.

Robert, B., Sassoon, D., Jacq, B., Gehring, W. J., and Buckingham, M. (1989). *Hox-7,* a mouse homeobox gene with a novel pattern of expression during embryogenesis. *EMBO J.,* 8, 91–100.

Ruiz i Altaba, A., and Melton, D. A. (1989). Bimodal and graded expression of the *Xenopus* homeobox gene *Xhox3* during embryonic development. *Development,* 106, 173–183.

Schughart, K., Kappen, C., and Ruddle, F. H. (1989). Duplication of large genomic regions during the evolution of vertebrate homeobox genes. *Proc. Natl. Acad. Sci. USA,* 86, 7067–7071.

Scott, M. P., Tamkun, J. W., and Hartzell, G. W., III. (1989). The structure and function of the homeodomain. *Biochim. Biophys. Acta,* 989, 25–48.

Shapiro, M. B., and Senapathy, P. (1987). RNA splice junctions of different classes of eukaryotes: Sequence statistics and functional implications in gene expression. *Nucleic Acids Res.,* 15, 7155–7175.

Smith, D. B., and Johnson, K. S. (1988). Single-step purification of polypeptides expressed in *Escherichia coli* as fusions with glutathione S-transferase. *Gene,* 67, 31–40.

Stern, C. D., and Canning, D. R. (1990). Origin of cells giving rise to mesoderm and endoderm in chick embryo. *Nature,* 343, 273–275.

Wedden, S. E., Pang, K., and Eichele, G. (1989). Expression pattern of homeobox-containing genes during chick embryogenesis. *Development,* 105, 639–650.

Chapter 11

The Molecular Biology and Genetic Control of Growth in Poultry

Christopher Goddard, Alexander Gray, Hazel Gilhooley, and Iris E. O'Neill

Summary. Improvements in growth rate and associated characteristics using modified selection programs or by a transgenic approach will depend on having a detailed knowledge of the genetic control of growth, an understanding of the biology of the regulatory factors, clones of the genes encoding these factors, and detailed knowledge of the regulation of their expression. A number of different approaches to identify and understand "trait genes" associated with growth have been attempted. One way is to analyze genes and gene products known to be associated with growth, such as growth hormone, the growth hormone receptor, and the insulin-like growth factors. This strategy is illustrated in this chapter by discussion of the somatotrophic axis in poultry. Other alternatives include the identification of the genes through their actions, for example, by dissection of growth into physiological components, or the assessment of the expression of a large number of genes and gene products at random and looking for those that are associated with growth. In order to use these other approaches, we have to modulate growth genetically and look for the response in expression of trait genes. One way of doing this is by using a transgenic biology, but since this is not well established in poultry, another approach is to use animals divergently selected for growth. We have used this strategy for the last few years and have interesting results at both the cellular and molecular levels. A number of examples, such as genetic differences in growth hormone, ornithine decarboxylase, and rates of skeletal muscle cell division have been demonstrated. The results of these studies suggest that a complete understanding of genotype and phenotype allowing identification of genes and gene products that regulate growth will enable the full potential of transgenic biology to be fulfilled in avian species as the methodology becomes fully developed.

Introduction

Successful manipulation of the avian genome resulting in differences in growth rate or food conversion efficiency would be a major step forward both scientifically and commercially. So far, progress in these traits has been made entirely by conventional genetic selection, and attempts to modify growth rate in poultry by a transgenic approach have not yet proven successful (Souza et al., 1984; Bosselman et al., 1989a, 1989b, 1990). Relatively little is known about the biological basis of the genetic control of growth, and it is not easy to identify the genes that could be manipulated to affect the trait. Before further improvements can be made using modified genetic selection programs or by the successful application of transgenic biology, a detailed knowledge of the genetic control of growth is required, the biology of the regulatory factors must be understood, the genes encoding them cloned, and the regulation of their expression clearly defined.

A number of different strategies can be used to answer some of these questions, and this chapter will focus on two of them. A conventional approach can be adopted in which genes and gene products known to be involved in the regulation of growth are studied. This is illustrated in the first section on the somatotrophic axis, focusing on growth hormone (GH), the GH receptor gene, the insulin-like growth factor-I (IGF-I) gene, and the type I IGF receptor gene.

The second section illustrates how progress might be made in future identification of trait genes. Selection of the avian phenotype for characteristics related to the rate of growth, which has resulted in great progress for the poultry industry in the last 25 years, may eventually lead us to the genes directly regulating growth and then to manipulation of the avian genome itself. We are analyzing strains of chicken with different rates of growth to determine the consequences of genetic selection at the cellular and molecular levels, and we have included some recent data that suggest that it may be a successful approach. Chickens are particularly useful for a number of reasons. Embryogenesis has been well characterized, there is easy access to all stages of development, and the biology of growth at the cellular and molecular levels suggests that results are applicable to other animals. The generation interval is shorter than for most other domestic livestock, which means that genetic selection experiments are also feasible, and finally, from a commercial point of view, the results obtained in the model system are potentially transferable into practice.

Somatotrophic Axis

Considerable progress has been made toward understanding the physiology of posthatch growth in the chicken, particularly the role of the somatotrophic axis. The definitions, concepts, and major components of the somatotrophic axis have been recently reviewed (Scanes, 1987; Scanes et al., 1990; Scanes and Vasilatos-Younken, 1991; De Pablo, 1991; Goddard and Boswell, 1991).

Growth Hormone

Growth hormone is clearly involved in the regulation of growth in poultry and has been a target gene for transgenesis (Souza et al., 1984; Bosselman et al., 1989a, 1989b, 1990). The key developments in GH biology in poultry are shown in Table 1.

Although GH activity was identified in the chicken before the 1970s, it was not until then that purification was achieved. This enabled Harvey and Scanes (1977) to develop a reliable radioimmunoassay that has provided the basis for much of the work on the physiology of GH for the last 15 years. A number of studies have investigated the developmental profile in plasma of

Table 1
Biology of Growth Hormone in Avian Species

Purification	Papkoff and Hayashida, 1972
	Farmer et al., 1974
	Harvey and Scanes, 1977
Radioimmunoassay	Harvey and Scanes, 1977
Developmental profiles	Harvey et al., 1979
	Burke and Marks, 1982
	Stewart and Washburn, 1983
	Lilburn et al., 1986
	Lauterio et al., 1986
	Goddard et al., 1988
GH cells in the embryo	Gasc and Sar, 1981
Amino acid content	Lai et al., 1984
	Leung et al., 1984a
Amino acid sequence	Souza et al., 1984
	Burke et al., 1987
cDNA sequence	Souza et al., 1984
	Burke et al., 1987
	Zvirblis et al., 1987
	Lamb et al., 1988
	Chen et al., 1988
	Foster et al., 1990
Chromosomal mapping	Shaw et al., 1990

chickens and in general have shown that GH increases in the circulation from day 17 *in ovo*, peaks during the rapid growth phase, and then declines to adult levels (Harvey et al., 1979; Burke and Marks, 1982; Stewart and Washburn, 1983; Lilburn et al., 1986; Lauterio et al., 1986; Goddard et al., 1988). The minor differences in these profiles can be attributed to the use of different strains of animal. The increase and decrease in GH concentration are due to changes in pulse amplitude. Gasc and Sar (1981) showed the presence of somatotrophs at 6 – 7 days of embryogenesis, and it may be worthwhile reexamining the timing of the appearance of GH in serum in the embryo using a sensitive assay, particularly in view of the reported growth-promoting effects of administration of GH at 11 and 18 days *in ovo* (Dean et al., 1990; Moore et al., 1990) and the increase in IGF-I concentration in chick embryo at day 15 (Robcis et al., 1991). The last 7 years have seen details emerge of the amino acid sequence (Lai et al., 1984; Leung et al., 1984a), nucleotide sequences of chicken (Souza et al., 1984; Burke et al., 1987; Zvirblis et al., 1987; Lamb et al., 1988), duck (Chen et al., 1988), and turkey GH (Foster et al., 1990), the production of recombinant DNA-derived GH, and the mapping of the GH gene to chromosome 1 in the chicken (Shaw et al., 1990).

There is little information on the structural gene in avian species. No convincing Southern blots have been published, although by analogy to the nucleotide and amino acid sequences of mammalian GH molecules it seems

Table 2
Manipulation of Plasma Growth Hormone Concentrations in
Intact Chickens

Exogenous GH	Scanes et al., 1986
	Leung et al., 1986
	Burke et al., 1987
	Vasilatos-Younken et al., 1988
	Peebles et al., 1988
	Cogburn et al., 1989
	Cravener et al., 1989
	Cravener et al., 1990
Exogenous TRH	Leung et al., 1984b
	Burke, 1987
Exogenous GRF	Baile et al., 1986
	Leung et al., 1986
Retroviral insertion of GH gene	Souza et al., 1984
	Bosselman et al., 1989a
	Bosselman et al., 1989b
	Bosselman et al., 1990
In ovo administration of GH	Dean et al., 1990
	Moore et al., 1990

reasonable to suggest that it contains 5 exons and 4 introns. Analogy to the GH locus in mammals (Chen et al., 1989) suggests it is unlikely that the chicken contains more than one pituitary-specific GH gene. By analogy, the signal peptide is probably coded for by exon I and the mature peptide by part of exon II and exons III to V inclusive. The regulation of GH gene expression and the structure of the promoter region are other areas requiring work. We know of no information within the public domain concerning the 5' upstream region of the GH gene likely to contain promoter sequences and responsive elements that mediate the effects of known secretagogues and inhibitors of GH secretion, such as growth hormone-releasing factor (GRF), somatostatin (SRIF), thyrotropin releasing-hormone (TRH), triiodothyronine (T_3), gluco-corticoids, and IGF peptides. There is no information on the regulation of GH gene expression by cell- and tissue-specific factors such as the homeobox protein PIT-1/GHF-1, which binds to cell-specific elements in the mammalian GH promoter (Fox et al., 1990). Much of this information will be required before the effective and correct expression of homologous GH is possible in transgenic chickens.

There have been many attempts to manipulate the GH concentration in chickens in order to affect body weight, growth rate, or lean tissue deposition (Table 2). Growth hormone has been administered to chickens of different ages, sexes, and strains by injection, continuous infusion, or pulsatile infusion (Scanes et al., 1986; Leung et al., 1986; Burke et al., 1987; Vasilatos-Younken et al., 1988; Peebles et al., 1988; Cogburn et al., 1989; Cravener et al., 1989, 1990). In general, GH administration has failed to alter growth rate in an analogous manner to that observed in pigs. The best responses in chickens have been in older animals (>11 weeks) in response to pulsatile administration. There has been no consistent alteration of growth when concentrations of endogenous GH have been increased by the use of TRH; in one study positive effects were noted (Leung et al., 1984b), whereas another reported no effects (Burke, 1987). No major changes were seen when GRF was used to manipulate GH concentrations, and the slight response was transitory (Leung et al., 1986).

Bosselman and colleagues have published some interesting and encouraging results using retroviral insertion of chicken GH cDNA that resulted in expression of GH in the embryo (Souza et al., 1984; Bosselman et al., 1989a, 1989b). Unfortunately, none of these transgenic embryos hatched, so we do not know if elevation of GH would have had any positive effects on growth rate or body composition (Bosselman et al., 1990). A number of recent reports have outlined the possibility of *in ovo* manipulation of growth hormone concentrations resulting in an increased growth rate (Dean et al., 1990; Moore et al., 1990). The data published so far have demonstrated some interesting although transient effects of injecting mammalian GH preparations at day 11 or day 18 of embryogenesis on the subsequent growth of the young hatched bird. These are exciting and potentially useful data, although we have not seen any results published using chicken GH.

Despite all these inconsistencies, the data are still encouraging enough to pursue a transgenic approach using GH DNA constructs. The next step must be to actually produce a germline transgenic chicken expressing extra copies of the GH gene, ideally controlled in such a way that the plasma concentrations reflect endogenous pulsatility. The problem may be that major effects of the transgene will not be observed until a later stage of development, for example, in animals over 10 – 12 weeks of age, which is of less use to the commercial broiler producers although of great importance scientifically.

Growth Hormone Receptor

Growth hormone binds with high affinity and specificity to the GH receptor in cell membranes and to a related glycoprotein in the serum. The GH receptor mediates the cellular response to GH. The cDNA sequence for the GH receptor has been reported for a number of mammalian species, and certain features have been conserved within the class Mammalia. The GH receptor contains an extracellular binding domain of about 210 – 250 residues, including seven cysteines, a transmembrane domain, and a cytoplasmic domain of variable length (Leung et al., 1987a). The specific GH binding protein has been isolated from sera of human, rabbit, and mouse (Leung et al., 1987a; Smith et al., 1989; Fuh et al., 1990). The liver receptor and serum-binding protein

Table 3
Growth Hormone Receptors in Avian Species

Study	Specific binding	Age	Tissue
Leung et al., 1984a	1.37 ± 0.31%	1 week	Liver
	3.03 ± 0.27%	8 weeks	Liver
Leung et al., 1986	6.60 ± 0.66%	9 weeks	Liver
	3.84 ± 0.77%	9 weeks	Liver
		(+ 25 μg/day GH)	
Leung et al., 1987b	<1.0%	6 weeks	Liver
	2.94 ± 0.25%	8 weeks	Liver
	17.0%	20 weeks	Liver
		(Hubbard broilers)	
	<0.1%	6 weeks	Liver
	0.40 ± 0.08%	8 weeks	Liver
	<0.1%	20 weeks	Liver
		(sex-linked dwarfs)	
Krishnan et al., 1989	18–27%	Adult	Liver
		(chicken and turkey)	
Attardo and Harvey, 1990	2.1 ± 0.41%	8 weeks	Hypothalamus
Houston et al., 1990	4.75%	11 weeks	Liver
Vasilatos-Younken et al., 1990	0.44	2 weeks	Liver
	1.25	24 weeks	Liver
	(pmol/mg)	(male turkey)	
Burnside and Cogburn, 1990	10-fold higher	35 weeks	Liver
		4 weeks	Liver

have the same N-terminal sequence and the same specificity for naturally occuring variants of GH; the serum-binding protein is thought to represent a truncated form of the receptor.

It has generally been difficult to measure GH receptors in young, rapidly growing chickens and turkeys by conventional ligand-binding techniques (Leung et al., 1987b). A limited number of studies has suggested that hepatic GH binding sites are present in low concentrations in young chickens and that their number increases throughout development at the same time that plasma concentration of GH decreases (Table 3) (Leung et al., 1987b; Krishnan et al., 1989; Vasilatos-Younken et al., 1990). This evidence is consistent with downregulation of the GH receptor in the liver by high plasma GH concentrations. Concern has been expressed because the increase in concentration of GH receptors also occurs at the same time as a decrease in the overall growth rate, but hepatic GH receptor concentrations may not be a reflection of the GH receptor level in other tissues (Vasilatos-Younken et al., 1990). The developmental profile of GH receptor in turkeys appears to be quite similar to that seen in the chicken, with a low specific binding of GH in early development that increases gradually up to 25 weeks of age (Vasilatos-Younken et al., 1990). Again this suggests a downregulation of the GH receptor by an increase in GH concentration. GH-binding proteins have recently been demonstrated in both the chicken and turkey (Jones et al., 1990). A clear pattern exists in animals in which the sex-linked dwarf gene has been introduced. Low or undetectable binding of GH is observed in hepatic membranes prepared from sex-linked dwarf animals at all ages, regardless of the genetic background to which the gene is introduced (Leung et al., 1987b).

Two laboratories have been working on the molecular biology of the chicken GH receptor, and Burnside et al. (1991) have recently published the sequence of a chicken GH receptor cDNA. They amplified sequences from a λgt11 chicken liver library by the polymerase chain reaction (PCR) using oligonucleotide primers complementary to highly conserved regions of mammalian GH receptors. The isolated DNA fragment was sequenced and found to be 55 – 60% homologous to the mammalian GH receptors, and was then used to screen a λZap chicken liver library to obtain full-length clones. The largest (2.4 kb) was sequenced, and a restriction map and predicted amino acid sequence were determined. The clone contained a single open reading frame of 608 amino acids with a single ATG codon 35 bp downstream of the start of the clone, which fullfilled the criteria for the initiation of translation.

We have also isolated a number of GH receptor cDNA clones using a different strategy. A chicken liver cDNA library primed with oligo dT was screened using a 426-bp rabbit GH receptor cDNA that contains a 280-bp extracellular domain sequence, the transmembrane domain and 96-bp intracellular domain, and we have isolated 16 clones. The largest of these (3.9 kb) was cloned into Bluescribe and the sequence was determined. The clone is a full-length cDNA containing both 5'- and 3'-untranslated sequences including a polyadenylation consensus sequence. The sequence encoding the signal and mature peptides is identical to the sequence reported by Burnside et al. (1991).

The protein structure predicted from the nucleotide sequence is similar to mammalian GH receptors. The data suggested a 16-amino acid signal peptide and a mature peptide of 592 amino acids containing a 24-amino acid transmembrane domain between amino acids 238 and 261. The extracellular domain of the chicken GH receptor appears to be shorter than the mammalian sequences, but despite this there are seven cysteine residues and four out of five potential glycosylation sites common to all species. The intracellular domain also contains five conserved cysteine residues. The overall amino acid homology between the chicken and a number of mammalian species is between 50% and 60%, but there are several domains in which the homology is much higher. For example, the extracellular region contains a 17-amino acid sequence that is almost completely identical to the corresponding region in mammalian GH receptors. This is closely followed by a sequence of 170 amino acids that is about 70% homologous to the mammalian sequence, which Burnside et al. (1991) have suggested could be the GH-binding domain. The transmembrane domain shows only a 46% structural homology, although the predicted tertiary structure is likely to be similar in all species. The region immediately 3' to the transmembrane domain is 76% homologous to other

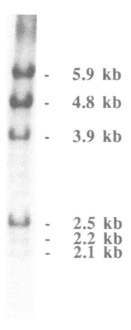

FIGURE 1. Southern blot analysis of *Eco*R1 digests of chicken genomic DNA (10 μg) probed with the 3.9-kb GH receptor cDNA. The size of each restriction fragment was determined by reference to a 1-kb ladder.

species, and this is followed by small alternating regions of both high and low homology. Since the signal transduction mechanism induced by GH binding to the receptor has not been identified, the functional significance of these domains is unknown.

Burnside et al. (1991) observed three bands when *Eco*R1-digested chicken genomic DNA was probed with a fragment of the chicken GH receptor cDNA, and two bands were observed when a cDNA probe corresponding to the intracellular domain was used. The estimated size of the gene indicated by the sum of the restriction fragment lengths was about 13 kb, which is considerably shorter than the 87 kb reported for the human GH receptor gene. We have used our 3.9-kb chicken GH receptor cDNA to probe *Eco*R1 digests of genomic DNA to identify at least six restriction fragments totaling about 22 kb (Figure 1). We have recently identified further *Eco*R1 restriction fragments of about 12 kb on extended Southern blots, which would suggest the gene is likely to be of a similar size to that of mammals.

Using reverse transcriptase coupled to PCR (RT-PCR), Burnside et al. (1991) have demonstrated GH receptor expression in a wide range of tissues (liver, skin, heart, lung, bursa, kidney, muscle, brain, testes, and adrenal) from an adult cockerel. Northern blot analysis of liver RNA identified a major transcript of 4.7 kb, a second transcript of about 4 kb, and a smaller one of

Goddard, Gray, Gilhooley, and O'Neill   **173**

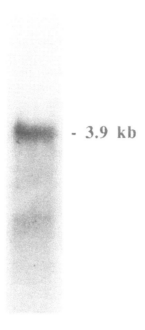

FIGURE 2. Northern blot analysis of total RNA (20 μg) prepared from the liver of a 6-day-old chick and probed with the 3.9-kb GH receptor cDNA. The major hybridizing species was 3.9 kb.

1 kb. There was also evidence for a modified form of the receptor from the RT-PCR study, although the size difference was smaller than that observed on the Northern blots. We have identified a single major transcript of about 3.9 kb on Northern blots of liver RNA isolated from a 5-day-old chick (Figure 2). This may be the major species at this stage, since there appears to be developmental regulation of the receptor (Burnside and Cogburn, 1990).

Comparison of Southern and Northern blots of normal and sex-linked dwarf animals from the same genetic background showed some very interesting results. The banding pattern on Southern blots of *Hind*III-digested DNA from animals carrying the sex-linked dwarf gene was different from that for their full-sized counterparts. Hepatic RNA prepared from sex-linked dwarfs also contained an aberrant-sized message of about 3 kb, consistent with a deletion of about 1–1.7 kb compared to the normal animals. These results suggest that the defective GH receptor protein is the primary lesion in the sex-linked dwarf that results in a reduction in growth rate, and the absence of hepatic GH-binding activity in these animals is not due to downregulation of the receptor by high serum GH concentrations (Burnside et al., 1991). The defect has not yet been precisely mapped, but these animals represent a good model for the correction of this defect by transgenesis.

Insulin-like Growth Factor-I

The structure of mammalian IGFs and the multiple forms of mRNA that arise from alternative splicing and promoter sites during gene transcription have been reviewed recently (Daughaday and Rotwein, 1989), along with the IGF-binding proteins (Ooi and Herington, 1988), IGF receptors (Rechler and Nissley, 1985), postbinding effector mechanisms (Roth, 1988), and biochemistry of the IGF peptides in poultry (Goddard and Boswell, 1991).

The literature on the molecular biology of IGF peptides in poultry is limited. Fawcett and Bulfield (1990) described the cloning, sequence analysis, and expression of putative IGF-I cDNAs from the chicken. They probed a cDNA library from the liver of a 5-week-old broiler with human IGF-IA cDNA (Jansen et al., 1983) and found evidence for at least two forms of IGF-I mRNA. The first was homologous to human IGF-IA cDNA, but the second contained another exon (1A) spliced between exons 1 and 2. The nucleotide sequences encoding the mature peptide of the chicken and human are 78% homologous. The amino acid sequence derived from the cDNA containing exon 1A predicted a 130-amino acid peptide comprising a 25-amino acid leader peptide, a putative mature chicken IGF-I (70 amino acids), and a peptide corresponding to the E-domain (35 amino acids). The cDNA without exon 1A contained an open reading frame encoding a peptide homologous to the leader peptide of human IGF-I (48 amino acids). There are eight substitutions between mature chicken and human IGF-I, four of which are chemically similar amino acids. A chicken IGF-I cDNA, cloned in Rotwein's laboratory (Kajimoto and Rotwein, 1989), encoded a 153-amino acid peptide precursor identical in length and structure to that predicted by Fawcett and Bulfield (1990) using the ATG start site in exon 1. This is the only methionine common among primary IGF-I translation products from different species.

Fawcett and Bulfield (1990) also isolated clones from a chicken genomic library. Southern blotting revealed that the gene was at least 23 kb in length. Kajimoto and Rotwein (1991) recently reported that the organization of the gene in the chicken might be simpler than in mammals and that comparative analysis could identify essential features common to vertebrates. They suggested that the gene contains four exons that are distributed over 50 kb of chromosomal DNA and that are transcribed and processed into a major RNA of 2.6 kb and a minor species of 1.9 kb.

Multiple IGF-I mRNA species, similar to the pattern of expression seen in mammals, were found using the full-length chicken IGF-I cDNA probe (Fawcett and Bulfield, 1990). The major transcript was seen as a diffuse band of 0.65–0.85 kb, with additional minor bands perhaps representing incompletely spliced precursors, in contrast to the major IGF-I mRNA transcript of 2.6 kb detected using a single-stranded antisense chicken IGF-I probe (Kajimoto and Rotwein, 1989). Expression of IGF-I mRNA was detected in liver from 7-week-old and in liver, brain, skeletal muscle, and heart from 1-week-old chickens using a solution hybridization assay. We have recently demonstrated IGF-I mRNA expression in skeletal muscle, liver, brain, kidney, ovary, adipose tissue, and heart of 6-week-old broiler chickens by RT-PCR using oligonucleotide primers derived from exons 2 and 3 of the chicken IGF-I sequence (Goddard and Boswell, 1991), and IGF-I mRNA expression has been demonstrated throughout embryogenesis using PCR (Serrano et al., 1990).

Our recent studies have focused on the role of IGF peptides in muscle development, and we have shown the importance of IGF-I and -II in the regulation of muscle satellite cell proliferation *in vitro* by the addition of exogenous peptides (Duclos et al., 1991). Northern blots of mRNA from proliferating and differentiated muscle satellite cells were probed with a riboprobe corresponding to the sequence encoding the mature chicken IGF-I peptide. This hybridized to a major transcript of about 2.6 kb in cells both before and after cellular differentiation. This is of similar size to the transcript identified by Kajimoto and Rotwein (1989), with little evidence of multiple bands in these Northern blots (Figure 3).

Despite the important role of IGF peptides in the regulation of growth and development in poultry, the major contribution of an autocrine/paracrine mechanism may make manipulation of the IGF system by transgenesis more difficult, although disruption of the IGF-II gene in transgenic mice has elegantly demonstrated its role in development of the embryo (DeChiara et al., 1990).

The importance of IGF binding proteins in the biological action of IGF peptides has clearly been established in mammals, and a number of these binding proteins have been cloned (Brewer et al., 1988; Brinkman et al., 1988; Wood et al., 1988), but it remains an area of avian growth and development that is underinvestigated. There are few publications and no data on genetic control at the molecular level. A complete understanding of avian growth and development requires additional information in this area.

The biological action of IGFs is initiated by their binding to specific receptors on the surface of cells in the target tissue. Distinct binding sites for both IGF-I and -II have been characterized in mammalian tissues, and two receptor types have been identified. The type I receptor is structurally related to the insulin receptor, usually binds IGF-I, IGF-II, and insulin with decreasing affinities, and is involved in the mediation of a variety of biological effects of IGF peptides (Rechler and Nissley, 1985). The type II receptor is structurally distinct, consists of a transmembrane molecule with a large extracellular domain, and is the cation-independent mannose 6-phosphate receptor (Morgan et al., 1987). It is not clear if the type II receptor is a transmembrane signalling protein or has perhaps a structural or transport function (Roth, 1988). The chick embryo has been used to investigate the role of IGF peptides during development, but no evidence has yet been found for a

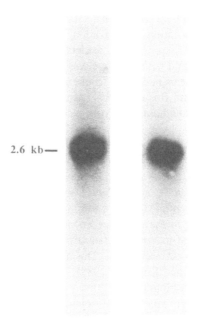

2.6 kb—

Proliferating   Differentiated

FIGURE 3. Northern blot analysis of total RNA (25 μg) prepared from proliferating and differentiated muscle satellite cells in culture. The probe was an antisense sequence prepared from a *Hind*III-*Eco*R1 fragment corresponding to the nucleotide sequence encoding the mature chicken IGF peptide.

type II receptor (Kasuga et al., 1982; Bassas et al., 1988). There is no evidence for binding of IGF-II to the chicken cation mannose 6-phosphate receptor (Canfield and Kornfeld, 1989; Clairmont and Czech, 1989), and we have recently demonstrated that biological actions of IGF peptides in the liver and muscle of young, rapidly growing chickens are mediated through the type I or a related receptor (Duclos and Goddard, 1990; Duclos et al., 1991) with complete absence of binding to a receptor with type II characteristics.

We are currently examining the IGF receptor in the chicken at the molecular level and have some potential IGF type I receptor clones isolated by hybridization to the human IGF type I receptor (Ullrich et al., 1986). These clones are currently being sequenced, and the initial data show up to 90% homology to the sequence of the human receptor α-subunit domain (D. G. Armstrong and C. Goddard, unpublished results)

Identification of Trait Genes

The second part of this chapter returns to the problem of identification of "trait genes" so that they may be manipulated by a transgenic approach.

Table 4

Bromodeoxyuridine Labeling Index in Tissues
from 5-Day-Old Broiler (Line 101) and Layer
(Line 71) Strain Chicks

	Labeling index (%)	
Tissue	Line 101	Line 71
Breast muscle	3.16 ± 0.62*	1.36 ± 0.36
Bone (tibia/femur epiphysial plate)	22.46 ± 1.40	19.06 ± 2.05

Values are means ± standard deviation ($n = 4$ animals).
*$p < 0.01$ versus the other group.

Bulfield (1989) divided potential solutions to the problem of identification of trait genes into three approaches. The first, illustrated in the section on the somatotrophic axis, is the analysis of genes and gene products known to be associated with a particular trait. The second is the dissection of a commercial trait into its physiological components, with the aim of identification of the genes through their actions. The third is to find a way of assessing the expression of a large number of genes and gene products at random and look for those that are associated with a particular trait. In order to use these approaches, we have to modulate commercial characteristics genetically and look for the response in expression of trait genes. One way of doing this is via a transgenic approach as demonstrated by Palmiter and Brinster (Palmiter et al., 1983), but since this is not yet routine in poultry, the second approach is to use animals divergently selected for a commercial trait. We have practised this strategy for the last few years and have some interesting results at both the cellular and molecular levels. A number of examples are illustrated below.

We have used chickens from a commercial strain of broiler derived from a White Cornish strain selected over 40 generations for an increase in body weight and breast muscle conformation (line 101) and compared them to a line derived from the same genetic stock except that the selection criteria were relaxed after 18 generations (line MK) and also to a closed population of White Leghorn-type layer chickens (line 71). The body weight differences and serum concentrations of a number of hormones have been published (Goddard et al., 1988). The major finding was that although there was an increase in GH concentrations during the period of most rapid growth in all three lines, there was a significant difference between them, with the slow-growing animals having the highest GH concentrations (Table 2). We have recently shown that this is due to a decrease in GH pulsatility, at least when comparing the commercial broiler strain to White Leghorns, and that these differences may be related to the relative difference in fatness between the two lines (Griffin et al., 1991).

These rapidly growing broilers have a higher rate of muscle growth than the slow-growing animals. Between hatch and maturity, the muscle DNA-to-protein ratio remains constant, regardless of genetic differences in muscle growth or age, implying that nuclear division is continuing in proportion to muscle mass posthatch (Tinch, 1990). The number of muscle fibers is fixed during development (Goldspink, 1977) and, although the nuclei within muscle fibers have lost the ability to divide (Stockdale and Holtzer, 1961), the DNA content per muscle fiber increases dramatically during postnatal growth (Moss, 1968). This is due to proliferation of satellite cells, which lie between the basement membrane surrounding each muscle fiber and the sarcolemma (Mauro, 1961). These cells fuse with adjacent fibers to increase the number of nuclei within them and enhance the synthesis and accumulation of muscle protein (Moss and Leblond, 1970). This has focused our attention on the proliferation of satellite cells, and we have used bromodeoxyuridine labeling to show that the rate of cell division is threefold greater in breast muscle sections taken from 5-day-old broiler chicks compared to White Leghorns (Table 4). Interestingly, in this preliminary experiment, the rate of cell division in the epiphysial plate of the tibia/femur was not significantly different between the lines, although a more extensive study is required to confirm this (C. Goddard, N. Loveridge, and C. Farquaharson, 1991, unpublished observations).

In the light of these results, it is interesting to note that at least one indicator of cell division, ornithine decarboxylase activity, is up to 20-fold higher in the breast muscle of broilers compared to the layer strain (Bulfield et al., 1988). Johnson and Bulfield have recently cloned and analyzed ornithine decarboxylase cDNA from a chicken cDNA library (R. Johnson and G. Bulfield, unpublished results). One of the interesting features was the presence of a polymorphism in *Hae*II digests of genomic DNA prepared from broilers that is completely absent from DNA of layers. It would be interesting to know if the polymorphism is linked to the difference in ornithine decarboxylase activity and the difference in proliferation of muscle satellite cells *in vivo* and, if so, how these differences are regulated by growth factors such as the IGF peptides. Gray and Tait have recently identified a polymorphism in the ornithine decarboxlase gene in replicate selection lines of mice associated with an increased growth rate and have strongly implicated the ornithine decarboxylase gene as a trait gene that has been modified during the selection process (A. Gray and A. Tait, unpublished observations). There seems to be no reason why other trait genes related to the regulation of growth rate cannot be identified in the same way and then used in transgenic approaches to alter growth rate genetically.

Conclusions

The production of transgenic laboratory and farm animals is becoming a routine procedure, and the appropriate technology should allow this to happen in avian species in the near future. An ideal use of this technology would be to manipulate genes controlling commercially important traits such as growth.

For this purpose, it is important to develop strategies to isolate and characterize these genes and their products. Some of the approaches to be taken have been outlined in this chapter, and the results of these studies suggest that a complete understanding of genotype and phenotype will enable the full potential of transgenic biology to be achieved in the bird for further understanding of processes controlling growth.

Acknowledgments

We would like to thank Ross Breeders, Newbridge, Midlothian, UK, for a generous and continued supply of animals.

References

Attardo, D., and Harvey, S. (1990). Growth hormone-binding sites in chicken hypothalamus. *J. Mol. Endocrinol.,* 4, 23–29.

Baile, C. A., Della-Fera, M. A., and Buonomo, F. C. (1986). The neurophysiological control of growth. In: *Control and Manipulation of Animal Growth,* edited by P. J. Buttery, D. B. Lindsay, and N. B. Haynes. Butterworths, London, pp. 105–118.

Bassas, L., Lesniak, M. A., Serrano, J., Roth, J., and de Pablo, F. (1988). Developmental regulation of insulin and type I insulin-like growth factor receptors and absence of type II receptors in chicken embryo tissues. *Diabetes,* 37, 637–644.

Bosselman, R. A., Hsu, R-Y., Boggs, T., Hu, S., Bruszewski, J., Ou, S., Kozar, L., Martin, F., Green, C., Jacobsen, F., Nicholson, M., Schultz, J. A., Semon, K. M., Rishell, W., and Stewart, R. G. (1989a). Germline transmission of exogenous genes in the chicken. *Science,* 243, 533–535.

Bosselman, R. A., Hsu, R-Y., Boggs, T., Hu, S., Bruszewski, J., Ou, S., Souza, L. A., Kozar, L., Martin, F., Nicholson, M., Rishell, W., Schultz, J. A., Semon, K. M., and Stewart, R. G. (1989b). Replication-defective vectors of reticuloendotheliosis virus transduce exogenous genes into somatic stem cells of the unincubated chicken embryo. *J. Virol.,* 63, 2680–2689.

Bosselman, R. A., Hsu, R-Y., Boggs, T., Hu, S., Nicholson, M., and Briskin, M. J. (1990). Insertion and expression of model genes in the chicken germline using a replication-defective REV vector. *Proceedings of the 4th World Congress on Genetics Applied to Livestock Production,* XVI, 94–96.

Brewer, M. T., Stetler, G. L., Squires, C. H., Thompson, R. C., Busby, W. H., and Clemmons, D. R. (1988). Cloning, characterisation, and expression of a human insulin-like growth factor binding protein. *Biochem. Biophys. Res. Commun.,* 152, 1289–1297.

Brinkman, A., Groffen, C., Kortleve, D. J., Geurts van Kessel, A., and Drop, S. L. S. (1988). Isolation and characterisation of a cDNA encoding the low molecular weight insulin-like growth factor binding protein (IBP-1), *EMBO J.,* 7, 2417–2423.

Bulfield, G. (1989). The biochemical control of quantitative traits. In: *Evolution and Animal Breeding*, edited by W. G. Hill, and T. F. C. Mackay. CAB International, Wallingford, Oxon, UK, chap. 32.

Bulfield, G., Isaacson, J. H., and Middleton, R. J. (1988). Biochemical correlates of selection for weight-for-age in chickens: Twenty-fold higher muscle ornithine decarboxylase levels in modern broilers. *Theor. Appl. Genet.*, 75, 432–437.

Burke, W. H. (1987). Influence of orally administered thyrotropin-releasing hormone on plasma growth hormone, thyroid hormones, growth, feed efficiency, and organ weights of broiler chickens. *Poultry Sci.*, 66, 147–153.

Burke, W. H., and Marks, H. L. (1982). Growth hormone and prolactin levels in nonselected and selected broiler lines of chickens from hatch to eight weeks of age. *Growth*, 46, 283–295.

Burke, W. H., Moore, J. A., Ogez, J. R., and Builder, S. C. (1987). The properties of recombinant chicken growth hormone and its effects on growth, body composition, feed efficiency and other factors in broiler chickens. *Endocrinology*, 120, 651–658.

Burnside, J., and Cogburn, L. A. (1990). Regulation of growth hormone (GH) receptor in broiler breeder males. *Poultry Sci.*, 69, Suppl. 1, 28.

Burnside, J., and Cogburn, L. A. (1991). Developmental profile of growth hormone receptor gene expression in broiler chickens. *J. Cell. Biochem.*, Suppl. 15E, 201, abstract CI 104.

Burnside, J., Liou, S. S., and Cogburn, L. A. (1991). Molecular cloning of the chicken growth hormone receptor complementary deoxyribonucleic acid; Mutation of the gene in sex-linked dwarfs. *Endocrinology*, 128, 3183–3192.

Canfield, W. M., and Kornfeld, S. (1989). The chicken liver cation independent mannose-6-phosphate receptor lacks the high affinity binding site for IGF-II. *J. Biol. Chem.*, 264, 7100–7103.

Chen, E. Y., Liao, Y. C., Smith, D. H., Barrera-Saldana, H. A., Gelinas, R. E., and Seeburg, P. H. (1989). The human growth hormone locus: Nucleotide sequence, biology, and evolution. *Genomics*, 4, 479–497.

Chen, H.-T., Pan, F.-M., and Chang, W.-C. (1988). Purification of duck growth hormone and cloning of the complementary DNA. *Biochem. Biophys. Acta*, 949, 247–251.

Clairmont, K. B., and Czech, M. P. (1989). Chicken and *Xenopus* mannose-6-phosphate receptors fail to bind insulin-like growth factor II. *J. Biol. Chem.*, 264, 16390–16392.

Cogburn, L. A., Liou, S. S., Rand, A. L., and McMurtry, J. P. (1989). Growth, metabolic and endocrine responses of broiler cockerels given a daily subcutaneous injection of natural or biosynthetic chicken growth hormone. *J. Nutr.*, 119, 1213–1222.

Cravener, T. L., Vasilatos-Younken, R., and Andersen, B. J. (1990). Hepatomegaly induced by the pulsatile, but not continuous, intravenous administration of purified chicken growth hormone in broiler pullets: Liver composition and nucleic-acid content. *Poultry Sci.*, 69, 845–848.

Cravener, T. L., Vasilatos-Younken, R., and Wellenreiter, R. H. (1989). Effect of subcutaneous infusion of pituitary-derived chicken growth hormone on growth performance of broiler pullets. *Poultry Sci.*, 68, 1133–1140.

Daughaday, W. H., and Rotwein, P. (1989). Insulin-like growth factors I and II. Peptide, messenger ribonucleic acid and gene structures, serum and tissue concentrations. *Endocrinol. Revs.*, 10, 68–91.

De Pablo, F. (1991). Insulin-like growth factor-I and insulin as growth and differentiation factors in chicken embryogenesis. *Poultry Sci.*, 70, 1790–1796.

Dean, C. E., Moore, R., Hargis, P. S., and Hargis, B. M. (1990). In ovo growth hormone administration affects embryonic tissue development and thyroid hormone function. *Poultry Sci.*, 69, Suppl. 1, 41.

DeChiara, T. M., Efstratiadis, A., and Robertson, E. J. (1990). A growth deficiency phenotype in heterozygous mice carrying an insulin-like growth factor II gene disrupted by targeting. *Nature*, 245, 78–80.

Duclos, M. J., and Goddard, C. (1990). Insulin-like growth factor receptors in chicken liver membranes: Binding properties, specificity, developmental pattern and evidence for a single receptor type. *J. Endocrinol.*, 125, 199–206.

Duclos, M. J., Wilkie, R. S., and Goddard, C. (1991). Stimulation of DNA synthesis in chicken muscle satellite cells by insulin and insulin-like growth factors: Evidence for exclusive mediation by a type-I IGF receptor. *J. Endocrinol.*, 128, 35–42.

Farmer, S. W., Papkoff, H., and Hayashida, T. (1974). Purification and properties of avian growth hormone. *Endocrinology*, 95, 1560–1565.

Fawcett, D. H., and Bulfield, G. (1990). Molecular cloning, sequence analysis and expression of putative chicken insulin-like growth factor I cDNAs. *J. Mol. Endocrinol.*, 4, 201–211.

Foster, D. N., Foster, L. K., Kim, S. U., and Enyeart, J. J. (1990). Turkey growth hormone complementary DNA nucleotide sequence. *Poultry Sci.*, 69, Suppl. 1, 52.

Fox, S. R., Jong, M. T. C., Casanova, J., Ye, Z.-S., Stanley, F., and Samuels, H. (1990). The homeodomain protein, Pit-1/GHF-1, is capable of binding to and activating cell-specific elements of both growth hormone and prolactin gene promoters. *Mol. Endocrinol.*, 4, 1069–1080.

Fuh, G., Mulkerrin, M. G., Bass, S., McFarland, N., Brochier, M., Bourell, J. H., Light, D. R., and Wells, J. A. (1990). The human growth hormone receptor. *J. Biol. Chem.*, 265, 3111–3115.

Gasc, J. M., and Sar, M. (1981). Appearance of LH-immunoreactive cells in the Rathke's pouch of the chicken embryo. *Differentiation*, 20, 77–80.

Goddard, C., and Boswell, J. M. (1991). The molecular biology and genetic control of growth in poultry. *CRC Crit. Rev. Poultry Biol.*, 3, 325–340.

Goddard, C., Wilkie, R. S., and Dunn, I. C. (1988). The relationship between insulin-like growth factor, growth hormone, thyroid hormone and insulin in chickens selected for growth. *Dom. Anim. Endocrinol.*, 5, 165–176.

Goldspink, G. (1977). The growth of muscles. In: *Growth and Poultry Meat Production*, edited by K. N. Boorman, and B. J. Wilson. British Poultry Science Ltd., Edinburgh, pp. 13–28.

Griffin, H. D., Windsor, D., and Goddard, C. (1991). Why are young broiler chickens fatter than layer-strain chicks? *Comp. Biochem. Physiol.*, 100A, 205–210.

Harvey, S., and Scanes, C. G. (1977). Purification and radioimmunoassay of chicken growth hormone. *J. Endocrinol.*, 73, 321–329.

Harvey, S., Davison, T. F., and Chadwick, A. (1979). Ontogeny of growth hormone and prolactin secretion in the domestic fowl (*Gallus domesticus*). *Gen. Comp. Endocrinol.* 39, 270–273.

Houston, B., O'Neill, I. E., Mitchell, M. A., and Goddard, C. (1990). Purification and characterisation of a single charge isomer of chicken growth hormone. *J. Endocrinol.*, 125, 207–215.

Jansen, M., van Schaik, F. M. A., Ricker, A. T., Bullock, B., Woods, D. E., Gabbay, K. H., Nussbaum, A. L., Sussenbach, J. S., and Van Den Brande, J. L. (1983). Sequence of cDNA encoding human insulin-like growth factor I precursor. *Nature*, 306, 609–611.

Jones, D. L., Anderson, B. J., Vasilatos-Younken, R., McMurtry, J. P., and Rosebrough, R. W. (1990). Identification and verification of the existence of circulating growth hormone (GH) binding activity in domestic poultry. *Poultry Sci.*, 69, Suppl. 1, 70.

Kajimoto, Y., and Rotwein, P. (1989). Structure and expression of a chicken insulin-like growth factor I precursor. *Mol. Endocrinol.*, 3, 1907–1913.

Kajimoto, Y., and Rotwein, P. (1991). Structure of the chicken insulin-like growth factor I gene reveals conserved promoter elements. *J. Biol. Chem.*, 266, 9724–9731.

Kasuga, M., Van Obberghen, E., Nissley, S. P., and Rechler, M. M. (1982). Structure of the insulin-like growth factor receptor in chicken embryo fibroblasts. *Proc. Natl. Acad. Sci. USA*, 79, 1864–1868.

Krishnan, K. A., Proudman, J. A., and Bahr, J. M. (1989). Avian growth hormone receptor assay: Use of chicken and turkey liver membranes. *Mol. Cell. Endocrinol.*, 66, 125–134.

Lai, P.-H., Duyka, D. R., Souza, L. M., and Scanes, C. G. (1984). Purification and properties of chicken growth hormone. *IRCS Med. Sci.*, 12, 1077–1078.

Lamb, I. C., Galehouse, D. M., and Foster, D. N. (1988). Chicken growth hormone cDNA sequence. *Nucleic Acids Res.*, 16, 9339.

Lauterio, T. J., Decuypere, E., and Scanes, C. G. (1986). Growth, protein synthesis and plasma concentrations of growth hormone, thyroxine and triiodothyronine in dwarf, control and growth selected strains of broiler type domestic fowl. *Comp. Biochem. Physiol.*, 83A, 627–632.

Leung, D. W., Spencer, S. A., Cachianes, G., Hammonds, R. G., Collins, C., Henzel, W. J., Waters, M. J., and Wood, W. I. (1987a). Growth hormone receptor and serum binding protein: Purification, cloning and expression. *Nature*, 330, 537–543.

Leung, F. C., Styles, W. J., Rosenblum, C. I., Lilburn, M. S., and Marsh, J. A. (1987b). Diminished hepatic growth hormone receptor binding in sex-linked dwarf broiler and leghorn chickens. *Proc. Soc. Exp. Biol. Med.*, 184, 234–238.

Leung, F. C., Taylor, J. E., Steelman, S. L., Bennett, C. D., Rodkey, J. A., Long, R. A., Serio, R., Weppelman, R. M., and Olson, G. (1984a). Purification and properties of chicken growth hormone and the development of a homologous radioimmunoassay. *Gen. Comp. Endocrinol.*, 56, 389–400.

Leung, F. C., Taylor, J. E., and Van Iderstine, A. (1984b). Effects of dietary thyroid hormones on growth and serum T3, T4 and growth hormone in sex-linked dwarf chickens. *Proc. Soc. Exp. Biol. Med.*, 177, 77–81.

Leung, F. C., Taylor, J. E., Wien, S., and Van Iderstine, A. (1986). Purified chicken growth hormone (GH) and a human pancreatic GH-releasing hormone increase body weight gain in chickens. *Endocrinology*, 118, 1961–1965.

Lilburn, M. S., Leung, F. C., Ngiam-Rilling, K., and Smith, J. H. (1986). The relationship between age and genotype and circulating concentrations of triiodothyronine, thyroxine and growth hormone in commercial meat strain chickens. *Proc. Soc. Exp. Biol. Med.*, 182, 336–343.

Mauro, A. (1961). Satellite cells of skeletal muscle fibres. *J. Biophys. Biochem. Cytol.*, 9, 493–495.

Moore, R., Dean, C. E., Hargis, B. M., and Hargis, P. S. (1990). In ovo endocrine manipulation of growth and development of chickens. *Poultry Sci.*, 69, Suppl. 1, 96.

Morgan, D. O., Edman, J. C., Standring, D. N., Fried, V. A., Smith, M. C., Roth, R. A., and Rutter, W. J. (1987). Insulin-like growth factor II receptor as a multifunctional binding protein. *Nature*, 329, 301–307.

Moss, F. P. (1968). The relationship between the dimensions of the fibers and the number of nuclei during normal growth of skeletal muscle in the domestic fowl. *Am. J. Anat.*, 122, 555–564.

Moss, F. P., and Leblond, C. P. (1970). Satellite cells as the source of nuclei in muscle of growing rats. *Anat. Rec.*, 170, 421–436.

Ooi, G. T., and Herington, A. C. (1988). The biological and structural characterisation of specific serum binding proteins for the insulin-like growth factors. *J. Endocrinol.*, 118, 7–18.

Palmiter, R. D., Norstedt, G., Gelinas, R. E., Hammer, R. E., and Brinster, R. L. (1983). Metallothionein-human GH fusion genes stimulate growth of mice. *Science*, 222, 809–814.

Papkoff, H., and Hayashida, T. (1972). Pituitary growth hormone from turtle and duck: Purification of and immunochemical studies. *Proc. Soc. Exp. Biol. Med.*, 140, 251–255.

Peebles, E. D., Burke, W. H., and Marks, H. L. (1988). Effects of recombinant chicken growth hormone in randombred meat-type chickens. *Growth Dev. Age*, 52, 133–138.

Rechler, M. M., and Nissley, S. P. (1985). The nature and regulation of the receptors for insulin-like growth factors. *Ann. Rev. Physiol.*, 47, 425–442.

Robcis, H. L., Caldes, T., and de Pablo, F. (1991). Insulin-like growth factor-I serum levels show a midembryogenesis peak in the chicken that is absent in growth-retarded embryos cultured ex ovo. *Endocrinology*, 128, 1895–1901.

Roth, R. A. (1988). Structure of the receptor for insulin-like growth factor II: The puzzle amplified. *Science*, 239, 1269–1271.

Scanes, C. G. (1987). The physiology of growth, growth hormone, and other growth factors in poultry. *CRC Crit. Rev. Poultry Biol.*, 1, 51–105.

Scanes, C. G., Arámburo, C., and Campbell, R. L. (1990). Hormonal involvement in avian growth and development: Growth hormone and insulin-like growth factor I. In: *Endocrinology of Birds: Molecular to Behavioural*, edited by M. Wada, S. Ishii, and C. G. Scanes. Japanese Scientific Societies Press, Tokyo/Springer-Verlag, Berlin, pp. 93–110.

Scanes, C. G., Duyka, D. R., Lauterio, T. J., Bowen, S. J., Huybrechts, L. M., Bacon, W. L., and King, D. B. (1986). Effects of chicken growth hormone, triiodothyronine and hypophysectomy in growing domestic fowl. *Growth*, 50, 12–31.

Serrano, J., Shuldiner, A. R., Roberts, C. T. Jr., LeRoith, D., and de Pablo, F. (1990). The insulin-like growth factor-I gene is expressed in chick embryos during early organogenesis. *Endocrinology*, 127, 1547–1549.

Shaw, E. M., Shoffner, R. N., Foster, D. N., and Guise, K. S. (1990). Mapping of the growth hormone gene on chicken chromosomes by in situ hybridisation. *Poultry Sci.*, 69, Suppl. 1, 122.

Smith, W. C., Kuniyoshi, J., and Talamantes, F. (1989). Mouse serum growth hormone (GH) binding protein has GH receptor extracellular and substituted transmembrane domains. *Mol. Endocrinol.*, 3, 984–990.

Souza, L. M., Boone, T. C., Murdock, D., Langley, K., Wypych, J., Fenton, D., Johnson, S., Lai, P.-H., Everett, R., Hsu, R.-Y., and Bosselman, R. (1984). Applications of recombinant DNA technology to studies on chicken growth hormone. *J. Exp. Zool.*, 232, 465–473.

Stewart, P. A., and Washburn, K. W. (1983). Variation in growth hormone, triiodothyronine (T3) and lipogenic enzyme activity in broiler strains differing in growth and fatness. *Growth*, 47, 411–425.

Stockdale, F. E., and Holtzer, H. (1961). DNA synthesis and myogenesis. *Exp. Cell Res.*, 24, 508–520.

Tinch, A. E. (1990). The genetics of muscle growth in chickens and mice. Ph.D. thesis, University of Edinburgh.

Ullrich, A., Gray, A., Tam, A. W., Yang-Feng, T., Tsubokawa, M., Collins, C., Henzel, W., Le Bon, T., Kathuria, S., Chen, E., Jacobs, S., Francke, U., Ramachandran, J., and Fujita Yamaguchi, Y. (1986). Insulin-like growth factor I receptor primary structure: Comparison with insulin receptor suggests structural determinants that define functional specificity. *EMBO J.*, 5, 2503–2512.

Vasilatos-Younken, R., Cravener, T. L., Cogburn, L. A., Mast, M. G., and Wellenreiter, R. H. (1988). Effect of pattern of administration on the response to exogenous, pituitary-derived chicken growth hormone by broiler-strain pullets. *Gen. Comp. Endocrinol.*, 71, 268–283.

Vasilatos-Younken, R., Gray, K. S., Bacon, W. L., Nestor, K. E., Long, D. W., and Rosenberger, J. L. (1990). Ontogeny of growth hormone (GH) binding in the domestic turkey: Evidence of sexual dimorphism and developmental changes in relationship to plasma GH. *J. Endocrinol.*, 126, 131–139.

Vasilatos-Younken, R., and Scanes, C. G. (1991). Growth hormone and IGFs in poultry growth: required, optimal or ineffective. *Poultry Sci.*, 70, 1764–1780.

Wood, W. I., Cachianes, G. C., Henzel, W. J., Winslow, G. A., Spencer, S. A., Hellmiss, R., Martin, J. L., and Baxter, R. C. (1988). Cloning and expression of the growth hormone-dependent insulin-like growth factor-binding protein. *Mol. Endocrinol.*, 2, 1176–1185.

Zvirblis, G. S., Gorbulev, V. G., Rubtsov, P. M., Karapetyan, R. V., Zuravlev, I. V., Fisinin, V. I., Skryabin, K. G., and Bayev, A. A. (1987). Genetic engineering of peptide hormones. I. Cloning and primary structure of chicken growth hormone cDNA. *Mol. Biol. (USSR)*, 21, 1620–1624.

Chapter 12

The Major Histocompatibility Complex in Chickens

Susan J. Lamont

Summary. The major histocompatibility complex (MHC) genes encode highly polymorphic cell-surface proteins of three classes. These antigens are markers of "self" and are important in regulation of cellular communication in the immune response. Restriction of cellular interactions by the MHC includes such phenomena as T-B cell cooperation in antibody production, generation of germinal centers, antigen presentation to T cells and T-cell cytotoxic reactions against virally infected cells. Via its role as a restriction element and via specific immune response genes, the MHC has profound effects on genetic control of immunoresponsiveness. The chicken MHC influences total immunoglobulin levels; antibody production to synthetic peptides, soluble antigens and cellular antigens; cell-mediated immunity; complement-mediated hemolytic activity; chemotaxis and activity of macrophages; and the percentage of T cells of specific subsets. Allelic diversity of the MHC has also been associated with differences in various production traits such as mortality, egg production, body weight, and feed efficiency.

The B-G antigens (class IV) are expressed on erythrocytes and some other cell types. The B-F antigens (class I), expressed on all cell types, consist of 39- to 43-kDa glycoproteins of three extracellular domains and are associated with $\beta2$ microglobulin. The B-L (class II) antigens consist of one monomorphic α chain associating with polymorphic β chains and are primarily expressed on cells of the immune system. Recent molecular cloning of B-complex genes has helped establish a molecular map of the chicken MHC. There are several striking differences between chicken and mammalian MHC molecular genetic organization. The chicken B-F and B-Lβ genes are interspersed rather than in separate class I and II regions, the class III genes have not yet been mapped to the MHC, and the introns in chicken MHC genes are much smaller than introns in mammalian MHC genes. The telomeric end of the MHC is adjacent to the nucleolar organizing region on a microchromosome. There are several non-MHC genes interspersed in the region, including gene C8.4, which has characteristics of the immunoglobulin superfamily, and gene C12.3, which bears homology to GTP-binding proteins.

Because of the important role of the chicken MHC in health and fitness, and rapid advances in the molecular definition of this system, the chicken MHC should provide useful genetic markers for breeding populations and should be a good candidate for fruitful modifications by genetic engineering.

Introduction

The major histocompatibility complex (MHC) is a group of genes that influences tissue graft acceptance or compatibility (Snell, 1953). The chicken was the second species in which the MHC was identified (Schierman and

Nordskog, 1961). Gene products of the MHC are cell-surface antigens that serve as markers of genetic identity. Mammalian MHC proteins have been divided, according to structural and functional properties, into three classes: class I, class II, and class III. The class I and class II homologs have been identified in the chicken, and are also designated B-F and B-L, respectively. The chicken possesses an additional unique class (class IV or B-G) of MHC proteins restricted in their distribution to red blood cells. Because of its genetic linkage with the B blood group (Briles et al., 1950) the chicken MHC has also been designated as the "B complex." The combination of all alleles that an individual possesses for its MHC genes is known as its MHC haplotype.

The regulation of immune response and disease resistance is a major function of proteins encoded by the MHC (Dorf, 1981). Examples of differences in immune response that have been associated with MHC allelic differences include the response to simple chemical antigens and to complex native antigens (Benedict et al., 1975; Pevzner et al., 1975, 1978). Numerous studies confirm that gene products that are encoded by the chicken MHC affect resistance to disease, including autoimmune diseases, and those induced by viruses, bacteria, and parasites (Bacon, 1987; Lamont, 1989). Many non-immune phenomena, including reproduction, are also affected by gene products encoded by the MHC (Edidin, 1983; Warner, 1986). Examples of MHC associations with growth and egg production in the chicken are available (Simonsen et al., 1982b; Gavora et al., 1986; Lamont et al., 1987b; Kim et al., 1987). Thus the MHC has major effects on the economically-important traits of disease resistance, immune response, growth, and reproduction.

Products of the Major Histocompatibility Complex Genes

B-G (Class IV) Antigens

The location of the erythrocyte antigen B (*Ea-B*) genes within the chicken MHC has allowed the use of blood typing with serological reagents to identify allelic differences in the chicken MHC by using hemagglutination. Alloantisera that potentially detect both B-G and B-F erythrocyte antigens are prepared from serum antibodies produced by immunizations with whole blood between birds that differ for their *Ea-B* blood groups (Briles and Briles, 1982). Alloantisera that are specific for either B-G or B-F antigens can be prepared by immunizations between birds of appropriate MHC-recombinant haplotypes or by immunizations with purified leukocytes or red blood cells and subsequent absorptions with the appropriate cell type. The MHC-recombinant haplotypes represent putative previous chromosomal crossover events, which resulted in a combination of segments of different standard MHC types into a new recombinant haplotype. A workshop held in 1981 in Innsbruck recommended a uniform nomenclature for standard and recombinant haplotypes and also designated the standard chicken strains to be used as the reference sources for each of 27 known *B* haplotypes (Briles et al., 1982).

The B-G (class IV) antigens of the chicken are apparently highly immunogenic in the mouse, and this has allowed the production of several monoclonal antibodies specific for one, some, or all B-G allelic products (Longenecker et al., 1979, Longenecker, 1982; Miller et al., 1982). Nomenclature for chicken MHC (B) antigens defined by monoclonal antibodies has been proposed by Longenecker and Mosmann (1981). Initially, it was hoped that a comprehensive panel of monoclonal antibodies could be developed for use as highly specific blood-typing reagents. This has only been realized for a few haplotypes, but the monoclonal antibodies have been extremely valuable in studies of B-G expression and structure.

Monoclonal antibodies directed against a widely shared B-G determinant were used to immunoprecipitate B-G antigens from erythrocyte membranes for analysis by using two-dimensional (2D) gel electrophoresis (Miller et al., 1984). The patterns generated varied among haplotypes from slight microheterogeneity to extremely complex patterns. This demonstrated a greater level of complexity for the locus than had been previously recognized. The high-resolution 2D gel electrophoresis technique provided biochemical confirmation of recombination in the subregion that had been originally identified by serological typing (Miller et al., 1988b).

Early reports on the number of polypeptide chains and on the molecular mass of B-G antigens showed discrepancies (Pink et al., 1977; Wolf et al., 1984; Miller et al., 1984; Salomonson et al., 1987). Recent work (Kaufman et al., 1990) has demonstrated that B-G molecules occur as homodimers and heterodimers with intrachain disulfide bonds, and no N-linked carbohydrate. Haplotype differences in size range from 30 to 55 kDa. Size variation is generated in the cytoplasmic region, which extends from the cell membrane into the cytoplasm. The structure of the cytoplasmic portion of each chain of the B-G dimer is that of an α-helical coil composed of heptad repeats. The variety of molecular sizes may be generated by proteolysis, biosynthetic modification, or alternative splicing. The alternative splicing is hypothesized to control the association of B-G chains with each other or with other cytoplasmic proteins. The immunological function of the B-G antigens is not yet understood, although they are highly immunogenic and appear to be necessary as an adjuvant for an efficient anti-B-F response (Hála et al., 1981). Their recent identification in the intestinal epithelium (Miller et al., 1990) suggests a role in antigen recognition, perhaps in a defensive function or as a receptor.

B-F (Class I) Antigens

The B-F (class I) antigen is expressed on the cell membranes of almost all somatic cells including nucleated erythrocytes. Ziegler and Pink (1975, 1976, 1978) described an MHC antigen that was located on both peripheral blood leukocytes and erythrocytes and consisted of a polymorphic heavy chain of 40 – 45 kDa associated with a smaller monomorphic polypeptide of about

12 kDa. Homology of chicken class I (B-F) antigen with that of mouse and human was shown by comparing partial amino acid sequences (Vitetta et al., 1977). Monoclonal antibodies to the chicken B-F antigen (Crone et al., 1985; Pink et al., 1985; Sgonc et al., 1987) were used to demonstrate that the expression of B-F antigen on chicken cells increases rapidly from hatch, when no cells express the antigen, to 9 days posthatch when 100% of bursal cells, red blood cells, and peripheral blood lymphocytes and 20% of thymus cells express B-F antigen (Sgonc et al., 1987). The B-F antigens continue to be expressed on these cells throughout life. Sequential immunoprecipitations of B-F antigens by monoclonal antibodies of different specificities indicated the existence of heterogeneity in the gene product (Crone et al., 1985).

The smaller peptide with which the B-F molecule is associated is the avian equivalent of mammalian β-2 microglobulin. A recent study (Skjoedt et al., 1986) characterizing the β-2 microglobulin molecule of chickens and turkeys revealed that the apparent molecular weight of this molecule is approximately 14,000. The 23 amino acids at the N-terminus of β2-microglobulin are identical in chickens and turkeys, and bear a strong homology to mammalian β2-microglobulin sequences.

B-L (Class II) Antigens

The B-L (class II) antigen was identified by Ziegler and Pink (1976) as an antigen that is precipitated by an anti-B antiserum, is 30 kDa, is not expressed on erythrocytes, and is not associated with β2-microglobulin. Genetic and serological evidence indicated, respectively, that the B-L antigens were on the same chromosome as *Ea-B* and that they were analogous to MHC class II Ia (immune-associated) antigens of mice (Ewert and Cooper, 1978; Ewert et al., 1980). Crone et al. (1981) demonstrated that the chicken B-L antigen is comprised of a nonpolymorphic α chain of 30–34 kDa and two polymorphic β chains (27–29 kDa). These chains are noncovalently bound on the cell surface. The α and β chains undergo complex N-glycosylation in the maturation process in the cytoplasm (Guillemot et al., 1986). The class II antigens are associated with invariant molecules (In) in the cytoplasm, but not on the cell surface (Guillemot and Auffray, 1989). The B-L antigens are expressed primarily on cells of the immune system: B cells, plasma cells, macrophages/monocytes in the circulation and tissues, and activated T cells (Peck et al., 1982; Ewert et al., 1984; Guillemot et al., 1984; Hála et al., 1984).

Functions of the Major Histocompatibility Complex Genes

Immune Response

One of the most crucial functions of the chicken major histocompatibility complex is the regulation of cellular communication in the immune response (Lamont and Dietert, 1990). The cell-surface protein products of the MHC, which are unique to each genetically unique individual, serve to distinguish "self" from "nonself". This allows immune reaction to occur against foreign antigens of an almost infinite spectrum to protect each organism's identity while simultaneously preserving the organism from self-destruction. The MHC cell-surface glycoproteins interact with both the antigen and with complementary structures of other immune cells. This communication generates an immune response that is specific for the inducing antigen.

Many cell types, cell-surface proteins, and soluble factors can be involved in the generation of different manifestations of immune response. In a generalized model of immune response, the inducing antigen is phagocytosed by a macrophage or other antigen-presenting cell, digested, and then expressed on the cell surface in association with MHC molecules. The T lymphocytes are activated by contact with the processed antigen and MHC molecules on the antigen-presenting cell surface. Some T cells differentiate to become cytotoxic cells that can lyse virally infected target cells. Other T cells differentiate into helper T cells that recognize specific antigens and stimulate production of antibodies by differentiated B lymphocytes (plasma cells).

Alloreactivity and MHC Restriction

Early studies of the *B* complex showed it to be linked to genes controlling skin graft rejection (Schierman and Nordskog, 1961), graft-versus-host (GVH) splenomegaly (Jaffe and McDermid, 1962; Simonsen et al., 1977) and mixed lymphocyte reactions (Miggiano et al., 1974). These studies confirmed that the chicken MHC controlled activation of T cells by allogeneic cells (cells from a different individual of the same species) in a manner similar to that known for the mouse and human.

Restriction of immune function by the MHC means that for some of the different cell types of the immune system to communicate effectively, they must share at least one MHC haplotype. Dependence on a common haplotype determines the level of interaction between mature cells in the immune response or in immune cell differentiation (Vainio and Toivanen, 1987; Vainio et al., 1987).

Normally, all cells in the immune system of an individual share their MHC haplotypes. To demonstrate the different levels at which MHC restriction affects immune function, specialized model systems were utilized. The cyclophosphamide (CP) model was used to elucidate the molecular requirements for T-B cell interaction in the chicken. Administration of CP to young chicks causes a permanent depletion of B lymphocytes. Transplantation with allogeneic B-cell precursors can then generate individuals with B cells and T cells of different MHC haplotypes. Antibody production by B cells to thymusdependent antigens can be restored only if T and B cells share at least one MHC class II haplotype (Toivanen et al., 1974; Vainio et al., 1983; 1984).

Combination in tissue culture of different cell types from birds of different MHC haplotypes also allows identification of the MHC antigen requirements that restrict function in the immune system. The interaction of antigen-presenting cells with T cells (Vainio et al., 1988) and the cytotoxic activity of T cells against some virus-infected and/or transformed cells was also shown to be restricted by a requirement for identity of at least one MHC haplotype (MacCubbin and Schierman, 1986; Weinstock and Schat, 1987). The latter may represent an important immune surveillance mechanism to limit proliferation of virally induced tumors in chickens.

Immunoresponsiveness

The MHC has profound effects on genetic control of immunoresponsiveness via its role as a restriction element and via specific immune response genes. The level of antibody production against a variety of antigens has been shown to be linked to the chicken MHC. The immune response to synthetic polypeptides such as (T, G) – A — L (Gunther et al., 1974), GAT (Benedict et al., 1975), and GT (Pevzner et al., 1979) is under genetic control of the chicken MHC. Antibody titers to other soluble (bovine serum albumin, BSA) and cellular (*Salmonella pullorum* bacterin) antigens are also associated with the chicken MHC (Pevzner et al., 1975). Additionally, total serum IgG levels were shown to be under genetic control, partially associated with the *B* complex, but more significantly associated with the *Ir-GAT* locus, which controls the level of antibodies produced after secondary immunization with the random polymer, glu-ala-tyr (Rees and Nordskog, 1981).

Assays of cell-mediated immunity such as the delayed wattle reaction (DWR) (Klesius et al., 1977) can also be used to evaluate genetic control of immunoresponsiveness. The antigen used to induce localized intradermal inflammation of the wattle can be a general activator of T cells, such as phytohemagglutinin (Taylor et al., 1987) or a specific antigen, such as *Staphylococcus aureus* (Cotter et al., 1987), which is an important pathogen of poultry. The DWR to *S. aureus* antigen is influenced by an interaction between *B* complex and sex. In males, the *B* genotype significantly influenced the maximum response level of the DWR; however, in females, no *B* genotype effect was detected. Production of the lymphokine, interleukin-2 (IL-2), has been shown to be at least partially under control of genes linked to the chicken MHC (Knudston et al., 1990).

Although the complement system does not exhibit the specificity and rapid, high-level secondary response of the humoral and cell-mediated immune system, complement proteins are extremely important in the immune reaction as a means to destroy invading pathogens by destroying bacterial cell walls. Differences in the concentrations of sera that result in complement-mediated lysis of erythrocytes have been associated with genetic variation in the MHC in chickens (Chanh et al., 1976).

Analysis of partially developed (two to five backcross generations) $15I_5$-B congenic lines showed no significant difference in immunoglobulin M or G levels or differential blood cell counts, but did show significantly different hematocrits and hemolytic complement levels (Shen et al., 1984). There were also B-genotype associated differences in antibody responses to sheep red blood cells, *Brucella abortus* antigen, and infectious bursal disease virus vaccine. The activity and recruitment to the peritoneal cavity of macrophages were also shown to differ among the B-congenic lines (Qureshi et al., 1986). All these examples illustrate the diverse facets of the immune response over which the MHC exerts genetic control.

Disease Resistance

The association of the MHC with resistance to disease has been one of the primary reasons for interest in both the basic biology and the practical application of knowledge about the chicken MHC. The influence of the MHC on disease resistance has been the subject of several recent reviews (Bacon, 1987; Lamont, 1989; Gavora, 1990). The earliest and most significant association of the MHC with resistance to a specific disease is that of Marek's disease, a virus-induced lymphoma. MHC differences are associated with incidence of tumor formation, mortality, and transient paralysis induced by Marek's disease virus. The B^{21} haplotype seems to convey a favorable response in a wide variety of genetic backgrounds. Genetic complementation of MHC alleles with other genes such as *Ir-GAT* may affect the susceptibility to Marek's disease (Figure 1, Steadham et al., 1987). Relative resistance or susceptibility to Marek's disease virus of different sublines that are identical at the B locus also illustrates the importance of non-MHC genes (Bacon, 1987). Other virally induced diseases influenced by the MHC include Rous sarcoma virus tumors (reviewed by Schierman and Collins, 1987), avian leukosis, and infectious bursal disease. Besides virally mediated diseases, other categories of disease have also been demonstrated to be, at least partially, under genetic control of the MHC (Gavora, 1990). These include spontaneous autoimmune thyroiditis in the Obese strain of chickens, coccidiosis, and fowl cholera (Lamont et al., 1987a). Thus, the variety of diseases influenced by the chicken MHC is extensive.

One obvious feature that emerges from these studies, however, is that there is not a single haplotype that performs optimally in all genetic backgrounds in response to all disease challenges. Given the many different immune mechanisms involved in resistance to disease and the variety of levels at which the MHC exerts genetic control, this is not surprising. Also, the associations of resistance with the MHC, but not consistently with one MHC haplotype, may suggest an important role for non-MHC genes, or currently unidentified MHC genes, located within the complex. The strong genetic associations indicate that there are genes of importance in disease response

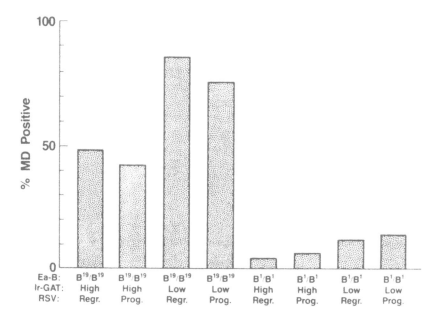

FIGURE 1. Incidence of Marek's disease (MD) in eight sublines of Iowa State University S1 chickens challenged with JM strain MD virus. Ea-B, erythrocyte alloantigen B; Ir-GAT, immune response to glutamic acid-alanine-tyrosine; RSV, response to Rous sarcoma virus-induced tumors, progressor (prog.) or regressor (regr.) of tumor after firmation. (From Steadham et al., 1987.)

within the MHC, but the lack of consistent association with certain haplotypes suggests that genes other than the currently identified MHC genes may have a major influence. Identification of specific resistance genes and utilization of genetic engineering technologies may allow, in the future, the construction of resistant haplotypes that have not yet appeared in conventional breeding approaches.

Production Traits

Genetic variation in production traits such as general mortality (Briles and McGibbon, 1948; Gilmour, 1959) has long been associated with genetic diversity of MHC. Growth and reproduction traits as diverse as fertilization rate, embryonic mortality, hatchability, juvenile and adult mortality, body weight, and egg production are influenced by the MHC genotype (Kim et al., 1987, 1989; reviewed in Bacon, 1987). MHC allelic frequencies have been significantly altered by selection for egg production traits and disease resistance (Simonsen et al., 1982b; Gavora et al., 1986; Lamont et al., 1987b).

The exact mechanism by which the MHC may influence these traits is not known. The effects may be indirect, via the MHC influence on immunocompetency and disease resistance, or the effects may be more direct and based on the role of the MHC in mediating cellular development and communication.

Molecular Genetic Structure of the Major Histocompatibility Complex: MHC Gene Isolation, Cloning, and Sequence Analysis

The complex nature of the chicken *B* system was not fully appreciated until the discovery of chromosomal recombinants in the chicken MHC. Hála et al. (1976) described a crossover between the MHC haplotypes of inbred lines CC and CB. This demonstrated, via genetic and serological analyses, that the chicken *B* system had a minimum of two loci, for which the designations *B-F* and *B-G* were proposed. The *B-F* locus was shown to determine histocompatibility and serologically determined antigens as well as controlling allogenic responses, and *B-G* was shown to determine only serological antigens (Vilhelmová et al., 1977). Ziegler and Pink (1975, 1976, 1978) combined biochemical characterization of MHC antigens with genetic analysis of *B* recombinants to develop a three-locus model of the chicken MHC (Pink et al., 1977). The application of molecular genetic approaches to analysis of the genes of the chicken MHC has led to recent progress in understanding the unique genomic organization of this complex.

B-G (Class IV)

The first *B-G* clone isolated was selected from an erythroid cell λgt11 expression library by using antiserum against purified B-G antigen (Goto et al., 1988). This cDNA clone, designated bg28, was confirmed as a *B-G* subregion clone by using it as a hybridization probe. In Northern hybridizations, it bound specifically with erythrocyte mRNA; in Southern hybridizations, it mapped to the MHC-bearing chromosome. The hybridization patterns also suggested that multiple loci exist within the *B-G* subregion. More recently, the techniques of chromosome walking and screening expression libraries with specific antibodies have enabled isolation of additional *B-G* subregions (reviewed in Kroemer et al., 1990a).

B-F (Class I)

Six *B-F* genes have been isolated from a cosmid library derived from a CB line chicken of the B^{12} haplotype (Guillemot et al., 1988). The longest *B-F* cDNA clone, designated F10, was sequenced. Based on inferred amino acid sequence, the chicken class I molecule was shown to be more divergent from mammalian class I molecules (equivalent avian and mammalian domains range from 24% to 52% amino acid sequence homology) than mammalian class I MHC molecules are from each other. Most residues involved in structural features of the antigen, however, were conserved. One of the six *B-F* genes, *B-FIV*, was identified as the transcribed gene (Kroemer et al., 1990c). It possesses regulatory elements typical of mammalian class II genes.

β2 Microglobulin

The cDNA coding for chicken β_2M was isolated from a λgt11 cDNA library from spleen and bone marrow cells, by screening with degenerate oligonucleotide probes based on mammalian β_2M sequences (Skjoedt et al., 1988). This gene is not linked with the MHC. The β_2M and class I MHC molecules associate at the cell surface.

B-L (Class II)

Several clones containing chicken *B-L* genes have been isolated (Béhar et al., 1988; Bourlet et al., 1988; Guillemot et al., 1988; Xu et al., 1989). Initial isolation of chicken class IIβ genomic clones was carried out by hybridizing with a human class II MHC probe under low-stringency conditions. The chicken class IIβ genes encoding specific domains showed 62–66% nucleotide homologies with the genes of corresponding domains from humans. The domains are the expected size (92–93 amino acids) and have the intrachain disulfide loop that is typical of the immunoglobulin supergene family (Bourlet et al., 1988). Subsequent isolations from libraries of other haplotypes (B^{12} and B^6) were done by screening with chicken class II (*B-L*) β exons (Guillemot et al., 1988; Xu et al., 1989). Sequence homologies among MHC class IIβ domains of chickens from different haplotypes range from 90% to 99% (Xu et al., 1989). A transcribed *B-L* gene of the B^{12} haplotype was sequenced (Zoorob et al., 1990). The 5' flanking region contains transcription-controlling sequence motifs analogous to those known in mammals.

Molecular Organization

Molecular mapping of a large segment of the chicken MHC has revealed details regarding the genomic organization of the *B* complex (Guillemot et al., 1988, 1989b; Béhar et al., 1988; Kroemer et al., 1990b). In contrast to mammalian MHCs, distances between genes of the chicken MHC are much smaller, as are the introns, which consist of about 100 bp each. The close proximity and interspersing of the individual *B-F* and *B-L* genes may account for the scarcity of recombinants identified between these two MHC subregions. The nucleolar organizer is tightly linked to the telomeric end of the *B* complex. Several genes of currently unknown function are also located within the *B* complex. The gene *C12.3* belongs to the same family as genes encoding G-protein β subunits and is linked to the MHC in chickens, but not in humans (Guillemot et al., 1989a). The G proteins are important in signal transduction and lymphocyte activation, and it is hypothesized that they may regulate susceptibility of T lymphocytes to virus-induced transformation, as in the case of Marek's disease (Guillemot et al., 1989a). The gene *8.4* is located between two *B-Lβ* genes. It codes for a cell membrane-bound protein with characteristics of the immunoglobulin supergene family (Kroemer et al., 1990a).

Despite unique features, the general structure of MHC genes and molecules are typical of those of the members of the immunoglobulin superfamily. This suggests that the chicken MHC has also arisen from a primordial gene that was the progenitor of all immune function-related genes, such as those encoding immunoglobulins, T-cell receptors, and MHC antigens.

Restriction Fragment Length Polymorphism Analysis of the MHC

The technique of restriction fragment length polymorphism (RFLP) analysis provides a molecular genotyping approach that can be used to characterize differences in MHC genes among individuals (Lamont et al., 1987b). Because the MHC has a well-established role in disease resistance, it should show RFLPs that are associated with traits of economic value. Therefore RFLP analysis of the MHC should be of value in a breeding program (Warner, 1986).

Chicken RFLPs have been generated with probes from the human MHC (Andersson et al., 1987) and from the chicken *B-L*, *B-F*, and *B-G* subregions (Miller et al., 1988a; Pitcovski et al., 1988; Lamont et al., 1987c, 1990; Chaussé et al., 1989, 1990; Warner et al., 1989;). Figure 2 shows *B-Lβ* RFLPs of the Iowa State University S1 line. The *B-Lβ* polymorphism is associated with *Ea-B* differences, but not *IrGAT* differences. This suggests that *IrGAT*, although linked to the MHC, is not controlled by a *B-Lβ* gene. *B-F* polymorphisms are also associated with *Ea-B*, but not *Ir-GAT*, differences in this line (Chen and Lamont, 1992). This RFLP is also associated with differences in disease resistance and immune response in this population (Figure 1, Lamont et al., 1987a; Steadham et al., 1987; Knudtson et al., 1990).

The use of RFLP analysis to characterize the MHC biochemically is a valuable addition to serological analysis of the MHC. Caution must be applied in analyzing restriction patterns, however, because fragment lengths indicate the distance between two restriction sites, but do not give any other information about the intervening sequences. The use of RFLP analysis is nonetheless valuable to examine subregions and genotypes for which the production of serological reagents has been difficult, to analyze MHC chromosomal recombinants, and to categorize newly identified genotypes. In the future, standard nomenclature should be defined for RFLP analysis of MHC haplotypes, as was previously done for serological analysis (Briles et al., 1982).

MHC Aneuploidy Effects

Chickens that are aneuploid (trisomic or tetrasomic) for the MHC-bearing chromosome have been used to map the MHC, determine linkages with the

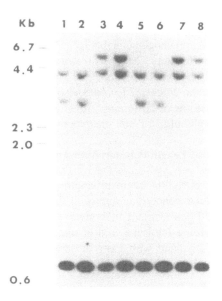

FIGURE 2. Autoradiogram of chicken DNA digested with *Pvu*II, separated by electrophoresis through agarose gel, blotted and hybridized with a chicken class II clone, p234. The B-G genotype, IrGAT type, and response to Rous sarcoma virus-induced tumors are: (1) B^1B^1, IrGAT^{Low}, progressor; (2) B^1B^1, IrGAT^{Low}, regressor; (3) $B^{19}B^{19}$, IrGAT^{High}, progressor; (4) $B^{19}B^{19}$, IrGAT^{Low}, regressor; (5) B^1B^1, IrGAT^{High}, regressor; (6) B^1B, IrGAT^{High}, progressor; (7) $B^{19}B^{19}$, IrGAT^{High}, regressor; (8) $B^{19}B^{19}$, IrGAT^{Low}, progressor (From Pitcovski et al., 1988.)

MHC, and examine effects of MHC gene dosage on several biological systems (Bloom et al., 1987, 1988). Linkage studies demonstrated that the chicken MHC (*B* complex) and the ribosomal ribonucleic acid (rRNA) genes (nucleolar organizer region, *NOR*) are linked on a microchromosome of approximately size 16 (Bloom and Bacon, 1985). The excess rRNA genes appear to be inactivated in aneuploid birds, but the MHC genes, on the same chromosome, appear to be expressed in a dose-dependent fashion. Based on both the mean hemagglutination titers of disomic, trisomic, and tetrasomic cells with polyclonal antiserum, and on the number of each cell type needed to reduce by 50% the ability of the antiserum to hemagglutinate disomic erythrocytes, the aneuploids have a greater concentration of MHC glycoproteins on their erythrocyte surfaces than do disomic chickens (Bloom et al., 1987). Flow cytometric analysis of bursal lymphocytes demonstrated a stepwise increase in the amount of B-L antigen on trisomic and tetrasomic cells (Delaney et al., 1988). The finding that extra MHC copies are not inactivated has important implications for the feasibility of producing chickens with increased copy numbers of the MHC genes to enhance health or production characteristics (Bloom et al., 1988; Lamont, 1989). Such birds could be produced by utilizing natural variation or genetic engineering techniques.

Industrial Applications of MHC Manipulations

Contemporary industrial use of the MHC in chickens is primarily in two areas. The first is in ascertaining pedigree purity. Because of the obvious separation of hens from eggs and progeny in commercial poultry breeding and the huge number of eggs and chicks that may be handled simultaneously at a hatchery, even a very small percentage of pedigree errors could result in large numbers of incorrectly pedigreed chicks. If genetic lines hold unique genetic markers, then these markers can be used to assure purity of lines. Blood group antigens, including the B blood group within the MHC, are excellent candidates to be used as genetic markers because they are quickly and economically identifiable from a small blood sample.

The second current use of MHC manipulation in the poultry industry is to alter MHC allelic frequencies to improve correlated traits such as disease resistance. Certain MHC alleles have been identified as beneficially associated with Marek's disease resistance in specific commercial genetic backgrounds. These MHC alleles can be either increased in frequency or fixed within one or more parent lines. The extreme polymorphism of the MHC (over 50 allelic forms in Leghorns, more in meat-type birds) allows many options for choice of specific alleles in parent lines and combinations in commercial birds.

New industrial applications of MHC manipulation should incorporate recently developed biotechnology procedures. With the use of specific probes, individuals can be examined for allelic forms of the MHC genes in the DNA. Direct analysis of class I and II genes will avoid the possibility of misclassifying allelic forms due to crossovers between serologically determined class IV genes and the class I and II genes. Modulation of gene copy number is an exciting possibility for future manipulation of the MHC. Bloom and co-workers (Bloom et al., 1988; Delaney et al., 1988) have shown that the gene products of the MHC are expressed in a dose-dependent fashion. That is, there is no downregulation within the cell to limit MHC expression to the usual diploid level, as is the case with some neighboring genes. This suggests that expanding the immune response repertoire of an individual by adding many additional gene copies may be feasible. This increase could be accomplished by genetic engineering or by identifying and utilizing naturally occurring variations in gene copy number. A single individual could then represent the entire spectrum of immune responsiveness normally seen in a population. Because most traits for disease resistance associated with the MHC are dominant traits (Bacon, 1987), the single individual with multiple gene copies might then be resistant to a greater variety of diseases.

References

Andersson, L., Lundberg, C., Rask, L., Gissel-Nielsen, B., and Simonsen, M. (1987). Analysis of class II genes of the chicken MHC (*B*) by use of human DNA probes. *Immunogenetics*, 26, 79–84.

Bacon, L. D. (1987). Influence of the MHC on disease resistance and productivity. *Poultry Sci.*, 66, 802–811.

Béhar, G., Bourlet, Y., Fréchin, N., Guillemot, F., Zoorob, R., and Auffray, C. (1988). Molecular analysis of chicken immune response genes. *Biochemie*, 70, 909–917.

Benedict, A. A., Pollard, L. W., Morrow, P. R., Abplanalp, H. A., Maurer, P. H., and Briles, W. E. (1975). Genetic control of immune responses in chickens. I. Responses to a terpolymer of poly $(Glu^{60}Ala^{30}Try^{10})$ associated with the major histocompatibility complex. *Immunogenetics*, 2, 313–324.

Bloom, S. E., and Bacon, L. D. (1985). Linkage of the major histocompatibility (B) complex and the nucleolar organizer in the chicken. *J. Heredity*, 76, 146–154.

Bloom, S. E., Briles, W. E., Briles, R. W., Delaney, M. E., and Dietert, R. R. (1987). Chromosomal localization of the major histocompatibility (B) complex (MHC) and its expression in chickens aneuploid for the major histocompatibility complex/ribosomal deoxyribonucleic acid microchromosome. *Poultry Sci.*, 66, 782–789.

Bloom, S. E., Delaney, M. E., Muscarella, D. M., and Dietert, R. R. (1988). Gene expression in chickens aneuploid for the MHC-bearing chromosome. In: *The Molecular Biology of the Major Histocompatibility Complex of Domestic Animal Species*, edited by C. M. Warner, M. F. Rothschild, and S. J. Lamont. Iowa State University Press, Ames, IA.

Bourlet, Y., Béhar, G., Guillemot, F., Fréchin, N., Billault, A., Chaussé, A.-M., Zoorob, R., and Auffray, C. (1988). Isolation of chicken major histocompatibility complex class II (B-L) β chain sequences: Comparison with mammalian β chains and expression in lymphoid organs. *EMBO*, 7, 1031–1039.

Briles, W. E., and Briles, R. W. (1982). Identification of haplotypes of the chicken major histocompatibility complex (B). *Immunogenetics*, 15, 449–459.

Briles, W. E., Bumstead, N., Ewert, D. L., Gilmour, D. G., Gogusev, J., Hála, K., Koch, C., Longenecker, B. M., Nordskog, A. W., Pink, J. R. L., Schierman, L. W., Simonsen, M., Toivanen, A., Toivanen, P., Vainio, O., and Wick, G. (1982). Nomenclature for the chicken major histocompatibility (B) complex. *Immunogenetics*, 15, 441–447.

Briles, W. E., and McGibbon, W. H. (1948). Heterozygosity of inbred lines of chickens at two loci effecting cellular antigens. *Genetics*, 33, 605 (Abstr.).

Briles, W. E., McGibbon, W. H., and Irwin, M. R. (1950). On multiple alleles affecting cellular antigens in the chicken. *Genetics*, 35, 663–652.

Chanh, T. C., Benedict, A. A., and Abplanalp, H. (1976). Association of serum hemolytic complement levels with the major histocompatibility complex in chickens. *J. Exp. Med.*, 144, 555–561.

Chaussé, A.-M., Coudert, F., Dambrine, G., Guillemot, F., Miller, M. M., and Auffray, C. (1989). Molecular genotyping of four chicken B complex haplotypes with B-Lβ, B-F, and B-G probes. *Immunogenetics*, 29, 127–130.

Chaussé, A.-M., Thoraval, P., Coudert, F., Auffray, C., and Dambrine, G. (1990). Analysis of B complex polymorphism in Rous sarcoma progressor and regressor chickens with B-G, B-F, and B-Lβ probes. *Avian Dis.*, 34, 934–940.

Chen, Y., and Lamont, S. J., (1992). Major histocompatibility complex class I restriction fragment length polymorphism analysis in highly inbred chicken lines and lines selected for major histocompatibility complex and immunoglobulin production. *Poultry Sci.*, 71, 999–1006.

Cotter, P. F., Taylor, R. L., Wang, T. L., and Briles, W. E. (1987). Major histocompatibility (B) complex-associated differences in the delayed wattle reaction to Staphylococcal antigen. *Poultry Sci.*, 66, 203–208.

Crone, M., Jensenius, J., and Koch, C. (1981). Evidence for two populations of B-L (Ia-like) molecules encoded by the chicken MHC. *Immunogenetics*, 13, 381–391.

Crone, M., Simonsen, M., Skjoedt, K., Linnet, K., and Olsson, L. (1985). Mouse monoclonal antibodies to class I and class II antigens of the chicken MHC. Evidence for at least two class I products of the B complex. *Immunogenetics*, 21, 181–187.

Delaney, M., Dietert, R. R., and Bloom, S. E. (1988). MHC-chromosome dosage effects: Evidence for increased expression of Ia glycoprotein and alteration of B cell subpopulations in neonatal aneuploid chickens. *Immunogenetics*, 27, 24–30.

Dorf, M. E., ed. (1981). *The Role of the Major Histocompatibility Complex in Immunobiology.* Garland Press, New York.

Edidin, M. (1983). MHC antigens and nonimmune functions. *Immunol. Today*, 4, 269.

Ewert, D. L., and Cooper, M. D. (1978). Ia-like alloantigens in the chicken: Serologic characterization and ontogeny of cellular expression. *Immunogenetics*, 7, 521–535.

Ewert, D. L., Gilmour, D. G., Briles, W. E., and Cooper, M. D. (1980). Genetics of Ia-like alloantigens in chickens and linkage with B major histocompatibility complex. *Immunogenetics*, 10, 169–174.

Ewert, D. L., Munchus, M. S., Chen, C.-L., and Cooper, M. D. (1984). Analysis of structural properties and cellular distribution of avian Ia antigen by using monoclonal antibody to monomorphic determinants. *J. Immunol.*, 132, 2524–2530.

Gavora, J. S. (1990). Disease genetics. In: *Poultry Breeding and Genetics*, edited by R. D. Crawford. Elsevier, Amsterdam, chap. 33.

Gavora, J. S., Simonsen, M., Spencer, J. L., Fairfull, R. W., and Gowe, R. S. (1986). Changes in the frequency of major histocompatibility haplotypes in chickens under selection for both high egg production and resistance to Marek's disease. *Z. Tierzucht Zuchtungsbiol.*, 103, 218–226.

Gilmour, D. G. (1959). Segregation of genes determining red cell antigens at high levels of inbreeding in chickens. *Genetics*, 44, 14–33.

Goto, R., Miyada, C. G., Young, S., Wallace, R. B., Abplanalp, H., Bloom, S. E., Briles, W. E., and Miller, M. M. (1988). Isolation of a cDNA clone from the *B-G* subregion of the chicken histocompatibility (*B*) complex. *Immunogenetics*, 102–109.

Guillemot, F., and Auffray, C. (1989). Molecular biology of the chicken major histocompatibility complex. *Poultry Biol.*, 2, 255–275.

Guillemot, F., Billault, A., and Auffray, C. (1989a). Physical linkage of a guanine nucleotide-binding protein-related gene to the chicken major histocompatibility complex. *Proc. Natl. Acad. Sci. USA*, 86, 4594–4598.

Guillemot, F., Billault, A., Pourquié, O., Béhar, G., Chaussé, A.-M., Zoorob, R., Kreibich, G., and Auffray, C. (1988). A molecular map of the chicken major histocompatibility complex: The class II β genes are closely linked to the class I genes and the nucleolar organizer. *EMBO*, 7, 2775–2785.

Guillemot, F. P., Furmel, P., Charron, D., LeDouarin, N., and Auffray, C. (1986). Structure, biosynthesis, and polymorphism of chicken MHC class II (B-L) antigens and associated molecules. *J. Immunol.*, 137, 1251–1257.

Guillemot, F., Kaufman, J. F., Skjoedt, K., and Auffray, C. (1989b). The major histocompatibility complex in the chicken. *Trends in Genet.*, 5, 300–304.

Guillemot, F. P., Oliver, P. D., Peault, B. M., and LeDouarin, N. M. (1984). Cells expressing Ia antigens in the avian thymus. *J. Exp. Med.*, 160, 1803–1819.

Gunther, E., Balcarová, J., Hála, K., Rude, E., and Hraba, T. (1974). Evidence for an association between immune responsiveness of chicken to (T, G)–A — L and the major histocompatibility system. *Eur. J. Immunol.*, 4, 548–553.

Hála, K., Plachý, J., and Schulmannová, J. (1981). Role of the B-G region antigen in the humoral immune response to the B-F region antigen of chicken MHC. *Immunogenetics*, 14, 393–401.

Hála, K., Vilhelmová, M., and Hartmanová, J. (1976). Probable crossing over in the *B* blood group system of chickens. *Immunogenetics*, 3, 97–103.

Hála, K., Wick, G., Boyd, R. L., Wolf, H., Bock, G., and Ewert, L. D. (1984). The B-L (Ia-like) antigens of the chicken. Lymphocyte plasma membrane distribution and tissue location. *Devel. Comp. Immunol.*, 8, 673–682.

Jaffe, W. P., and McDermid, E. M. (1962). Blood groups and splenomegaly in chick embryos. *Science*, 137, 984–985.

Kaufman, J., Salomonsen, J., Skjoedt, K., and Thorpe, D. (1990). Size polymorphism of chicken major histocompatibility complex-encoded B-G molecules is due to length variation in the cytoplasmic heptad repeat region. *Proc. Natl. Acad. Sci. USA*, 87, 8277–8281.

Kim, C. D., Lamont, S. J., and Rothschild, M. F. (1987). Genetic associations of body weight and immune response with the major histocompatibility complex in White Leghorn chicks. *Poultry Sci.*, 66, 1258–1263.

Kim, C. D., Lamont, S. J., and Rothschild, M. F. (1989). Associations of major histocompatibility complex haplotypes with body weight and egg production traits in S1 White Leghorn chickens. *Poultry Sci.*, 68, 464–469.

Klesius, P., Johnson, W., and Kramer, T. (1977). Delayed wattle reaction as a measure of cell-mediated immunity in the chicken. *Poultry Sci.*, 56, 249–256.

Knudtson, K. L., Kaiser, M. G., and Lamont, S. J. (1990). Genetic control of interleukin-2-like activity is distinct from that of mitogen response in chickens. *Poultry Sci.*, 69, 65–71.

Kroemer, G., Bernot, A., Béhar, G., Chaussé, A.-M., Gastinel, L.-N., Guillemot, F., Park, I., Thoraval, P., Zoorob, R., and Auffray, C. (1990a). Molecular genetics of the chicken MHC: Current status and evolutionary aspects. *Immunol. Rev.*, 113, 119–145.

Kroemer, G., Guillemot, F., and Auffray, C. (1990b). Genetic organization of the chicken MHC. *Immunol. Rev.*, 9, 8–19.

Kroemer, G., Zoorob, R., and Auffray, C. (1990c). Structure and expresion of a chicken MHC class I gene. *Immunogenetics*, 31, 405–409.

Lamont, S. J. (1989). The chicken major histocompatibility complex in disease resistance and poultry breeding. *J. Dairy Sci.*, 72, 1328–1333.

Lamont, S. J., Bolin, C., and Cheville, N. (1987a). Genetic resistance to fowl cholera is linked to the major histocompatibility complex. *Immunogenet.*, 25, 284–289.

Lamont, S. J., and Dietert, R. R. (1990). Immunogenetics. In: *Poultry Breeding and Genetics*, edited by R. D. Crawford. Elsevier, Amsterdam, chap. 22.

Lamont, S. J., Gerndt, B. M., Warner, C. M., and Bacon, L. D. (1990). Analysis of restriction fragment length polymorphisms of the major histocompatibility complex of 15I$_5$-B-congenic chicken lines. *Poultry Sci.*, 69, 1195–1203.

Lamont, S. J., Hou, Y. H., Young, B. M., and Nordskog, A. W. (1987b). Differences in major histocompatibility complex gene frequencies associated with feed efficiency and laying performance. *Poultry Sci.*, 66, 1064–1066.

Lamont, S. J., Warner, C. M., and Nordskog, A. W. (1987c). Molecular analysis of the chicken major histocompatibility complex genes and gene products. *Poultry Sci.*, 66, 819–824.

Longenecker, B. M. (1982). Preparation and properties of monoclonal antibodies against cell surface polymorphic or allelic determinants: Their use in blood typing and the study of cell differentiation. *An. Bld. Grp. Biochem. Polymorph.*, 13, 225–238.

Longenecker, B. M., and Mosmann, T. R. (1981). Nomenclature for chicken MHC (B) antigens defined by monoclonal antibodies. *Immunogenetics*, 13, 25–28.

Longenecker, B. M., Mosmann, T. R., and Shiogawa, C. (1979). A strong preferential response of mice to polymorphic antigenic determinants of the chicken MHC analyzed with mouse hybridoma (monoclonal) antibodies. *Immunogenetics*, 9, 137–147.

MacCubbin, D. L., and Schierman, L. W. (1986). MHC-restricted cytotoxic response of chicken T cells: Expression, augmentation, and clonal characterization. *J. Immunol.*, 136, 12–16.

Miggiano, V. C., Birgen, J., and Pink, J. R. L. (1974). The mixed leukocyte reaction in chickens. Evidence for control by the major histocompatibility complex. *Eur. J. Immunol.*, 4, 397–401.

Miller, M. M., Abplanalp, H., and Goto, R. (1988a). Genotyping chickens for the *B-G* subregion of the major histocompatibility complex using restriction fragment length polymorphisms. *Immunogenetics*, 28, 374–379.

Miller, M. M., Goto, R., and Abplanalp, H. (1984). Analysis of the B-G antigens of the chicken MHC by two dimensional gel electrophoresis. *Immunogenetics*, 20, 373–385.

Miller, M. M., Goto, R., and Briles, W. E. (1988b). Biochemical confirmation of recombination within the B-G subregion of the chicken major histocompatibility complex. *Immunogenetics*, 27, 127–132.

Miller, M. M., Goto, R., and Clark, S. D. (1982). Structural characterization of developmentally expressed antigenic markers on chicken erythrocytes using monoclonal antibodies. *Devel. Biol.*, 94, 400–414.

Miller, M. M., Goto, R., Young, S., Liu, J., and Hardy, J. (1990). Antigens similar to major histocompatibility complex B-G are expressed in the intestinal epithelium in the chicken. *Immunogenetics*, 32, 45–50.

Peck, R., Murthy, K. K., and Vainio, O. (1982). Expression of B-L (Ia-like) antigens on macrophages from chicken lymphoid organs. *J. Immunol.*, 129, 4–5.

Pevzner, I. Y., Nordskog, A. W., and Kaeberle, M. L. (1975). Immune response and the B blood group locus in chickens. *Genetics*, 80, 753–759.

Pevzner, I. Y., Trowbridge, C. L., and Nordskog, A. W. (1978). Recombination between genes coding for immune response and the serologically determined antigens in the chicken B system. *Immunogenetics*, 7, 25–33.

Pevzner, I. Y., Trowbridge, C. L., and Nordskog, A. W. (1979). B-complex genetic control of immune response to HSA, (T, G)–A — L, GT and other substances in chickens. *Immunogenetics*, 6, 453–460.

Pink, J. R. L., Droege, W., Hála, K., Miggiano, V. C., and Ziegler, A. (1977). A three locus model for the chicken major histocompatibility complex. *Immunogenetics*, 5, 203–216.

Pink, J. R., Vainio, O., and Rijnbeek, A.-M. (1985). Clones of B lymphocytes in individual follicles of the bursa of Fabricius. *Eur. J. Immunol.*, 15, 83–87.

Pitcovski, J., Lamont, S. J., Nordskog, A. W., and Warner, C. M. (1988). Analysis of *B-G* and immune response genes in the Iowa State University S1 chicken line by hybridization of sperm DNA with a major histocompatibility complex class II probe. *Poultry Sci.*, 68, 94–99.

Qureshi, M. A., Dietert, R. R., and Bacon, L. D. (1986). Genetic variation in the recruitment and activation of chicken peritoneal macrophages. *Proc. Soc. Exp. Biol. Med.*, 181, 560–568.

Rees, M. J., and Nordskog, A. W. (1981). Genetic control of serum immunoglobulin G levels in the chicken. *Immunogenetics*, 8, 425–431.

Salomonson, J., Skjoedt, K., Crone, M., and Simonsen, M. (1987). The chicken erythrocyte-specific MHC antigen. Characterization and purification of the B-G antigen by monoclonal antibodies. *Immunogenetics*, 25, 373–382.

Schierman, L. W., and Collins, W. M. (1987). Influence of the major histocompatibility complex on tumor regression and immunity in chickens. *Poultry Sci.*, 66, 812–818.

Schierman, L. W., and Nordskog, A. W. (1961). Relationship of blood type to histocompatibility in chickens. *Science*, 134, 1008–1009.

Sgonc, R., Hála, K., and Wick, G. (1987). Relationship between the expression of class I antigen and reactivity of chicken thymocytes. *Immunogenetics*, 26, 150–154.

Shen, P. F., Smith, E. J., and Bacon, L. D. (1984). The ontogeny of blood cells, complement, and immunoglobulins in 3- to 12-week-old $15I_5$-*B* congenic White Leghorn chickens. *Poultry Sci.*, 63, 1083–1093.

Simonsen, M., Crone, M., Koch, C., and Hála, K. (1982a). The MHC haplotypes of the chicken. *Immunogenetics*, 16, 513–532.

Simonsen, M., Hála, K., and Vilhelmová, M. (1977). The use of GVH methods in analysis of the B-complex. *Folia Biol. (Praha)*, 23, 402–410.

Simonsen, M., Kolstad, N., Edfors-Lilja, I., Liledahl, L. E., and Sorensen, P. (1982b). Major histocompatibility genes in egg-laying hens. *Am. J. Reprod. Immunol.*, 2, 148–152.

Skjoedt, K., Andersen, R., and Kaufman, J. (1988). Chicken B-F, B-G, and β_2M cDNA clones. In: *The Molecular Biology of the Major Histocompatibility Complex of Domestic Animal Species*, edited by C. M. Warner, M. F. Rothschild, and S. J. Lamont. Iowa State University Press, Ames, Ia.

Skjoedt, K., Welinder, K. G., Crone, M., Verland, S., Salomonsen, J., and Simonsen, M. (1986). Isolation and characterization of chicken and turkey beta 2-microglobulin. *Molec. Immunol.*, 23, 1301–1309.

Snell, G. D. (1953). The genetics of transplantation. *J. Natl. Cancer Inst.*, 14, 691–703.

Steadham, E. M., Lamont, S. J., Kujdych, I., and Nordskog, A. W. (1987). Association of Marek's disease with Ea-B and immune response genes in subline and F2 populations of the Iowa State S1 Leghorn line. *Poultry Sci.*, 66, 571–575.

Taylor, R. L., Jr., Cotter, P. F., Wing, T. L., and Briles, W. E. (1987). Major histocompatibility (B) complex and sex effects on the phytohemagglutinin wattle response. *Anim. Genet.*, 18, 343–350.

Toivanen, A., and Toivanen, P. (1977). Histocompatibility requirements for cellular cooperation in the chicken: Generation of germinal centers. *J. Immunol.*, 118, 431–436.

Toivanen, P., Toivanen, A., and Vainio, O. (1974). Complete restoration of bursa-dependent immune system after transplantation of semi-allogeneic stem cells into immunodeficient chicks. *J. Exp. Med.*, 139, 1344–1349.

Vainio, O., Koch, C., and Toivanen, A. (1984). B-L antigens (class II) of the chicken major histocompatibility complex control T-B cell interaction. *Immunogenetics*, 19, 131–140.

Vainio, O., Peck, R., Koch, C., and Toivanen, A. (1983). Origin of peripheral blood macrophages in bursa cell-reconstituted chickens: Further evidence for MHC-restricted interactions between T and B lymphocytes. *Scand. J. Immunol.*, 17, 193–199.

Vainio, O., and Toivanen, A. (1987). Cellular cooperation in immunity. In: *Avian Immunology: Basis and Practise*, edited by A. Toivanen and P. Toivanen. CRC Press, Boca Raton, FL, pp. 1–12.

Vainio, O., Toivanen, P., and Toivanen, A. (1987). Major histocompatibility complex and cell cooperation. *Poultry Sci.*, 66, 795–801.

Vainio, O., Veromaa, T., Eerola, E., Toivanen, P., and Ratcliffe, M. (1988). Antigen presenting cell-T cell interaction in the chicken is MHC class II antigen restricted. *J. Immunol.*, 140, 2864–2868.

Vilhelmová, M., Miggiano, V. C., Pink, J. R. L., Hála, K., and Hartmanová, J. (1977). Analysis of the alloimmune properties of a recombinant genotype in the major histocompatibility complex of the chicken. *Eur. J. Immunol.*, 7, 674–679.

Vitetta, E. S., Uhr, J. W., Klein, J., Pazderka, F., Moticka, E. J., Ruth, R. F., and Capra, J. D. (1977). Homology of murine (H-2) and (human) HLA with a chicken histocompatibility complex. *Nature*, 270, 535–536.

Warner, C. M. (1986). Genetic manipulation of the major histocompatibility complex. *J. Anim. Sci.*, 62, 279–287.

Warner, C. M., Gerndt, B., Xu, Y., Bourlet, Y., Auffray, C., Lamont, S., and Nordskog, A. (1989). Restriction fragment length polymorphism analysis of major histocompatibility complex class II genes from inbred chicken lines. *Anim. Genet.*, 20, 225–231.

Weinstock, D., and Schat, K. A. (1987). Virus specific syngeneic killing of reticuloendotheliosis virus transformed cell line target cells by spleen cells. In: *Avian Immunology*, edited by W. T. Weber and D. L. Ewert. *Prog. Clin. Biol. Res.*, 238, 253–264.

Wolf, H., Hála, K., Boyd, R. L., and Wick, G. (1984). MHC- and non-MHC-encoded surface antigens of chicken lymphoid cells and erythrocytes recognized by polyclonal xeno-, allo-, and monoclonal antibodies. *Eur. J. Immunol.*, 14, 831–839.

Xu, Y., Pitcovski, J., Peterson, L., Auffray, C., Bourlet, Y., Gerndt, B. M., Nordskog, A. W., Lamont, S. J., and Warner, C. M. (1989). Isolation and characterization of three class II MHC genomic clones from the chicken. *J. Immunol.*, 142, 2122–2132.

Ziegler, A., and Pink, J. R. L. (1975). Characterization of major histocompatibility complex (B) antigens of the chicken. *Transplantation*, 20, 523–527.

Ziegler, A., and Pink, J. R. L. (1976). Chemical properties of two antigens controlled by the major histocompatibility complex of the chicken. *J. Biol. Chem.*, 251, 5391–5396.

Ziegler, A., and Pink, J. R. L. (1978). Structural similarity of major histocompatibility antigens on leukocytes and erythrocytes. *Immunochemistry*, 15, 515–516.

Zoorob, R., Béhar, G., Kroemer, G., and Auffray, C. (1990). Organization of a functional chicken class II *B* gene. *Immunogenetics*, 31, 179–187.

Chapter 13

Synthesis and Deposition of Egg Proteins

Roger G. Deeley, Ruth A. Burtch-Wright, Caroline E. Grant,
Pamela A. Hoodless, Aimee K. Ryan, and Timothy J. Schrader

Introduction

The objective of this chapter is not to provide a comprehensive review of the regulation of egg protein synthesis and deposition, but rather to consider aspects of these processes that are of interest from both a biotechnological standpoint as well as that of basic science. Of necessity, some of the features of this model system will be dealt with somewhat superficially, while selected areas will be presented in more detail, particularly those involving regulation of expression of the major yolk protein genes.

Over the last 25 to 30 years, the regulation of egg protein synthesis has provided an exceptional model system for physiological and biochemical studies on developmental and hormonal regulation of gene expression (Clemens, 1974; Johnson, 1986). Some of the genes specifying major egg proteins have served as paradigms for detailed analyses of the molecular mechanisms involved. Figure 1 provides a brief summary of the major proteins that have been studied most extensively.

Synthesis of the egg-white proteins takes place in the tubular gland cells of the oviduct. Primary activation of the egg-white protein genes can only be elicited by estrogen, and it occurs concomitantly with the hormonally induced proliferation and differentiation of the oviduct (Oka and Schimke, 1969; O'Malley et al., 1969; Palmiter, 1972). Once this differentiation has been initiated, expression of the egg-white protein genes can be maintained, or induced to a limited degree, by other steroid hormones including progesterone, glucocorticoids, and androgens (Palmiter et al., 1978; LeMeur et al., 1981; Chambon et al., 1984). Interestingly, some of the egg-white protein genes are expressed in other tissues, where they are not regulated by estrogen despite the fact that the tissue contains estrogen receptor. For example, conalbumin is synthesized in the liver, where it is more typically referred to as transferrin. In the liver, conalbumin mRNA levels are controlled by iron, not by estrogen (Lee et al., 1978). Lysozyme is also synthesized by macrophages, again in an estrogen-independent fashion (Hauser et al., 1981).

Yolk protein synthesis takes place in the parenchymal cells of the liver. Some of the yolk proteins, such as apolipoprotein B (apoB) and retinol-binding protein, are normal constituents of the serum, and the onset of vitellogenesis simply results in an increase in their rate of production, typically of the order of 10-fold (Kirchgessner et al., 1987; DurgaKumari and Adiga, 1986). Other

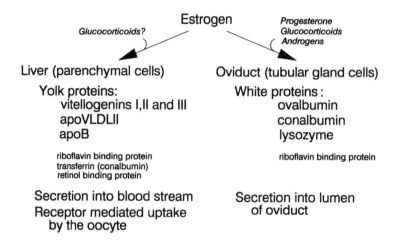

FIGURE 1. Synthesis and deposition of egg proteins. This figure summarizes the hormonal regulation, site of synthesis, and mode of deposition of the major egg-yolk and egg-white proteins. The hormones in italics are those capable of maintaining gene expression or eliciting a limited response following primary activation by estrogen.

yolk protein genes, such as those specifying the vitellogenins and apo very-low-density lipoprotein II (apoVLDLII), are completely dormant in immature birds. Expression of these genes can be induced by exogenous estrogens (Wiskocil et al., 1980). Unlike the egg-white protein genes, they can be activated without a requirement for DNA synthesis, not only in female chicks, but also in roosters, male chicks, and embryos of either sex at defined developmental stages (Wiskocil et al., 1980; Chan et al., 1976; Elbrecht et al., 1981; Bergink et al., 1974). Again in contrast to the egg-white protein genes, expression of this group of yolk protein genes cannot be maintained or induced by other steroid hormones, even in birds that have been primed with estrogen. However, it is less certain whether estrogen and other steroids can act synergistically to maximize their levels of expression. As in the case of the egg-white protein genes, some of the yolk protein genes are expressed in several tissues. ApoB, for example, is synthesized in the intestine and kidney, but is estrogen responsive only in the liver (Kirchgessner et al., 1987). Riboflavin-binding proteins can be found in both the yolk and the white, and are the products of the same gene (reviewed in White and Merrill, 1988).

Uptake of Yolk Proteins by the Oocyte

Egg-white proteins are secreted directly from the tubular gland cells into the lumen of the oviduct, while the yolk proteins, following secretion from the liver, are transported in the blood and then enter the oocyte via receptor-mediated mechanisms (Jones et al., 1975). All of the major yolk proteins are secreted from the liver as lipoprotein particles. Vitellogenin II, which has been studied more intensively than the two minor vitellogenins, I and III, has been shown to circulate as a dimer and is present in the high-density lipoprotein (HDL) fraction of serum (Deeley et al., 1975). ApoB and apoVLDLII are found in the very-low-density lipoprotein fraction (VLDL), where they are associated with the same lipoprotein particle (Williams, 1979).

At the peak of growth, oocytes incorporate 1–1.5 g of yolk protein per day (Johnson, 1986; Gilbert, 1981), most, and possibly all, of which enters the oocyte by receptor-mediated mechanisms. It is now apparent from the studies of Schneider and colleagues that a single receptor is responsible for the internalization of both the vitellogenins and VLDL (Stifani et al., 1990a). The oocyte receptor is functionally and genetically distinct from the somatic receptor that is responsible for the uptake of circulating low-density lipoprotein (LDL) by somatic cells. It appears to be the lack of the oocyte receptor that is responsible for the defect in restricted ovulator hens, while the somatic receptor in these birds is normal (Nimpf et al., 1989; Jones et al., 1975; McGibbon, 1977).

The somatic receptor is presumed to be the functional equivalent of the mammalian LDL receptor, the ligands for which have been shown to be apoB and a second apolipoprotein, apoE (Hayashi et al., 1989). ApoB functions as a ligand for the chicken receptor. However, birds do not produce a homolog of mammalian apoE, and the mammalian protein fails to bind to the chicken somatic receptor. The vitellogenins also do not bind to the somatic receptor, although they are certainly ligands for the oocyte receptor.

The region of the vitellogenin polypeptide that interacts with the receptor has been localized to the amino-proximal portion of the molecule. When the vitellogenins enter the yolk, they are subjected to limited proteolysis to generate the yolk proteins known as lipovitellins, phosvitins, and the phosvettes (Deeley et al., 1975; Bergink and Wallace, 1974). In the case of vitellogenin II, the lipovitellin moiety was predicted to constitute the amino-terminal portion of vitellogenin (Deeley et al., 1975), and this has now been confirmed by sequencing of the vitellogenin II gene (Byrne et al., 1984). It is the lipovitellin portion of the molecule that is the ligand for the receptor, but precisely which part of lipovitellin is involved is not yet known (Stifani et al., 1990b).

Unlike vitellogenin, which is recognized only by the oocyte receptor, apoB functions as the apolipoprotein ligand in VLDL that is recognized by both somatic and oocyte receptors (Hayashi et al., 1989). Surprisingly, mammalian apoE, although not recognized by the chicken somatic receptor, will function as a ligand for the oocyte receptor (Steyrer et al., 1990). That the chicken oocyte receptor may be more closely related to the mammalian somatic receptor than is the chicken somatic receptor is also suggested by the

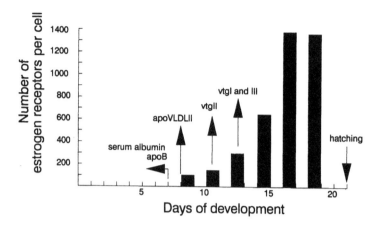

FIGURE 2. Accumulation of estrogen receptor during embryogenesis following treatment with estrogen. This figure shows the accumulation of estrogen receptors in the liver during embryogenesis following treatment of the embryo with estrogen. The bars represent the level of nuclear estrogen receptor 48 h after treatment. Estrogen receptor levels were determined as described in Haché et al. (1987). The time at which expression of the major egg-yolk protein genes can be induced in response to estrogen (indicated by the arrows) was determined by Northern blot analysis (described in Colgan et al., 1982; Evans et al., 1988).

fact that the oocyte receptor cross-reacts with anti-bovine LDL receptor antibodies, whereas the somatic receptor does not (Hayashi et al., 1989).

These observations raise very interesting questions relating to the origins of the mammalian LDL receptor and the evolutionary relationship between the ligands that these receptors recognize. They may also have very practical implications that can be exploited for the selective targeting of proteins produced in transgenic birds to the oocyte. As is described in more detail below, we have been investigating the possibility of using regulatory regions of the gene specifying the second major VLDL component, apoVLDLII, to provide highly efficient and tissue-specific expression of vectors that might be used as part of such technology.

Developmental Regulation of Expression of the Yolk Protein Genes

One of the reasons that we have pursued regulation of the apoVLDLII gene, rather than the other major yolk protein genes, is illustrated in Figure 2. The bar graph indicates the number of nuclear estrogen receptors per hepatocyte at various stages of embryogenesis, 48 h after receptor synthesis has been induced by injecting the egg with estrogen (Haché et al., 1987). As is apparent from the graph, the earliest stage at which receptors can be detected in the liver is approximately days 7–8 of incubation. Thereafter, the number of receptors induced in response to the hormone increases until hatching, at which time it reaches 30–40% of adult levels. The main reason for presenting these data is to draw attention to the relationship between receptor levels and the developmental staging of the ability to activate various yolk protein genes. Both serum albumin and apoB, which are synthesized in both sexes, are already detectable prior to day 7 (Elbrecht et al., 1984; Evans et al., 1988). The liver itself begins to differentiate between days 3 and 5, and it is possible that these genes are switched on at a comparable stage. On the other hand, the ability to activate those genes that are absolutely dependent on estrogen is staggered over a period of at least 4–5 days (Evans et al., 1988). Activation of the apoVLDLII gene can be detected as early as days 7–8, followed by vitellogenin II, at days 10–11, and the minor vitellogenins, I and III, at days 12–13 (Colgan et al., 1982; Elbrecht et al., 1984; Evans et al., 1988). This developmental sequence is recapitulated to some degree if the genes are activated by administering estrogen to a male bird, although the differences in the response times are compressed into several hours rather than days (Deeley et al., 1977; Evans et al., 1988). It has been suggested that the timing with which the genes can be activated may reflect the relative sensitivities of their regulatory regions to activation by estrogen. Alternatively, the relatively delayed response of the vitellogenin genes may indicate a requirement for factors that are not necessary for activation of the apoVLDLII gene. It is for these reasons, combined with the high efficiency with which the gene is expressed, that we have focused on dissecting the controlling regions of the apoVLDLII rather than the vitellogenin gene. Before summarizing some of our findings, we will review the organization of key elements involved in the regulation of the vitellogenin II and ovalbumin genes, in order to provide a background of information that can be compared with the results of studies on apoVLDLII.

Regulatory Regions of the Vitellogenin and Ovalbumin Genes

Like the apoVLDLII gene, both the vitellogenin and ovalbumin genes are dormant when estrogen and its receptor levels are below a critical threshold and are expressed at very high levels when this threshold is exceeded. However, the organization of their regulatory elements differs markedly.

The upper half of Figure 3 shows a schematic of approximately 1 kb flanking the 5' end of the vitellogenin II gene, indicating the location of sites of known or potential regulatory importance. When this gene is activated in the liver, a specific set of nuclease hypersensitive sites appear (indicated by the arrowheads) that extend approximately 900 nucleotides upstream (Burch and Weintraub, 1983; Philipsen et al., 1988; Saluz et al., 1988). This suggests that some reorganization of chromatin structure in this region is associated with activation of the gene. In addition, a site for the restriction enzyme *Msp*1, located approximately 650 nucleotides upstream, becomes demethylated

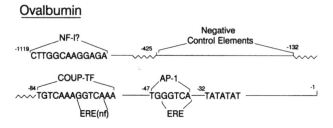

FIGURE 3. Comparison of the hormone response elements of the chicken vitellogenin II and ovalbumin genes. Approximately 1 kb of the 5' flanking regions of the chicken vitellogenin II and ovalbumin genes are illustrated. Potential binding sites for the estrogen receptor (ERE), glucocorticoid receptor or progesterone receptor (GR/PRE), nuclear factor 1 (NF-1?), chicken ovalbumin upstream promoter transcription factor (COUP-TF), and activator protein 1 (AP-1) have been indicated. Lowercase letters have been used to indicate differences between the putative response elements and their consensus sequence. In addition, nuclease hypersensitive sites (arrowheads) and the location of the *Msp*1 restriction enzyme site, which is demethylated in the 24-h period following estrogen stimulation, have been shown for the vitellogenin II gene. The putative TATA box, a region containing a negative control element and a nonfunctional half-ERE (EREnf), have been indicated for the ovalbumin gene.

during the 24 h following stimulation with estrogen (Wilks et al., 1982; Burch and Weintraub, 1983). The *Msp*1 site overlaps a sequence corresponding to a canonical estrogen response element (ERE) (Burch et al., 1988). Such elements typically have the sequence GGTCA spaced 3 nucleotides from the reverse complement TGACC. They serve as binding sites for homodimers of the estrogen receptor (Kumar and Chambon, 1988). A single, complete element such as this has been shown to be capable of conferring estrogen inducibility on genes that normally do not respond to the hormone (reviewed by O'Malley, 1990). The element located at −650 appears to be the major determinant of the estrogen inducibility of the vitellogenin II gene, despite its distance from the transcriptional initiation site. Although demethylation has been implicated as one of the mechanisms associated with conversion of inactive to active chromatin, it is not a prerequisite for the binding of estrogen receptor to this site, nor for the activation of the gene. Instead, it occurs during the 24 h following receptor binding.

Based on short-term transfection studies carried out by Burch and colleagues, the canonical ERE acts synergistically with a second imperfect ERE located approximately 300 nucleotides closer to the gene (Burch et al., 1988; Binder et al., 1990). A third potential ERE, even closer to the promoter and also near a nuclease hypersensitive site, is apparently nonfunctional (Seal et al., 1991). In addition, a perfect glucocorticoid, or progesterone, response element is located just a few nucleotides downstream from the major ERE. These two response elements have been shown to act synergistically when placed upstream from the thymidine kinase promoter (Cato et al., 1988). However, in the past we have not observed such synergism between glucocorticoids and estrogen *in vivo*. Our own studies using apoVLDLII constructs with extensive 5′ flanking regions indicate that glucocorticoids have a dominant negative, rather than positive, effect.

The efficiency and tissue specificity with which the vitellogenin gene is expressed clearly does not depend solely on interaction between the estrogen receptor and its response elements (Seal et al., 1991). Studies on the promoter regions of *Xenopus* vitellogenins, and more recently of chicken vitellogenin II, have identified binding sites for several liver-specific, or liver-enriched, trans-acting factors, that are important for determining both the efficiency and tissue specificity with which the genes are expressed (Kaling et al., 1991; Wahli et al., 1989; Corthésy et al., 1989). Most of the interactions characterized so far are located in the proximal promoter region extending approximately 100 nucleotides upstream from the transcriptional initiation site, although sites farther upstream may play as yet undefined roles. For example, one such interaction has been described for the avian gene by AB and colleagues that involves a site between 800 and 900 nucleotides upstream (Philipsen et al., 1988). This site appears to be occupied in an estrogen-dependent fashion by a member of the nuclear factor I (NF-1) family of trans-acting factors. Members of this family may also play a role in regulation of egg-white protein genes, such as lysozyme and ovalbumin, as well as a variety of mammalian genes (Borgmeyer et al., 1984; Bradshaw et al., 1988).

As described above, the hormonal dependency of the vitellogenin II gene appears to be dictated by at least two hormone response elements located several hundred nucleotides upstream from the transcriptional initiation site. This contrasts quite markedly with the organization of similar elements regulating the ovalbumin gene. The extreme hormonal dependence of the ovalbumin gene that is displayed *in vivo* has proven to be very difficult to reproduce *in vitro*. It has required a fine structure/function analysis of the promoter region simply to identify an estrogen response element.

Based on the sequence of the proximal promoter, i.e., the region extending approximately 80 nucleotides upstream from the major transcriptional initiation site, there are two elements, each of which corresponds to the conserved pentanucleotide that constitutes one half of a canonical ERE (Figure 3). From studies in other model systems, half-response elements like this appear to be inactive in isolation, but they can cooperate with other half-elements over limited distances to form a functional unit. In the case of the ovalbumin promoter, such cooperation does not take place and only one of the half-elements (the one located just upstream from the TATA sequence) appears to function as a response element, and this element confers a low degree of estrogen induction of approximately five- to sixfold (Tora et al., 1988). However, recent data by Sassone-Corsi and co-workers indicate that the half-ERE with the addition of two nucleotides forms part of a functional AP-1 site (Gaub et al., 1990). AP-1 sites are recognized by a heterodimer of the products of the proto-oncogenes *jun* and *fos*. These two proteins belong to the class of leucine zipper DNA-binding proteins and are typically involved in cellular responses to a number of mitogens (Curran and Franza, 1988). In this instance, *jun* and *fos* act synergistically with the estrogen receptor to increase markedly the inducibility of the promoter. One of the most intriguing aspects of these observations is the fact that the synergism does not require the DNA binding domain of the estrogen receptor and so is independent of direct interaction of the receptor with its response element. The second half-ERE forms part of a binding site for another factor named COUP-TF (chicken ovalbumin upstream promoter transcription factor), originally described by O'Malley and collegues (Sagami et al., 1986). This factor, like the estrogen receptor, is a member of the zinc finger family of trans-acting factors that includes steroid and thyroid receptors (Wang et al., 1989). COUP-TF is found in many tissues, and there are binding sites for the protein in both vitellogenin II and apoVLDLII promoters. It is not limited to birds, and it is currently unknown what, if any, ligand binds to the factor.

All of the regulatory features described so far have involved elements that have been identified as binding sites for positively acting factors. Studies on the ovalbumin promoter and, very recently, the vitellogenin II promoter, have also identified regions that suppress expression (Gaub et al., 1987; Sanders and McKnight, 1988). One such region is found in the vitellogenin II promoter, very close to the transcriptional initiation site (Seal et al., 1991). Its function appears to be to suppress expression of the gene in nonhepatic tissues. The negative control region upstream from the ovalbumin gene functions in a somewhat different fashion, by suppressing expression in the absence of steroid receptors occupied by the appropriate ligand. This suppression is relieved by the addition of hormone (Gaub et al., 1987; Sanders and Mc-Knight, 1988). However, the precise sequences involved have not been identified. The negative regulatory region appears to operate in conjunction with a positive control region between −730 to −900 (Schweers and Sanders, 1991). The notion that the extreme induction ratios displayed by some hormonally regulated genes is achieved by a combination of both positive and negative elements has been supported by studies on several model systems. This is also a feature of the regulation of the apoVLDLII gene, where we know more about the negative regulatory sequences involved.

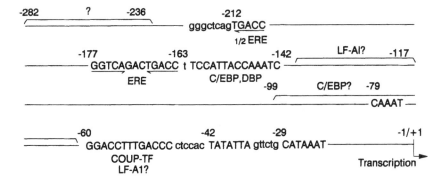

FIGURE 4. Elements in the apoVLDLII proximal promoter. The figure illustrates features of the apoVLDLII promoter from the major start of transcription (+1) to –282 bp. The sequence of the canonical estrogen response element (ERE), the major determinant of the estrogen responsiveness, and the imperfect estrogen response element (1/2-ERE) are shown. The elements indicated contain homologies to known recognition sequences for the transcription factors DBP, C/EBP, COUP-TF and LF-A1. *In vivo* footprinting experiments have shown that these sequences and those regions indicated by the brackets bind potential regulatory factors (Wijnholds et al., 1991). The locations of the putative TATA boxes (–49 and –29) and the CAAT box (–79) are also illustrated.

Organization of Regulatory Regions of the apoVLDLII Gene

We have examined the functional importance both of sites within the apoVLDLII promoter, such as those shown in Figure 4, as well as regions extending several kilobases upstream (Figure 5). The functional assay that we have used has involved short-term transfection of expression vectors using regulatory regions of the apoVLDLII gene to drive expression of the chloramphenicol acetyl transferase (CAT) gene in primary cultures. These cultures have been prepared from several different embryonic tissues, although the studies have been carried out predominantly in hepatocytes from day 20 embryos. A series of apoVLDLII expression vectors was constructed by introduction of various segments of the apoVLDLII promoter, first exon and upstream 5' flanking region into the plasmid pSV$_2$CAT following excision of the SV40 early promoter and enhancer region.

Figure 4 illustrates the major features of the promoter of the apoVLDLII gene (summarized from data in Wijnholds et al., 1988, 1991). One possible reason that the apoVLDLII gene responds as efficiently as it does is that it combines features of both the vitellogenin II and ovalbumin genes that are important determinants of the efficiency with which they respond to estrogen. We have used both deletion and point mutations to demonstrate that the major determinant of estrogen responsiveness of this gene, as in the case of vitellogenin II, is a canonical ERE. However, this element is in the proximal promoter itself, immediately adjacent to a binding site for what may be a

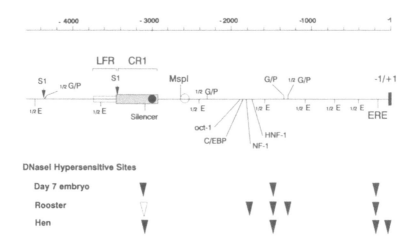

FIGURE 5. Selected features of the 5' flanking region of the apoVLDLII gene. The figure illustrates a number of potential regulatory features in the 5'-flanking region of the apoVLDLII gene (−4500 bp to −1/+1 bp). Locations of sequence matches to several steroid hormone responsive elements or half-elements are indicated (ERE, 1/2-ERE, G/P, 1/2G/P). The boxes represent the middle repetitive element which is a member of the chicken repeat 1 family, CR1 (stippled), and the adjacent lower frequency repeat, LFR (open). The closed circle within the CR1 element indicates the location of a protein-binding site identified by O'Malley and co-workers (Sanzo et al., 1984). A 9/9 nucleotide match to a known silencer element is indicated. The open circle shows the location of the protein-binding site associated with the *Msp*1 site that undergoes a developmentally regulated demethylation event. S1 hypersensitive sites (S1) are indicated by the arrowheads. The lower half of the figure shows the pattern of DNaseI hypersensitive sites in day 7 embryos, roosters, and laying hens. The closed triangles indicate constitutive hypersensitive sites, while the open triangle indicates a hypersensitive site that is induced in the rooster following estrogen treatment.

member of the family of trans-acting factors that include C/EBP, DBP and LAP (Johnson et al., 1987; Mueller et al., 1990; Descombes et al., 1990). These factors have been shown previously to bind to sites in the mammalian serum albumin promoter and to have a major effect on the efficiency of transcription.

In addition to the canonical ERE, an imperfect ERE is located approximately 35 nucleotides upstream. On the basis of its location, this element might be expected to be capable of cooperating with the canonical element. However, our short-term transfection studies indicate that this site contributes relatively little to the estrogen inducibility of the promoter. Furthermore, point mutations of each element indicate that the effects of the two sites are additive rather than synergistic. Additional binding sites induced by estrogen treatment have also been identified by *in vivo* footprinting by AB and colleagues (Wijnholds et al., 1988, 1991). Based on their sequence and other indirect data, these sites may be targets for COUP-TF and possibly for liver specific factors such as LF-AI (HNF-4) (Sladek et al., 1990; Ramji et al., 1991) as well as C/EBP and DBP (Johnson et al., 1987; Mueller et al., 1990).

Previous analyses of the 5' flanking region of the apoVLDLII gene have revealed alterations in nuclease hypersensitivity (Kok et al., 1985; Haché and Deeley, 1988) and methylation (Colgan et al., 1982) that occur during development or upon activation of the gene. These occur at locations extending more than 3 kb upstream from the sites of transcriptional initiation (Figure 5). Several specific protein interactions have been identified at upstream locations within this region (Hoodless et al., 1990). In addition, based solely on DNA sequence, the region contains potential binding sites for factors that have been implicated in regulation of several genes expressed in mammalian liver, such as HNF-1, NF-1, LF-A1, C/EBP, and DBP (reviewed by Johnson, 1990). There are also several potential glucocorticoid response elements located between −1.2 kb and −4.5 kb.

In order to examine the effects of upstream sequences, a series of exonuclease III deletion mutants was produced in which sites of potential regulatory importance (spanning the entire region from −4495 to −16) were sequentially removed. Constructs containing the entire flanking region are extremely estrogen dependent. Induction ratios as high as 200-fold have been obtained in hepatocytes following co-transfection with an estrogen receptor expression vector. The results of an extensive series of analyses with these constructs are summarized below, in relation to the schematic presented in Figure 5.

Constructs containing all mapped protein-binding sites of the proximal promoter region, shown in Figures 4 and 5, display extremely low levels of estrogen-independent expression and in the presence of hormone and receptor are almost as active as pSV_2CAT. Elimination of any of these binding sites markedly reduces expression. Inclusion of the entire region extending to approximately 4 kb upstream increases both the activity of the constructs five- to sixfold) and the tissue specificity of their expression. Such constructs typically are expressed 50- to 60-fold more in hepatocytes than in fibroblasts. This increase in activity and tissue specificity cannot be attributed to a single region, but is the result of a summation of both positively and negatively acting regions.

One region of particular interest is located approximately 3 kb upstream. It contains a middle repetitive element that is a member of the chicken repeat 1 (CR1) family (Stumph et al., 1981). This element acts as a negative control region (Baniahmad et al., 1987; T. J. Schrader, A. K. Ryan, R. Burtch Wright, and R. G. Deeley, unpublished results). However, in hepatocytes, inclusion of the adjacent upstream region overrides the negative effect of the CR1 element. This second region is also repeated in the chicken genome, albeit at a 10- to 20-fold lower frequency than the CR1 element (Haché and Deeley, 1988). Interestingly, the junction between the low-frequency repeat and the CR1 sequence displays hypersensitivity to S1 nuclease. This sensitivity is dependent on the degree of supercoiling of the DNA. Based on this behavior and the primary sequence of the region, we have suggested that it may be a recombinational hot spot (Haché and Deeley, 1988). We have identified two regions in the CR1 element that are responsible for the silencing activity by functional assays, protein footprinting and gel retardation studies T. J. Schrader, A. K. Ryan, R. Burtch Wright, and R. G. Deeley, unpublished results). One of these regions coincides with the binding site of a protein initially identified by O'Malley and co-workers (Sanzo et al., 1984). The other is located approximately 150 nucleotides upstream and, based on footprinting data, matches an element that has also been shown to function as a silencer in mammals (Salier et al., 1990; Savagner et al., 1990; Wang and Brand, 1990).

CR1 sequences have been shown very recently to be highly conserved in many avian species (Chen et al., 1991). They have also been found in a predominantly conserved orientation, flanking several avian genes in regions of transitional nuclease sensitivity (Stumph et al., 1983, 1984). This suggests that they are preferentially located close to the termini of transcriptional domains, and a number of functional roles have been postulated for them. Perhaps one of the most intriguing is that of nuclear matrix attachment sites. Although we have expended considerable effort investigating this possibility using the CR1 element flanking the apoVLDLII gene, we have not obtained convincing evidence in favor of the suggestion. However, we are attempting to exploit the CR1 element and its adjacent repeat as a means of increasing both the integration frequency of various constructs into the chicken genome and their levels of expression.

A second region that has been of interest to us for some time involves a site in the 5' flanking region that displays developmentally programmed demethylation. This site is located 2.6 kb upstream from the start of the gene and includes a sequence recognized by the restriction enzyme, *Msp*l. Demethylation of this site coincides with the time at which the liver becomes competent to express the apoVLDLII gene in response to estrogen. We have characterized and, more recently, purified and cloned proteins that bind to this site. The sequence of the binding site is very similar to sites recognized by C/EBP. However, the proteins that we have characterized are clearly distinct from C/EBP, DBP, and LAP and show a much higher degree of sequence specificity. Transfection experiments indicate that the site may increase the efficiency of expression approximately twofold. However, these results have to be viewed with some caution, since gel retardation assays indicate that the amount of proteins that recognize the site decreases very rapidly when hepatocytes are placed in culture (P. A. Hoodless, A. K. Ryan, and R. G. Deeley, unpublished results). This raises the possibility that we are underestimating the regulatory importance of this region because of the limitations of the assay system, or that its role may be limited primarily to earlier stages of development. The latter possibility is consistent with the fact that the level of proteins binding to this site is highest in the liver at early stages of development and decreases sharply between days 7 and 9 of embryogenesis (Hoodless et al., 1990). This decrease in binding activity is followed by a discrete shift in the mobility of protein/DNA complexes, detectable in gel retardation assays, that occurs between days 11 and 13 of embryogenesis (Hoodless et al., 1990). The timing of these alterations suggests a role for the proteins in acquisition of the ability to express the apoVLDLII, and possibly other yolk protein genes, in the liver.

The proteins involved are present at extremely low levels, which has precluded extensive biochemical characterization of them. However, we have now isolated two cDNA clones from an expression library constructed from embryonic liver mRNA that encode proteins binding to the site. One of them is a previously unidentified member of the C_2-H_2 family of zinc-finger proteins (Mitchell and Tjian, 1989). The other is also novel. The only structural similarity detected with other DNA-binding proteins is to the DNA-binding domain of the family of interferon regulatory factors (Driggers et al., 1990). We are currently determining whether the cloned proteins are identical to those that we have characterized during previous biochemical studies.

Hormone Specificity of the apoVLDLII Promoter

As described above, once activated by estrogen, the egg-white protein genes respond to several steroid hormones. Data have also been presented that estrogen and glucocorticoids may synergize in the regulation of expression of the vitellogenin II gene. In view of these observations, we have investigated both the hormone specificity of the native apoVLDLII gene and the possibility of increasing its expression by mutation of its regulatory elements.

Several glucocorticoid response elements present upstream from the gene are identical with functional elements in the mouse mammary tumor virus long terminal repeat, suggesting that glucocorticoids may have a positive effect on expression. In order to ensure that adequate receptors were present, the hepatocytes were co-transfected with a glucocorticoid receptor (GR) expression vector. No positive response was detected. However, we were able to show, using the native apoVLDLII promoter and flanking regions, that very low levels of the GR were sufficient to antagonize the ability of estrogen to activate the constructs, suggesting that glucocorticoids may be potent, dominant negative regulators of apoVLDLII expression. This contrasts with the data referred to above, which were obtained using a chimeric construct in which an upstream portion of the vitellogenin II gene was fused to the promoter of the thymidine kinase gene and transfected into human cells (Cato et al., 1988). In view of the coordinate regulation of the vitellogenin II and apoVLDLII genes, it will be of interest to determine whether glucocorticoids have a similar effect on the native vitelloginin II promoter in avian hepatocytes.

In an attempt to modify the responsiveness of the gene, we have also mutated each ERE in the proximal promoter region to different steroid-responsive elements. These experiments strongly suggest that other factors involved in the activation of the apoVLDLII promoter display a marked preference for the transactivation domain of the estrogen receptor. We are continuing to manipulate the structure of the apoVLDLII promoter region in order to obtain maximal levels of expression while maintaining its high degree of selectivity for estrogen as the activating hormone.

Conclusion

The results of the studies described above indicate that the regulatory regions of the apoVLDLII gene provide an excellent starting point for the construction of highly efficient expression vectors designed to produce recombinant proteins in a tissue-specific fashion. With recent advances in the production of transgenic birds, it should be possible in the near future to use these vectors to produce fusion proteins designed for export from the liver and for selective delivery to the oocyte.

References

Baniahmad, A., Muller, M., Steiner, Ch., and Renkawitz, R. (1987). Activity of two different silencer elements of the chicken lysozyme gene can be compensated by enhancer elements. *EMBO J.*, 6, 2297–2303.

Bergink, E. W., and Wallace, R. A. (1974). Precursor-product relationship between amphibian vitellogenin and the yolk proteins lipovitellin and phosvitin. *J. Biol. Chem.*, 249, 2897–2903.

Bergink, E. W., Wallace, R. A., Van de Berg, J. A., Bos, E. S., Gruber, M., and AB, G. (1974). Estrogen-induced synthesis of yolk proteins in roosters. *Am. Zool.*, 14, 1177–1193.

Binder, R., MacDonald, C. C., Burch, J. B. E., Lazier, C. B., and Williams, D. L. (1990). Expression of endogenous and transfected apolipoprotein II and vitellogenin II genes in an estrogen responsive chicken liver cell line. *Mol. Endocrinol.*, 4, 201–208.

Borgmeyer, U., Nowock, J., and Sippel, A. E. (1984). The TGGCA-binding protein: A eukaryotic nuclear protein recognizing a symmetrical sequence on double-stranded linear DNA. *Nucleic Acids Res.*, 12, 4295–4311.

Bradshaw, M. S., Tsai, M.-J., and O'Malley, B. W. (1988). A far upstream ovalbumin enhancer binds nuclear factor-1-like factor. *J. Biol. Chem.*, 263, 8485–8490.

Burch, J. B. E., and Weintraub, H. (1983). Temporal order of chromatin structural changes associated with activation of the major chicken vitellogenin gene. *Cell*, 33, 65–76.

Burch, J. B. E., Evans, M. I., Friedman, T. M., and O'Malley, P. J. (1988). Two functional estrogen response elements are located upstream of the major chicken vitellogenin gene. *Mol. Cell. Biol.*, 8, 1123–1131.

Byrne, B. M., Van het Schip, A. D., Van de klundert, J. A. M., Arnberg, A. C., Gruber, M., and AB, G. (1984). Amino acid sequence of phosvitin derived from the nucleotide sequence of part of the chicken vitellogenin gene. *Biochemistry*, 23, 4275–4279.

Cato, A. C. B., Heitlinger, E., Ponta, H., Klein-Hitpass, L., Ryffel, G. U., Bailly, A., Rauch, C., and Milgrom, E. (1988). Estrogen and progesterone receptor-binding sites on the chicken vitellogenin II gene: Synergism of steroid hormone action. *Mol. Cell. Biol.,* 8, 5323–5330.

Chambon, P., Dierich, A., Gaub, M.-P., Jakowlev, S., Jongstra, J., Krust, A., LePennec, J.-P., Oudet, P., and Reudelhuber, T. (1984). Promoter elements of genes coding for proteins and modulation of transcription by estrogens and progesterone. *Recent Prog. Hormone Res.,* 40, 1–42.

Chan, L., Jackson, R. L., O'Malley, B. W., and Means, A. R. (1976). Synthesis of very low density lipoproteins in the cockerel. *J. Clin. Invest.,* 58, 368–379.

Chen, Z., Ritzel, R. G., Lin, C. C., and Hodgetts, R. B. (1991). Sequence conservation in avian CR1, an interspersed repetitive DNA family evolving under functional constraints. *Proc. Natl. Acad. Sci. USA,* 88, 5814–5818.

Clemens, M. J. (1974). The regulation of egg yolk protein synthesis by steroid hormones. *Prog. Biophys. Mol. Biol.,* 28, 71–107.

Colgan, V., Elbrecht, A., Goldman, P., Lazier, C. B., and Deeley, R. G. (1982). The avian apoprotein II very low density lipoprotein gene. *J. Biol. Chem.,* 257, 14453–14460.

Corthésy, B., Cardinaux, J.-R., Claret, F.-X., and Wahli, W. (1989). A nuclear factor I-like activity and a liver-specific repressor govern estrogen-regulated in vitro transcription from the *Xenopus laevis* vitellogenin B1 promoter. *Mol. Cell. Biol.,* 9, 5548–5562.

Curran, T., and Franza, B. R., Jr. (1988). *Fos* and *Jun:* The AP-1 connection. *Cell,* 55, 395–397.

Deeley, R. G., Mullinix, K. P., Wetekam, W., Kronenberg, H. M., Meyers, M., Eldridge, J. D., and Goldberger, R. F. (1975). Vitellogenin synthesis in the avian liver: Vitellogenin is the precursor of the egg yolk phosphoproteins. *J. Biol. Chem.,* 250, 9060–9066.

Deeley, R. G., Udell, D. S., Burns, A. T. H., Gordon, J. I., and Goldberger, R. F. (1977). Kinetics of avian vitellogenin messenger RNA induction. *J. Biol. Chem.,* 252, 7913–7915.

Descombes, P., Chojkier, M., Lichtsteiner, S., Falvey, E., and Schibler, U. (1990). LAP, a novel member of the C/EBP gene family, encodes a liver-enriched transcriptional activator protein. *Genes Devel.,* 4, 1541–1551.

Driggers, P. H., Ennist, D. L., Gleason, S. L., Mak, W.-H., Marks, M. S., Levi, B.-Z., Flanagan, J. R., Appella, E., and Ozato, K. (1990). An interferon γ-regulated protein that binds the interferon-inducible enhancer element of major histocompatibility complex class I genes. *Proc. Natl. Acad. Sci. USA,* 87, 3743–3747.

DurgaKumari, B., and Adiga, P. R. (1986). Estrogen modulation of retinol-binding protein in immature chicks: Comparison with riboflavin carrier protein. *Mol. Cell. Endocrinol.,* 46, 121–130.

Elbrecht, A., Lazier, C. B., Protter, A. A., and Williams, D. L., (1984). Independent developmental programs for two estrogen-regulated genes. *Science,* 225, 639–641.

Elbrecht, A., Williams, D. L., Blue, M.-L. and Lazier, C. B. (1981). Differential ontogeny of estrogen responsiveness in the chick embryo liver. *Can. J. Biochem.,* 59, 606–613.

Evans, M. I., Silva, R., and Burch, J. B. E. (1988). Isolation of chicken vitellogenin I and III cDNAs and the developmental regulation of five estrogen-responsive genes in the embryonic liver. *Genes Devel.,* 2, 116–124.

Gaub, M.-P., Bellard, M., Scheuer, I., Chambon, P., and Sassone-Corsi, P. (1990). Activation of the ovalbumin gene by the estrogen receptor involves the fos-jun complex. *Cell,* 63, 1267–1276.

Gaub, M.-P., Dierich, A., Astinotti, D., Touitou, I., and Chambon, P. (1987). The chicken ovalbumin promoter is under negative control which is relieved by steroid hormones. *EMBO J.,* 6, 2313–2320.

Gilbert, A. B. (1981). The ovary. In: *Physiology and Biochemistry of the Domestic Fowl,* Vol. 3, edited by D. J. Bell and G. M. Freeman. Academic Press, London and New York, pp. 1163–1208.

Haché, R. J. G., and Deeley, R. G. (1988). Organization, sequence and nuclease hypersensitivity of repetitive elements flanking the chicken apoVLDLII gene: Extended sequence similarity to elements flanking the chicken vitellogenin gene. *Nucleic Acids Res.*, 16, 97–113.

Haché, R. J. G., Tam, S.-P., Cochrane, A., Nesheim, M., and Deeley, R. G. (1987). Long-term effects of estrogen on avian liver: Estrogen-inducible switch in expression of nuclear, hormone-binding proteins. *Mol. Cell. Biol.*, 7, 3538–3547.

Hauser, H., Graf, T., Beug, H., Geiser, W. I., Lindenmaier, W., Grez, M., Land, H., Giesecke, K., and Schutz, G. (1981). In: *Haematology and Blood Transfusion*, Vol. 26, edited by R. Neth, R. C. Gallo, T. Graf, K. Mannweiter, and K. Winkler. Springer Verlag, Berlin, pp. 175–178.

Hayashi, K., Nimpf, J., and Schneider, W. J. (1989). Chicken oocytes and fibroblasts express different apolipoprotein-B-specific receptors. *J. Biol. Chem.*, 264, 3131–3139.

Hoodless, P. A., Roy, R. N., Ryan, A. K., Haché, R. J. G., Vasa, M. Z., and Deeley, R. G. (1990). Developmental regulation of specific protein interactions with an enhancerlike binding site far upstream from the avian very-low-density apolipoprotein II gene. *Mol. Cell. Biol.*, 10, 154–164.

Johnson, A. L. (1986). Reproduction in the female. In: *Avian Physiology*, 4th ed., edited by P. D. Sturkie. Springer-Verlag, New York, Berlin, Heidelberg, Tokyo, pp. 403–431.

Johnson, P. F. (1990). Transcriptional activators in hepatocytes. *Cell Growth & Differentiation*, 1, 47–52.

Johnson, P. F., Landschulz, W. H., Graves, B. J., and McKnight, S. L. (1987). Identification of a rat liver nuclear protein that binds to the enhancer core element of three animal viruses. *Genes Devel.*, 1, 133–146.

Jones, D. G., Briles, W. E., and Schjeide, D. A. (1975). A mutation restricting ovulation in chickens. *Poultry Sci.*, 54, 1780–1783.

Kaling, M., Kugler, W., Ross, K., Zoidl, C., and Ryffel, G. U. (1991). Liver-specific gene expression: A-activator-binding site, a promoter module present in vitellogenin and acute-phase genes. *Mol. Cell. Biol.*, 11, 93–101.

Kirchgessner, T. G., Heinzmann, C., Svenson, K. L., Gordon, D. A., Nicosia, M., Lebherz, H. G., Lusis, A. J., and Williams, D. L. (1987). Regulation of chicken apolipoprotein B: Cloning, tissue distribution and estrogen induction of mRNA. *Gene*, 59, 241–251.

Kok, K., Snippe, L., AB, G., and Gruber, M. (1985). Nuclease-hypersensitive sites in chromatin of the estrogen-inducible apoVLDLII gene of chicken. *Nucleic Acids Res.*, 13, 5189–5202.

Kumar, V., and Chambon, P. (1988). The estrogen receptor binds tightly to its responsive element as a ligand-induced homodimer. *Cell*, 55, 145–156.

Lee, D. C., McKnight, G. S., and Palmiter, R. D. (1978). The action of estrogen and progesterone on the expression of the transferrin gene. *J. Biol. Chem.*, 253, 3494–3503.

LeMeur, M., Glanville, N., Mandel, J. L., Gerlinger, P., Palmiter, R., and Chambon, P. (1981). The ovalbumin gene family: Hormonal control of X and Y gene transcription and mRNA accumulation. *Cell*, 23, 561–571.

McGibbon, W. H. (1977). Evidence that the restricted ovulator gene (ro) in the chicken is sex-linked. *Genetics*, 86, S43 (Abstr.)

Mitchell, P. J., and Tjian, R. (1989). Transcriptional regulation in mammalian cells by sequence-specific DNA binding proteins. *Science*, 245, 371–378.

Mueller, C. R., Maire, P., and Schibler, U. (1990). DBP, a liver-enriched transcriptional activator, is expressed late in ontogeny and its tissue specificity is determined postranscriptionally. *Cell*, 61, 279–291.

Nimpf, J., Radosavljevic, M. J., and Schneider, W. J. (1989). Oocytes from the mutant restricted ovulator hen lack receptors for very low density lipoprotein. *J. Biol. Chem.*, 264, 1393–1398.

Oka, T., and Schimke, R. T. (1969). Interaction of estrogen and progesterone in chick oviduct development. *J. Cell. Biol.*, 43, 123–137.

O'Malley, B. W. (1990). The steroid receptor superfamily: More excitement predicted for the future. *Mol. Endocrinol.*, 4, 363–369.

O'Malley, B. W., McGuire, W. L., Kohler, P. O., and Korenman, S. G. (1969). *Recent Prog. Hormone Res.*, 25, 105–160.

Palmiter, R. D. (1972). Regulation of protein synthesis in chick oviduct. *J. Biol. Chem.*, 247, 6450–6461.

Palmiter, R. D., Mulvihill, E. R., McKnight, G. S., and Senear, A. W. (1978). Regulation of gene expression in the chick oviduct by steroid hormones. *Cold Spring Harbor Symp. Quant. Biol.*, 42, 639–647.

Philipsen, J. N. J., Hennis, B. C., and AB, G. (1988). *In vivo* footprinting of the estrogen-inducible vitellogenin II gene from chicken. *Nucleic Acids Res.*, 16, 9663–9676.

Ramji, D. P., Tadros, M. H., Hardon, E. M., and Cortese, R. (1991). The transcription factor LF-A1 interacts with a bipartite recognition sequence in the promoter regions of several liver-specific genes. *Nucleic Acids Res.*, 19, 1139–1146.

Sagami, I., Tsai, S. Y., Wang, H., Tsai, M.-J., and O'Malley, B. W. (1986). Identification of two factors required for transcription of the ovalbumin gene. *Mol. Cell. Biol.*, 6, 4259–4267.

Salier, J.-P., Hirosawa, S., and Kurachi, K. (1990). Functional characterization of the 5'-regulatory region of human factor IX gene. *J. Biol. Chem.*, 265, 7062–7068.

Saluz, H. P., Feavers, I. M., Jiricny, J., and Jost, J. P. (1988). Genomic sequencing and *in vivo* footprinting of an expression-specific DNase I-hypersensitive site of avian vitellogenin II promoter reveal a demethylation of a mCpG and a change in specific interactions of proteins with DNA. *Proc. Natl. Acad. Sci. USA*, 85, 6697–6700.

Sanders, M. M., and McKnight, G. S. (1988). Positive and negative regulatory elements control the steroid-responsive ovalbumin promoter. *Biochemistry*, 27, 6550–6557.

Sanzo, M., Stevens, B., Tsai, M.-J., and O'Malley, B. W. (1984). Isolation of a protein fraction that binds preferentially to chicken middle repetitive DNA. *Biochemistry*, 23, 6491–6498.

Savagner, P., Miyashita, T., and Yamada, Y. (1990). Two silencers regulate the tissue-specific expression of the collagen II gene. *J. Biol. Chem.*, 265, 6669–6674.

Schweers, L. A., and Sanders, M. M. (1991). A protein with a binding specificity similar to NF-κB binds to a steroid-dependent regulatory element in the ovalbumin gene. *J. Biol. Chem.*, 266, 10490–10497.

Seal, S. N., Davis, D. L., and Burch, J. B. E. (1991). Mutational studies reveal a complex set of positive and negative control elements within the chicken vitellogenin II promoter. *Mol. Cell. Biol.*, 11, 2704–2717.

Sladek, F. M., Zhong, W., Lai, E., and Darnell, J. E. (1990). Liver-enriched transcription factor HNF-4 is a novel member of the steroid hormone receptor superfamily. *Genes Devel.*, 4, 2353–2365.

Steyrer, E., Barber, D. L., and Schneider, W. J. (1990). Evolution of lipoprotein receptors. The chicken oocyte receptor for very low density lipoprotein and vitellogenin binds the mammalian ligand apolipoprotein E. *J. Biol. Chem.*, 265, 19575–19581.

Stifani, S., Barber, D. L., Nimpf, J., and Schneider, W. J. (1990a). A single chicken oocyte plasma membrane protein mediates uptake of very low density lipoprotein and vitellogenin. *Proc. Natl. Acad. Sci. USA*, 87, 1955–1959.

Stifani, S., Nimpf, J., and Schneider, W. J. (1990b). Vitellogenesis in *Xenopus laevis* and chicken: Cognate ligands and oocyte receptors. *J. Biol. Chem.*, 265, 882–888.

Stumph, W. E., Baez, M., Beattie, W. G., Tsai, M.-J., and O'Malley, B. W. (1983). Characterization of deoxyribonucleic acid sequences at the 5' and 3' borders of the 100 kilobase pair ovalbumin gene domain. *Biochemistry*, 22, 306–315.

Stumph, W. E., Hodgson, C. P., Tsai, M.-J., and O'Malley, B. W. (1984). Genomic structure and possible retroviral origin of the chicken CR1 repetitive DNA sequence family. *Proc. Natl. Acad. Sci. USA*, 81, 6667–6671.

Stumph, W. E., Kristo, P., Tsai, M.-J., and O'Malley, B. W. (1981). A chicken middle-repetitive DNA sequence which shares homology with mammalian ubiquitous repeats. *Nucleic Acids Res.*, 9, 5383–5397.

Tora, L., Gaub, M.-P., Mader, S., Dierich, A., Bellard, M., and Chambon, P. (1988). Cell-specific activity of a GGTCA half-palindromic oestrogen-responsive element in the chicken ovalbumin gene promoter. *EMBO J.*, 7, 3771–3778.

Wahli, W., Martinez, E., Corthesy, B., and Cardinaux, J.-R. (1989). *Cis*- and *trans*-acting elements of the estrogen-regulated vitellogenin gene B1 of *Xenopus laevis*. *J. Steroid Biochem.*, 34, 17–32.

Wang, L. H., Tsai, S. Y., Cook, R. G., Beattie, W. G., Tsai, M.-J., and O'Malley, B. W. (1989). COUP transcription factor is a member of the steroid receptor superfamily. *Nature*, 340, 163–166.

Wang, T. C. and Brand, S. J. (1990). Islet cell-specific regulatory domain in the gastrin promoter contains adjacent positive and negative DNA elements. *J. Biol. Chem.*, 265, 8908–8914.

White, H. B., and Merrill, A. H. (1988). Riboflavin-binding proteins, *Ann. Rev. Nutr.*, 8, 279–299.

Wijnholds, J., Muller, E., and AB, G. (1991). Oestrogen facilitates the binding of ubiquitous and liver-enriched nuclear proteins to the apoVLDL II promoter *in vivo*. *Nucleic Acids Res.*, 19, 33–41.

Wijnholds, J., Philipsen, J. N. J., and AB, G. (1988). Tissue-specific and steroid-dependent interaction of transcription factors with the oestrogen-inducible apoVLDLII promoter *in vivo*. *EMBO J.*, 7, 2757–2763.

Wilks, A. F., Cozens, P. J., Mattaj, I. W., and Jost, J.-P. (1982). Estrogen induces a demethylation at the 5' end region of the chicken vitellogenin gene. *Proc. Natl. Acad. Sci. USA*, 79, 4252–4255.

Williams, D. L. (1979). Apoproteins of avian very low density lipoprotein: Demonstration of a single high molecular weight apoprotein. *Biochemistry*, 18, 1056–1063.

Wiskocil, R., Bensky, P., Dower, W., Goldberger, R. F., Gordon, J. I., and Deeley, R. G. (1980). Coordinate regulation of two estrogen-dependent genes in avian liver. *Proc. Natl. Acad. Sci USA*, 77, 4474–4478.

Chapter 14

Poultry Breeding Technologies in the Twentieth Century: Where Have We Come from and Where Are We Going?

Robert N. Shoffner

Summary. R. N. (Bob) Shoffner was the keynote speaker at the banquet during the Colloquium on the Manipulation of the Avian Genome. He brought to the audience a living history of the changes and evolution of the poultry breeding industry that spanned 50 years of participation, observation, contribution, and leadership. He has served as a Research Assistant, Instructor, Assistant Professor, Associate Professor, Professor, Acting Head, and Professor Emeritus at the University of Minnesota, has been a Fulbright Scholar in Australia, and was a Visiting Professor at the University of Texas. He has served as a consultant to many of the public and private poultry breeding organizations in the United States and to academic and research institutions in Canada, France, India, and Brazil. He is the recipient of the Ranelious Award, the National Turkey Federation Award, and the Merck Research Award, has served as President of the Poultry Science Association, and is a Fellow of the Poultry Science Association and the American Association for the Advancement of Science. This chapter is a transcript of the banquet address.

Introduction and Review

I suppose that a keynote speaker may take several options for an address. I choose to be extremely informal, foregoing formal citation, with no bibliography. I will mention some names that come easily to mind. No discrimination is intended for others — it is just my poor memory. Much of what I review tonight can be found somewhere in the newly published *Poultry Breeding and Genetics*, Roy Crawford, editor, published by Elsevier Press. This is a superb compilation with 41 chapters and 37 authors covering the history and state of the art in all aspects of poultry genetics. My qualifications, such as they are, come from having lived through and practiced classical genetics, quantitative genetics, cytogenetics, and molecular genetics. I was fortunate to be at the right place at the right time during the changes in research emphasis of the last five decades.

There have been distinct surges and minor pulses of enthusiastic activity in poultry genetics and breeding. The first of these was a long period when the descendants of the Southeast Asian Red Jungle Fowl *Gallus bankiva* spread, over centuries, to the rest of the world. Chickens became differentiated into characteristic phenotypes through migration, isolation, mutation, and selection. There was a long period of domestication when chickens were valued for food, fighting, religion, and art. Husbandry practices eventually resulted in pure breed development with emphasis on plumage color and morphological traits. Poultry shows were popular, and breed characteristics were important selection criteria. There was a misconception that breed standard superiority was also related to performance in economic traits. However, in the development of breeds, some such as the Leghorns were generally superior in egg production, while the large-bodied breeds produced more meat. It should be recognized that the first evidence that Mendel's laws applied to animals was demonstrated by Bateson and Punnett using chickens as early as 1902, followed by other studies in subsequent years, in particular demonstrating sex-linked inheritance. In the 1920s, active research in poultry genetics expanded in universities and government agencies. The trapnest was invented, allowing accurate pedigreeing, and selection began to be effective for quantitatively inherited economic traits. By 1940 there was a great deal of documentation about the inheritance of phenotypic traits and determination of linkage groups. The 1932, 1940, and 1950 M. A. Jull's editions of *Poultry Breeding* (John Wiley & Sons) and F. B. Hutt's *Genetics of the Fowl*, 1949 (McGraw-Hill), effectively summarized poultry genetics and breeding up to that period. In addition to what was known about inheritance, breeding plans and selection schemes were proposed. This is about where I come in, as these were the poultry genetics "bibles" of the time.

The U.S. National Poultry Improvement Plan (NPIP) was introduced as a national systematized breeding plan. Performance data were available to prospective buyers. Similar plans were developed in Canada and elsewhere. Several hundred breeding establishments, both large and small, participated in the plan, developing a real enthusiasm for improvement in economic traits. The NPIP was handicapped by retention of "niggling" selection criteria, such as absence of side sprigs or stubs on shanks. Because the NPIP did not provide for performance recognition of crossbred or inbred line crosses, the Random Sample Test was instituted. Hatching eggs were collected in a random manner from a breeder's supply flock of a commercial stock by a disinterested agent and delivered to a central testing facility. Performance data were collected from this sample and results published. These tests were reasonably discriminating, and they brought about rapid and drastic culling within the industry, as the not-quite-so-good breeders lost customers. The broiler enterprise got off to a fast start when the "Chicken of Tomorrow" contest began in the late 1940s. Dr. Benjamin Pierce was instrumental in developing this program for improvement of the meat qualities of chicken fryers or broilers. A model chicken was made, and he persuaded a prominent grocery chain to sponsor "The Chicken of Tomorrow" contest for the broiler that most nearly matched the model. These tests stimulated much enthusiasm, resulting in the forerunner stocks of the modern broiler.

The technique of artificial insemination (AI) developed by Burrows and Quinn has proven to be an exceedingly useful tool in poultry breeding. It is used extensively in producing basic breeding stock of layers and some broiler stocks. AI is a way of life to the turkey producer, as almost all commercial turkeys result from this technique. Neither chicken nor turkey semen store well in the frozen state. However, while frozen semen is not commercially practical, even low levels of recovery are useful for germ plasm preservation.

The application of quantitative genetic theory to poultry breeding research began about 1940 at research institutions. Gradual adoption of quantitative genetic principles by poultry breeders began at about the same time. The theories of R. A. Fisher and Sewall Wright were popularized by J. Lush and his colleagues. Other names, such as Dickerson, Falconer, Robertson, Gowe, Bohren, Hazel, Nordskog, Warren, Abplanalp, Crow, Comstock, and Lerner, come to mind as some of the contributors to the great fund of knowledge about the genetics of poultry generated during the 20-year period from 1940 to 1960. There was debate about Lerner's genetic homeostasis theory, wherein intense selection for a single trait would drastically change the adaptive genotype, resulting in adverse effects on other traits. Furthermore, with relaxed selection, the genotype would revert to its original state. The reality and cause of genetic plateaus, and the prevalence of overdominance, was also of much interest. As a neophyte and even later, it was always stimulating and informative to attend the various meetings just to hear the big "bulls" rattle their horns.

A rapid accumulation of estimates of genetic parameters followed. Estimates of heritability, genetic correlation of traits, correlation of part – whole records, and heterosis became well known. Genetic and economic values were combined to optimize selection index accuracy. Effects of relaxed selection were studied. Breeding systems were developed, with various crossing and selection methods aided by development of computer analysis programs. Most of the breeding theory had been put into practice by 1960, and there was little research opportunity without extensive facilities. The breeding industry settled on *complementation* crossing schemes utilizing lines that differed in trait expression, for example, larger egg size in one parent line and greater egg numbers in another, coupled with heterosis resulting in a product with much of the attributes of both parental lines. Highly productive egg layer, broiler, and turkey stocks have been developed, but even so, breeders are constantly "dressing up" desired traits and/or minimizing undesirable ones. In this interval, meetings and symposiums of Poultry Science, World Poultry Science, Poultry Breeders Roundtable, Genetics Society, World Congress of Genetics Applied to Livestock Production, Symposia on Cytogenetics, along with local and regional meetings, were effective in spreading the word about poultry genetics and breeding.

The underlying operational dogma for breeding companies can be expressed as $P\acute{o}^2 = G\acute{o}^2 + E\acute{o}^2 + GE\acute{o}^2$, where $P\acute{o}^2$ is the phenotypic variance, $G\acute{o}^2$ is the genetic variance, $E\acute{o}^2$ is the environmental variance, and $GE\acute{o}^2$ is the covariance of the genetic and environmental components. Breeders have devoted their efforts to maximizing accuracy in estimates of $G\hat{\sigma}^2$ with reduction in confounding effects of $E\hat{\sigma}^2$. Selection methods and breeding schemes have very effectively caused continued improvement. Improved nutrition, management of the photoperiod, vaccination, and elimination of infectious organisms have greatly reduced the confounding effects of environmental variation. Consequently, selection efficiency has at least been maintained if not improved. If mutation occurs at the rate I think it does, genetic variance will not diminish much in the foreseeable future.

There were smaller blips of research activity and curiosity directed to specific topics. The development and use of inbred lines of chicken was investigated to a considerable extent, due in part to the success of inbred line crosses in corn. The emphasis at the outset of the North Central Regional Poultry Breeding Project was directed to investigation of inbreeding. A number of inbred lines were developed at the Regional Poultry Disease Laboratory (now known as the Avian Disease and Oncology Laboratory) for resistance or susceptibility to avian leukosis, resulting in several very specific lines. On the whole, inbred line development directed toward economic improvement was not thoroughly researched. Most lines were more or less just inbred without selective direction, hoping that diversity would be generated. On the other hand, several breeding companies, with selective direction, produced inbred line crosses commercially for some years. Other selection schemes have supplanted the purely inbred line crosses.

The contribution of cytogenetics has been largely descriptive and informative and to date has not contributed much to breeding practice. The karyotype of the chicken was determined with a chromosome number of 78, with ZZ sex chromosomes for the male and ZW for the female. The microchromosomes are real chromosomes and not aberrant elements. The chicken is tolerant to reciprocal translocations, inversions, trisomy, and triploidy. However, chromosomal aberrants contribute to early abortion, as shown by Bloom at Cornell and Fechheimer and colleagues at Ohio State. Dosage compensation for the Z sex chromosome appears to be absent, and seems not to be a major factor for autosomal expression in either trisomy or triploidy. Species karyotypes have been a useful differentiation criterion in avian systematics, and more than 300 species have been karyotyped.

An interesting concept called ''canalization'' was first developed by Waddington at Edinburgh when he discovered that crossveinless could be revealed in *Drosophila melanogaster* by lowering temperature during development. Upon subsequent selection after a few generations, a line of crossveinless flies was developed. This should not be confused with inheritance of acquired characters, but as uncovering hidden genetic variation. A practical application of the concept to use environmental variation to uncover heretofore concealed genetic variation has been the use of photoperiod control to change the oviposition interval of the chicken. Reproduction is controlled by both long-running and circadian light rhythms. The chicken normally takes 24 h or more to manufacture an egg. When the interval is longer than 24 h, one or more days are skipped in the cycle, with a consequent reduction in egg production. Several laboratories attempted to identify hens with a shorter oviposition interval. John Morris in Australia subjected laying females to continuous light, and this experiment has been continued by Bruce Sheldon, resulting in a significant reduction in laying. Trevor Morris at the University of Reading, England, and Harold Biellier at Missouri University have also selectively reduced the laying interval to below 24 h by exposing laying hens to a continuous noncycling light environment. If this genotype becomes incorporated into commercial laying stacks, egg numbers per hen will be significantly increased.

Dr. Elwood Briles was instrumental in the development of chicken blood group immunogenetics and deserves a great deal of credit for his contribution in this field and his cooperation with others. There are 11 loci for chicken blood groups. The most important one is the *Ea-B* group, which is intimately associated with the major histocompatibility complex (MHC) of the chicken. Certain *Ea-B* alleles confer either resistance or susceptibility to specific organisms. The graft-versus-host technique used by John Morris in Burnett's Lab, University of Melbourne, and later exploited by Schierman and Nordskog, led to the discovery of the association of this phenomenon with the MHC. The *Ea-B* alleles have been used in the selection of line-crossing combinations, and, together with other loci, have been used for parental identification.

Recombinant DNA and Related Technology

The DNA story began with the unequivocal demonstration by Avery and co-workers, in 1944, that DNA is the hereditary messenger material. Shortly afterward, the transformation and transduction phenomena were worked out and understood. The one-gene, one-enzyme hypothesis of Beadle and Tatum was exciting stuff. Molecular genetic technology has developed in an exponential way involving numerous Nobel Prize winners. The Watson-Crick description of the double helix and the triplet code gave the clue to the DNA messenger capability. Jacob and Monod's description of the codon-anticodon regulation circuitry for getting the message from the chromosome to the cell via mRNA with rRNA and tRNA aiding in assembly of proteins further increased our understanding of genetic control. The discovery of reverse transcriptase by Temin and polymerase by Kornberg were major contributions that eventually benefited recombinant technology. Another was the discovery and cataloguing of restriction enzymes (nucleases) that provide precise cutting tools to give specific DNA fragments. The development of methods for identifying, isolating, and sequencing DNA and RNA fragments made possible the identification of genes and their products.

Progress in DNA technology is proceeding by leaps and bounds, due in large part to thrust from the private sector. This is especially true for automation robotics, and for the availability of a variety of equipment and biologicals such as sequencing kits, restriction enzymes, and oligonucleotide kits. Plasmid, liposome, and retroviral vectors are being improved for integration efficiency, as are microinjection, electroporation, and ballistic techniques. Several of these delivery routes are being developed for integration of genes into appropriate totipotent stem cells. Antisense RNA is a means to block translation. It is relatively untried, but it may prove to be useful in the future for mitigating unwanted expression.

Genome Mapping

Functional regions of chicken chromosomes, including centromeres, G, C, and R bands, synaptonemal complexes, and the nuclear organizer region are well defined. The existing gene map of the chicken is incomplete for location of genes to respective chromosomes. The mapping of expressed genes is important for: (1) location of genes to chromosomes; (2) location of existing independent synteny groups to respective chromosomes; (3) providing reference points on chromosomes for positioning the location of polymorphic variable regions; (4) determining spatial relationship of gene insertions to homologous resident genes; and (5) determination of *cis*-trans relationships.

Gene mapping has been accomplished by classical recombination studies, rearrangement chromosome markers, and somatic cell hybrids that allow the assignment of genes to a specific chromosome but not to a specific location on the chromosome. The *in situ* hybridization technique is a direct method for accurately determining the locus of a gene. A promising mapping method to determine the location of a gene sequence is to recover DNA from slides by microisolation of either whole chromosomes or specific chromosome regions. With clever use of oligonucleotide primers and the polymerase chain reaction (PCR), recovered DNA sequences are amplified, allowing the construction of specific chromosome and subchromosome libraries. It is my opinion that determining the location of "real" genes, i.e., coding genes, is extremely important, not only to provide reference points, but also to facilitate detailed study of expression of these genes in normal biological function. Incidentally, the U.S. Department of Energy appears to have the same opinion.

Segments of DNA in the noncoding regions may be identified by several techniques, including fingerprinting, and detection of restriction fragment length polymorphisms (RFLP). These marker segments may be used to locate, through linkage, genes that are commonly referred to as quantitative trait loci (QTL). In theory, a map of these markers spaced throughout the chromosome and ordered by association (synteny) with precisely located reference genes would provide a practically useful map of the genome.

Transgenesis in Chickens

Gene transfer in chickens has been accomplished with plasmid, liposome, and retroviral vectors, although with very low efficiency. It is handicapped by the difficulty in reaching recipient totipotent cells. One of the problems in pursuing a program of transgenesis in chickens is the difficulty of identifying genes that should be introduced to give additional and favorable expression, to provide increased genetic variation, or to provide models for investigation of gene regulation and expression. The integration of the growth hormone gene has not been profitable, as the chicken system apparently ignores the presence of additional growth hormone. Integration into the delicate and complicated feedback circuitry of the hormonal system may be too much of a perturbation. I suggest we should be looking for genes in less complicated pathways and feedback systems. For example, in order to intensify the brown egg shell pigmentation desired in some commercial laying stocks, there is the possibility of modifying the one-to-one relationship between liver protoporphyrin pigment production and the recipient uterine gland pigment deposition system. Protoporphyrin pigment production is regulated by two liver mitochondrial enzymes, amino-levulonic acid (ALA) and ferric chelatase. Greater levels of ALA and lower levels of ferric chelatase result in a higher production of protoporphyrin and hence a greater deposition of dark pigment on the egg shell. Insertion of an ALA sequence may possibly increase the amount of protoporphyrin pigment without disturbing other functions. Modified single genes for resistance to infectious organisms that confer altered cell membrane receptors can be expected to be profitable approaches for "improvement" or for study of regulation.

The line between basic and applied research is often obscure. A better term is "adaptive" research, using whatever approach or combination of attack is required. In almost every issue of *Science, Nature,* or *Bio/Technology,* there are examples of research with benefits ranging from possible to actual. One should never run out of stimulating ideas. In fact, one could get thoroughly distracted if not well disciplined. There is an optimistic future for genetic engineering in poultry. However, limited funds and facilities for combining molecular genetic technology and avian biology, and the replacement of collegiality with the "corporate mentality" in universities, where secrecy and patent protection is hampering free exchange of information and materials, are somewhat discouraging. Attracting excellent students is a twofold dilemma. One is securing initial financial support with adequate facilities, and the other is concern for the well-being of trained scientists in the rat race for support. These down-side remarks are not real barriers to progress; they only represent some of the impediments common to the grass-roots researcher.

Several conditions need to be resolved in developing useful transgenic chickens for the poultry breeding industry:

1. Improved communication between the basic biologist, the molecular geneticist, and the breeder for identifying useful genes.
2. Improved vector transfection efficiency and totipotent cell manipulation.
3. Increased effort, throughout the world, in well-funded centers of excellence. The variability in expression of transgenic animals suggests that many birds may have to be produced and evaluated to finally secure one that is useful in industry.
4. Cost of development and introduction into breeding structure may be limiting. However, the poultry breeding structure makes it relatively straightforward to evaluate a transgenic bird and put it into distribution in minimum time.

Finally, and very importantly, there will be an increased thrust in the application of molecular technology for understanding the many facets of the basic biology of avian species. This knowledge is essential for further progress in genetic engineering by providing some insight into what genes may be useful for making changes in function and development.

Chapter 15

Genetic Control of Disease and Disease Resistance in Poultry

Jan S. Gavora*

Summary. Avian diseases in the broad sense include health problems arising from adverse physical environment or inadequate nutrition, developmental disorders, and infectious and parasitic diseases. Genetic improvement is of particular importance in the reduction of losses from the latter two groups of diseases. There is considerable knowledge of genetic mechanisms underlying many developmental disorders, mainly where major genes are involved, but their molecular basis is largely unknown. Ample evidence exists that genetic factors play an important role in resistance to infectious diseases. Such resistance appears to be controlled by polygenes, and major genes underlying the resistance have been identified in only a few instances. Knowledge of molecular bases of resistance mechanisms, such as the major histocompatibility system, immune response, and pathogen receptors, is rapidly increasing. Also, the molecular structure of many pathogen genomes is known and can be used in devising approaches to the improvement of host resistance. Molecular genetic information may provide tools for genetic improvement of resistance by selection using markers that are either linked to or are themselves components of resistance genes. Such selection does not require exposure to pathogens and, therefore, is desirable from the animal-welfare point of view. As well, DNA fingerprinting can improve the efficiency of transfer of genetic resistance by crossbreeding and backcrossing. For industrial applications, molecular gene transfer is justified primarily to introduce new resistance mechanisms, exemplified by pathogen-mediated resistance or antisense RNA, that do not exist in the avian species. Improvements of existing resistance mechanisms by gene transfer should await perfection of methods such as the culture of avian embryonic stem cells and homologous recombination. Progress in molecular mapping of avian genomes and further elucidation of molecular bases of resistance will play an important role in future improvements of genetic resistance to diseases in poultry.

Introduction

Health is one of the most important factors determining profitability of all systems of poultry production. Health status influences not only the average performance of a flock but also its variability (Gavora et al., 1983). Despite this importance, health-status considerations are often omitted in analyses of experimental data. Such considerations are of particular importance in studies of genotype by environment interactions, where diseases may be responsible for differences in the performance of stocks at various locations or under various management or nutritional regimes. Similarly, health status would be

*   For the Department of Agriculture (Agriculture Canada), Government of Canada.

expected to influence heterosis. Nevertheless, it is often ignored. For example, Barlow (1981), reviewing experimental evidence for interaction between heterosis and environment in animals, did not take into consideration health status, as most studies of genotype-environment interaction pay no or only cursory attention to the incidence of disease.

Compared to meat or egg production, relatively little attention has been paid by breeders to selection for resistance to diseases. Most often, the selection practised deals with total mortality, which has low heritability. Nevertheless, advances in the viability of commercial poultry stocks have been achieved (Hartmann 1985; McMillan et al., 1990), and it is likely that further progress in the improvement of disease resistance by genetic means will be made in the future.

An important reason for genetic improvement of disease resistance is that such improvements are cumulative and could be viewed as "one-time investment," provided that the pathogen for resistance to which selection was practised does not change its virulence. In this sense, it appears more advantageous to improve genetic resistance than to invest in vaccination and medication programs that have to be repeated every generation. Genetic improvement of disease resistance is desirable also from the points of view of environmental safety and sustainability of agricultural practices, as it has the potential to reduce or eliminate the utilization of undesirable inputs such as drugs or biologicals. Improvement of genetic disease resistance should be an integral part of the overall disease control programs and is beneficial in combination with all other means of disease control such as vaccination, medication, or eradication of pathogens (Gavora and Spencer, 1978).

Types of Health Problems

Various types of health problems are listed in Table 1, along with suggestions of what means are most suitable for their improvement. Problems of physical environment and trauma are best eliminated or reduced by removal of environmental causes. However, all environmental hardships cannot be completely eliminated, and some stress seems to be inevitable in modern poultry production systems. Effects of stress could be reduced by both environmental means and genetic approaches. A phenomenon that significantly influences the well-being of poultry in many climates is heat stress. The current knowledge of the molecular basis of the mechanisms through which organisms deal with heat stress, including information concerning the genes coding for heat-shock proteins (e.g., Catelli et al., 1985; Morimoto, et al., 1986; Sargan et al., 1986), may be applied in selection programs to improve resistance of poultry to heat. Nutritional disorders, similar to other negative environmental effects, should be dealt with primarily by improvement of the nutritional status of flocks, although selection for high productivity on low-quality rations shows some promise (Hutt and Nesheim, 1966; Sorensen, 1985).

Table 1
Health Problems and Means for Their Improvement

Problem	Environmental change	Means of improvement	
		Genetic	
		Conventional[a]	Gene transfer
Physical environment and trauma	+	o	o
Stress	+	+	o
Nutritional	+	o	o
Developmental and physiological disorders			
Major genes	o	+	o
Polygenes	o	+	o
Infectious and parasitic diseases			
Major genes	+	+	+
Polygenes	+	+	o

[a]   Including marker-assisted selection.
+   Most desirable approach to improvement.
o   Less important with regard to improvement.

Developmental and physiological disorders could be due to major gene effects or may be caused by the interaction of many genes. Under either circumstance, conventional selection would appear to be the best method to alleviate the problems.

Infectious and parasitic diseases form the largest and most important group of poultry health problems, and both environmental and genetic means have to be employed for their reduction (Table 1). Because of their importance, the remainder of this review will concentrate on the pathogen-host relationship that forms the basis of infectious and parasitic diseases, with emphasis on the use of molecular genetic methods in the improvement of resistance to disease.

Mechanisms of Resistance and Their Potential for Molecular Manipulation

Resistance to disease can be inherited as a polygenic trait or can behave as a major gene inherited in a Mendelian fashion. It is likely that, of the multiplicity of genes underlying polygenic inheritance, there are a few that are responsible for a significant portion (say, 60–70%) of its genetic variation, while the rest of the variation is due to a large number of genes with small effects. For example, immunoresponsiveness to natural antigens was shown to depend primarily on 2 to 15 loci (Biozzi et al., 1979).

Over the past decade, the understanding of disease-resistance mechanisms has been rapidly increasing, and this will have a positive effect on future developments of disease control in poultry. For example, the major histocompatibility complex (MHC) is recognized as a major factor and a central point in the immune system of the bird (see Chapter 12), and is now being analyzed at the molecular level (Auffray, 1990). Many other aspects of immune response are being successfully studied, and substantial advances have been made in the understanding of the differentiation markers of blood cells, immunoglobulins, T-cell receptors, and others (Lamont and Dietert, 1990). Another example is that the inheritance of receptors for avian leukosis viruses is well understood (Payne, 1985), although the molecular structure of the receptor substances is not known.

Of particular interest is "nonhost" resistance, a conglomerate of mechanisms that make, for example, a chicken resistant to a potato or mammalian virus. It was recognized long ago that certain biochemical changes in either the host or the pathogen may render a host resistant to a specific pathogen (Haldane, 1949), but our understanding of nonhost resistance is only fragmentary. Exploration of this resistance may provide suggestions on how to genetically engineer the host for resistance to specific pathogens. One way of obtaining some insight into nonhost resistance is to study the reverse phenomenon, i.e., the adaptation of pathogens, such as viruses, to nonhost environments. Thus, in studies recently conducted with Rous sarcoma virus, it was found that viral envelope glycoproteins of this virus, which normally infects only chickens, were modified during adaptation of the virus to infection of mammalian cells by multiple tissue culture passages (Kawai et al., 1989). Similar adaptation to duck cells also resulted in changes in the viral envelope coding region that appeared to be a combination of an exogenous and endogenous virus structure (Ryndich et al., 1990).

The rapidly increasing understanding of the molecular structure and physiology of pathogens led to the design of recombinant DNA containing genetic information from pathogens that can be introduced by gene transfer and confer resistance to the pathogens (Salter et al., 1987; Crittenden et al., 1989), as discussed in Chapter 9.

No known natural defence mechanisms exist against "selfish" DNA, certain types of repetitive DNA sequences that may be associated with reduced performance. In chickens, this type of DNA seems to be widely spread and consists of at least two families of genes: (1) RAV-type elements (Smith, 1986), and (2) EAV-elements (Dunwiddie et al., 1986). It has been shown that at least the RAV-type elements are associated with reduced performance of chickens in production traits (Kuhnlein et al., 1989a, 1989b; Gavora et al., 1991).

In this context, it is of interest to note findings concerning the effects on cellular function and growth of persistent, nontransforming viruses. After infection of BALB/C 3T3 murine cells in culture with an RNA virus (reovirus), some cells underwent a limited lytic phase. The cells that survived appeared morphologically normal under light microscopy but were found persistently infected and shed the virus. These infected cells exhibited a 70 – 90% decrease in receptor number for epidermal growth factor, while their insulin receptors were threefold increased in number and serum- and insulin-stimulated DNA synthesis was comparable to control uninfected cells (Verdin et al., 1986). Hence, such persistent infections may lead to major alterations in the control of cell growth by specific growth factors. The finding that the presence of endogenous virus-producing genes in chickens negatively affects egg production to a similar extent as subclinical infections with related exogenous lymphoid leukosis viruses indicates the possibility that alterations of cell function, as described above (Verdin et al., 1986), could underlie the observed macro effects (Gavora et al., 1980, 1991; Gavora and Spencer, 1985).

Possible mechanisms amenable to genetic manipulation are regulatory phenomena mediated by promoters and enhancers of gene function (de Villiers, 1986; Paquette et al., 1990), including viral long terminal repeats (Ju and Cullen, 1985), through which increased expression of resistance genes may be introduced.

From the point of view of general resistance to disease, it is interesting to note that work in laboratory mice indicates the existence of resistance genes of relatively broad spectrum. The gene *Bcg*, involved in resistance against mycobacterial infections (Gros et al., 1981), was found to be identical to genes *Ity* and *Lsh*, controlling innate resistance to infection with *Salmonella typhimurium* and *Leishmania donovani*, respectively (Skamene et al., 1982), and is believed to be involved in general resistance of macrophages to bacterial attack.

Examples of other general mechanisms of resistance to viral diseases are interferon-induced antiviral systems, such as the protein kinase system that is active in resistance to vaccinia virus and a reovirus. In contrast, the product of the *Mx* gene specifically inhibits primary transcription of influenza viral genes but does not confer resistance to any other virus (Joklik, 1990). Recent studies involving the transfer of the *Mx1* gene to chicken cells (Garber et al., 1991) show promise for the engineering of resistance against influenza viruses in avian species.

Putting a cloned gene in a backward orientation to its promoter results in transcription of anti-sense RNA that can inactivate the normal messenger RNA, thus preventing gene expression (Izant and Weintraub, 1985). Anti-sense RNA may be another potential disease protection mechanism against viruses that can be introduced by transgenesis.

Application of Molecular Genetics to the Improvement of Disease Resistance

It is generally recognized that heritability of total mortality is low. Total mortality may include deaths from causes ranging from trauma to many types of infectious diseases. Nevertheless, through a combination of environmental

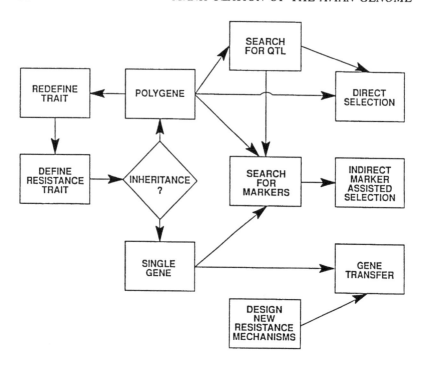

FIGURE 1. A strategy for research involving disease-resistance genetics.

and genetic efforts, there has been some improvement in the overall viability of commercial poultry. Evidence for such improvement in chickens selected for high egg production was provided by Hartmann (1985) and by McMillan et al. (1990).

A generalized strategy for the study of disease resistance and development of methods for improvement of resistance (Figure 1) shows the possible ways of dealing with disease resistance by conventional methods and by methods using molecular genetics. It is generally believed that, although useful in livestock improvement, molecular genetics itself will never replace traditional methods of improvement. Instead, conventional and molecular genetics approaches should be integrated for maximum improvement in the economic efficiency of domesticated animals and birds (Lande and Thompson, 1990).

Marker-Assisted Selection

Marker-assisted selection is particularly suitable for the improvement of disease resistance, as it provides an opportunity to replace disease challenge tests by indirect selection using molecular markers, in single or multiple stages (Cunningham, 1975). Disease challenge tests are undesirable from the animal-welfare point of view and because of the risk of spread of pathogens. In general, the efficiency of selection using only markers, relative to phenotypic selection of the same intensity, is

$$\sqrt{p/h^2}$$

where p is the total proportion of the additive genetic variance explained by the markers and h^2 is the heritability of the character under selection. Thus, indirect selection based on markers only can be more efficient than phenotypic selection if the proportion of additive genetic variance explained by the markers exceeds the heritability of the trait. As some disease-resistance traits have low heritability, marker-assisted selection may be particularly suitable for their improvement. However, molecular markers can only rarely be expected to have major direct effects on characters of economic importance, including disease resistance. Marker-assisted selection must, therefore, rely on linkage between the marker loci and the quantitative trait loci, although such linkage may be continually eroded by recombination. The most potent mechanism for generating such linkage is, therefore, an occasional hybridization of genetically differentiated lines (Lande and Thompson, 1990).

While molecular markers are expected to be based mostly on single-locus probes, complicated DNA fingerprint patterns could be used for genomic selection (Hillel et al., 1990). In this approach, DNA fingerprint loci that are detectable by minisatellite probes are used in introgression programs that rely on crossbreeding and backcrossing to select individuals with maximal genomic similarity to the recipient line and minimal similarity to the donor line. This genomic selection can significantly reduce the required number of backcross generations in an introgression breeding program and would appear particularly suitable for introgression of resistance genes from low-performance native breeds into modern commercial breeding stock. More detailed discussion of the use of DNA fingerprints and DNA markers in poultry improvement can be found in Chapter 16. The emergence of the RAPD technique for the analysis of DNA polymorphisms shows promise for application in routine breeding (Williams et al., 1990) because it could likely be at least partly automated and thus provide information on many samples, as may be required in a poultry breeding program.

Gene Transfer

An interesting approach to the demystification of genetic engineering was provided by Koshland (1989), who suggests "that if a molecular biologist comes to a geneticist with the suggestion that he would like to engineer a wolf into a dog, the geneticist will likely reply 'it's been done,' because over evolutionary time the friendliest of wolves learned how to earn a living 'delivering slippers' rather than hunting caribou in the wilds." Although poultry breeding will continue to be done by conventional means, genetic engineering by direct DNA modification will likely become, with improved technology, a rather commonplace tool in the poultry breeders' workshop. Gene transfer is a logical extension of the many years in which animal breeders practised genetic engineering of livestock by conventional means (Dickerson and Willham, 1983).

Gene transfer has been accomplished with varying degrees of success in all livestock species, and the efficiency of production of transgenic animals will improve continuously. However, accurate assessment of the consequences of transgene expression is necessary before the use of specific transgenic animals for production purposes (Pursel et al., 1989). The current most popular methods of gene transfer — namely, microinjection and the use of viral vectors — are relatively crude and do not provide any control over the site into which the exogenous DNA is inserted in the new host genome. Because of such uncertainty, extensive testing of the transgenic animals is needed (Smith et al., 1987; Pursel et al., 1989). However, there is a new approach that may dramatically improve the rate of success of targeted gene transfer, namely, homologous recombination in embryonic stem cell lines (Capecchi, 1989). With these methods, discussed by others in this volume, new horizons will be opened. Transgenesis could then be more widely and freely applied to the improvement of domestic animals and poultry, including creation of animals suitable for the production of important pharmaceuticals.

Transgenic technology will be of value in general studies of disease resistance in the bird. However, until the efficiency of production of specifically engineered genetic modifications improves, application of transgenic techniques to commercial birds will probably be restricted to introducing disease-resistance mechanisms that do not exist in the species being manipulated. The introduction into commercial strains of chickens of resistance genes from within the species should be, at least for the time being, performed by crossing and backcrossing, possibly using molecular markers for detection of the desired gene to be transferred and employing genomic selection for the more rapid return of the crossbred animals to the type of the desirable original parent.

Future Prospects

In the near future, large efforts and funds will be invested in the mapping of the poultry genome. A considerable effort in this direction is being mounted and coordinated from the U.S. Department of Agriculture's Agricultural Research Service, as well as similar projects being conducted in other parts of the world. The gradual assembly of the molecular map of the chicken genome will result from the work of multinational research teams. Improved knowledge of the chicken genome will allow the gradual elucidation of the molecular basis of resistance to many diseases and, eventually, should lead to the identification of many resistance genes. This process will be enhanced by the desirability of improved disease resistance and by the undesirability of exposure of poultry to pathogens that is required using current selection programs for disease resistance.

References

Auffray, C. (1990). Molecular biology of the chicken major histocompatibility complex. *Crit. Rev. Poultry Biol.*, 2, 255–275.

Barlow, R. (1981). Experimental evidence for interaction between heterosis and environment in animals. *Animal Breeding Abstr.*, 49, 715–737.

Biozzi, G., Mouton, D., Santa Anna, O. A., Passos, H. C., Gennari, M., Reis, Mitt., Ferreira, V. C. A., Heumann, A. M., Bouthillier, Y., Ibanez, O. M., Stiffel, C., and Siqueira, M. (1979). Genetics of immunorespusiveness to natural antigens in the mouse. *Current Topics Microbiol. Immunol.*, 85, 31–98.

Capecchi, M. R. (1989). Altering the genome by homologous recombination. *Science*, 244, 1288–1292.

Catelli, M. G., Binart, N., Feramisco, J. R., and Helfman, D. M. (1985). Cloning the chick hsp90 CDNA in expression vector. *Nucleic Acids Res.*, 13, 6035–6047.

Crittenden, L. B., Salter, D. W., and Federspiel, M. J. (1989). Segregation, viral phenotype, and proviral structure of 23 avian leukosis virus inserts in germ line of chicken. *Theor. Appl. Genet.*, 77, 505–515.

Cunningham, E. P. (1975). Multi-stage index selection. *Theor. Appl. Genet.*, 46, 56–61.

de Villiers, J. (1986). Eukaryotic transcriptional enhancer elements. *South Afr. J. Sci.*, 82, 479–482.

Dickerson, G. E., and Willham, R. L. (1983). Quantitative genetic engineering of more efficient animal production. *J. Anim. Sci.*, 57 (Suppl. 2), 248–264.

Dunwiddie, C. T., Resnick, R., Boyce-Jacino, M., Alegre, J. N., and Faras, A. J. (1986). Molecular cloning and characterization of *gag, pol,* and *env* related gene sequences in the ev' chicken. *J. Virol.*, 59, 669–675.

Garber, E. A., Chute, H. T., Condra, J. H., Gotlib, L., Colonno, R. J., Mills, E. O., Hancock, J., Hreniuk, D., and Smith, R. G. (1991). Avian cells expressing the murine Mx1 protein are resistant to influenza virus infection. *J. Cell. Biochem.*, Suppl. 15E, 204.

Gavora, J. S., and Spencer, J. L. (1978). Breeding for genetic resistance: Specific or general. *World's Poultry Sci. J.*, 34, 137–148.

Gavora, J. S., and Spencer, J. L. (1985). Effects of lymphoid leukosis virus infection on response to selection. In: *Poultry Genetics and Breeding,* edited by W. G. Hill, J. M. Mason, and D. Hewitt. British Poultry Science Ltd., Longman Group, Harlow.

Gavora, J. S., Chesnais, J., and Spencer, J. L. (1983). Estimation of variance components and heritability in populations affected by disease: Lymphoid leukosis in chickens. *Theor. Appl. Genet.*, 65, 317–322.

Gavora, J. S., Kuhnlein, U., Crittenden, L. B., Spencer, J. L., and Sabour, M. P. (1991). Endogenous viral genes: Association with reduced egg production rate and egg size in White Leghorns. *Poultry Sci.*, 70, 618–623.

Gavora, J. S., Spencer, J. L., Gowe, R. S., and Harris, D. L. (1980). Lymphoid leukosis virus infection: Effects on production, mortality and consequences in selection for high egg production. *Poultry Sci.,* 59, 2165–2178.

Gros, P., Skamene, E., and Forget, A. (1981). Genetic control of natural resistance to *Mycobacterium bovis* (Bcg) in mice. *J. Immunol.,* 127, 2417–2421.

Haldane, J. B. S. (1949). Disease and evolution. *La Ricerca Scientifica Supp.,* 19, 68–75.

Hartmann, W. (1985). The effect of selection and genetic factors on resistance to disease in fowls — A review. *World's Poultry Sci. J.,* 41, 20–35.

Hillel, J., Schaap, T., Haberfeld, A., Jeffreys, A. J., Plotzky, Y., Cahaner, A., and Lavi, U. (1990). DNA fingerprints applied to gene introgression in breeding programs. *Genetics,* 124, 783–789.

Hutt, F. B., and Nesheim, N. C. (1966). Changing the chick's requirement of arginine by selection. *Can. J. Genet. Cytol.,* 8, 251–259.

Izant, J. G., and Weintraub, H. (1985). Constitutive and conditional suppression of exogenous and endogenous genes by anti-sense RNA. *Science,* 229, 346–352.

Joklik, W. K. (1990). Interferons. In: *Virology,* edited by B. N. Fields, D. M. Knipe, et al., Raven Press, New York, pp. 383–409.

Ju, G., and Cullen, B. R. (1985). The role of avian retroviral LTRs in the regulation of gene expression and viral replication. *Adv. Virus Res.,* 30, 179–223.

Kawai, S., Nishizawa, M., Shinno-Kohno, H., and Shiroki, K. (1989). A variant Schmidt-Rupin Strain of Rous sarcoma virus with increased affinity to mammalian cells. *Jpn. J. Cancer Res.,* 80, 1179–1185.

Koshland, D. E., Jr. (1989). The engineering of species. *Science,* 244, 1233.

Kuhnlein, U., Gavora, J. S., Spencer, J. L., Bernon, D. E., and Sabour, M. P. (1989a). Incidence of endogenous viral genes in two strains of White Leghorn chickens selected for egg production and susceptibility and resistance to Marek's disease. *Theor. Appl. Genetics,* 77, 26–32.

Kuhnlein, U., Sabour, M., Gavora, J. S., Fairfull, R. W., and Bernon, D. E. (1989b). Influence of selection for egg production and Marek's disease resistance on the incidence of endogenous viral genes in White Leghorns. *Poultry Sci.,* 68, 1161–1167.

Lamont, S. J., and Dietert, R. R. (1990). Immunogenetics. In: *Poultry Breeding and Genetics,* edited by R. D. Cranford. Elsevier, Amsterdam.

Lande, R., and Thompson, R. (1990). Efficiency of marker-assisted selection in the improvement of quantitative traits. *Genetics,* 124, 743–756.

McMillan, I., Fairfull, R. W., Gowe, R. S., and Gavora, J. S. (1990). Evidence for genetic improvement of layer stocks of chickens during 1950–1980. *World's Poultry Sci. J.,* 46, 235–245.

Morimoto, R. I., Hunt, C., Huang, S.-Y., Berg, K. L., and Banerji, S. S. (1986). Organization, nucleotide sequence and transcription of the chicken hsp-70 gene. *J. Biol. Chem.,* 261, 12692–12699.

Paquette, Y., Kay, D. G., Rassart, E., Robitaille, Y., and Jolicoeur, P. (1990). Substitution of the U3 long terminal repeat region of the neurotropic Cas-Br E retrovirus affects its disease inducing potential. *J. Virol.,* 64, 3742–3752.

Payne, L. N. (1985). Genetics of cell receptors for avian retroviruses. In: *Poultry Genetics and Breeding,* edited by G. W. Hill, J. M. Mason, and D. Hewitt. British Poultry Science Ltd., Longman Group, Harlow, pp. 1–16.

Pursel, V. G., Pinkert, C. A., Miller, K. F., Bolt, D. J., Campbell, R. G., Palmiter, R. D., Brinster, R. L., and Hammer, R. E. (1989). Genetic engineering of livestock. *Science,* 244, 1284–1288.

Ryndich, A. V., Kashuba, V. I., Kavsan, V. M., Zubak, S. V., Dostalora, V., and Hlozanek, I. (1990). Molecular principles of retrovirus adaptation: Nucleated sequence of Raus sarcoma virus adapted to duck cells. *Genetika,* 26, 389–398.

Salter, D. W., Smith, E. J., Hughes, S. H., Wright, S. E., and Crittenden, L. B., (1987). Transgenic chickens: Insertion of retroviral genes into the chicken germ lines. *Virology,* 157, 236–240.

Sargan, D. R., Tsai, M.-J., and O'Malley, B. W. (1986). Hsp-108, a novel heat shock inducible protein of chicken. *Biochemistry,* 25, 6252–6258.

Skamene, E., Gros, P., Forget, A., Kongshavn, P., St. Charles, C., and Taylor, B. (1982). Genetic regulation of resistance to intracellular pathogens. *Nature,* 297, 506–509.

Sorensen, P. (1985). Influence of diet on response of selection for growth and efficiency. In: *Poultry Genetics and Breeding,* edited by W. G. Hill, J. M. Mason, and D. Hewitt. British Poultry Science Ltd., Longman Group, Harlow, pp. 85–95.

Smith, C., Menwissen, T. H. E., and Gibson, J. (1987). On the use of transgenes in livestock improvement. *Animal Breeding Abstr.,* 55, 1–10.

Smith, E. J. (1986). Endogenous avian leukemia viruses. In: *Avian Leukosis,* edited by G. F. de Boer. Martinus Nijhoff, Boston, pp. 101–120.

Verdin, E. M., Maratos-Flier, E., Carpentier, J.-L., and Kahn, C. R. (1986). Persistent infection with a nontransforming RNA virus leads to impaired growth factor receptors and response. *J. Cell. Physiol.,* 128, 457–465.

Williams, J. G. K., Kubelik, A. R., Livak, K. J., Rafalski, J. A., and Tingley, S. V. (1990). DNA polymorphisms amplified by arbitrary primers are useful as genetic markers. *Nucleic Acids Res.,* 18, 6531–6535.

Chapter 16

Multilocus DNA Markers: Applications in Poultry Breeding and Genetic Analyses

Jossi Hillel, E. Ann Dunnington, Alon Haberfeld, Uri Lavi, Avigdor Cahaner, Orit Gal, Yoram Plotsky, H. L. Marks, and Paul B. Siegel

Summary. Population genetic theory has facilitated considerable progress in development of superior poultry breeding stocks through the use of selection and heterosis. The development of techniques for DNA fingerprint analyses provides tools that may be used to complement present breeding programs through direct identification of genotype. Based on the fact that the information obtained in these procedures is derived from many highly polymorphic loci inherited in a simple co-dominant Mendelian manner, five potential uses are described in this chapter:

1. **Identification** — A wide range of identifications is discussed: individual, family, genetic stocks, and breeding lines. The level of similarity between DNA fingerprint patterns of closely related populations as a relative estimate for genetic distance between them is also considered.

2. **Evolution** — Populations divergently selected for quantitative traits were shown to have specific DNA fingerprint patterns. These genomic differences are attributable to cumulative mutations and genetic drift through the process of many generations of selection, or to selection where alleles of the loci coding for the selected trait are linked to alleles of the DNA fingerprint loci. Two replicates of each of two Japanese quail lines divergently selected for 4-week body weight and one unselected line were used to demonstrate how these forces could generate line-specific DNA fingerprint patterns. The relative weight of these three evolutionary forces, namely, mutation, genetic drift, and selection, combined with a linkage disequilibrium state, were estimated to be 16%, 59%, and 25%, respectively.

3. **Marker-assisted selection** — An efficient approach to identify linkage between DNA fingerprint bands as genetic markers and quantitative trait loci (QTL) was based on a study of the paternal distribution of a quantitative trait combined with a comparison of DNA fingerprint patterns of DNA or blood mixes of individuals from the two extremes of the distribution of the trait in the population. This approach was demonstrated by an experiment in which a DNA fingerprint band was shown to be linked to abdominal fat deposition in broilers.

4. **Genomic selection** — Introgression of a given heritable trait from one line to a well-developed commercial line requires large numbers of backcross generations. To reduce the required number of backcross generations, an approach is presented for the faithful recovery of the recipient

genome that involves selection based on an assessment of the level of DNA fingerprint similarity between progeny of the first and second backcross generations to the DNA fingerprint pattern of the recipient line (or minimum similarity to the donor line). This presentation is based on theoretical probabilities and on a brief description of experimental data.

5. **Heterosis** — Hybrid vigor is expected when lines involved in cross-breeding or the individual parents of a given cross are genetically distant. The larger the genetic distance, the larger the heterosis expected. An approach to predict hybrid vigor through selection of lines or individuals of minimum similarity between their DNA fingerprint patterns is discussed.

Introduction

Since the advent of plant and animal domestication, people have consistently tried to modify the phenotypic expression of animals and plants through artificial selection. Breeding programs for poultry improvement during this century have been based on Mendelian principles and quantitative genetic theory. Large breeding populations, intense selection, and utilization of heterosis have contributed much to the development of superior germplasm. Development of biotechnological methodologies during the past decade has created the possibility for additional genetic improvement of economically important traits in poultry breeding programs.

Currently, the genotype or breeding value of an individual with respect to a specific trait is assessed either by its own phenotypic performance or by the average of its relatives' performances. The phenotype, however, may be a poor estimator for breeding value of an individual (e.g., traits with low heritability), may be expensive or difficult to measure (e.g., disease resistance), may not be expressed (e.g., sex-limited traits), or can be measured only late in life or postmortem (e.g., egg production, carcass characteristics). Introgression of a monogenic trait from one stock to another is inefficient because numerous backcross generations are required to restore the genome of the recipient line. Finally, programs using heterosis suffer from the difficulty in predicting combining ability. In this chapter we discuss the potential use of multilocus molecular genetic markers, especially those resulting from a variable number of tandem repeats (VNTR) of DNA sequences, in overcoming these difficulties.

Molecular Genetic Markers

Recent developments in molecular genetics have demonstrated polymorphic DNA sequences that, if linked to relevant gene loci, may aid in assessment and identification of these gene loci and may complement the use of phenotypic measurements. Their usefulness for linkage analyses depends on: (1) assessment at the DNA level rather than at the gene product level;

(2) availability of measures for analysis at any developmental stage or age, in any tissue, and in both sexes; (3) distinguishing heterozygotes due to co-dominant inheritance; (4) high rates of polymorphism, entailing high levels of heterozygosity; (5) large numbers of polymorphic marker loci dispersed throughout the genome; and (6) ease of detection. The first three conditions may be met by use of classical restriction fragment length polymorphism (RFLP) markers (based mainly on the presence or absence of restriction endonuclease recognition sites). To fulfill the remaining conditions, a system involving higher levels of polymorphism is necessary. The recently discovered hypervariable regions (HVR) (Jeffreys et al., 1985a) provide a highly poly-morphic system, unrestricted by the shortcomings of classical RFLPs. Poly-morphism in HVRs is generated mainly by VNTR (Nakamura et al., 1987) of core sequences consisting of 10 to 16 base pairs (minisatellite loci) (Jeffreys et al., 1985a) or 2 to 3 bp (microsatellite loci) (Epplen, 1988). Hence, the number of alleles at a locus is essentially limitless, and one probe can hybridize to numerous HVR loci dispersed throughout the genome.

Endonuclease-digested, electrophoresed, and denatured DNA, hybridized to a multilocus DNA probe, will result in a multiband pattern, the "DNA fingerprint," which is absolutely individual-specific (Jeffreys et al., 1985b). Linkage analysis and allelic relationships among DNA fingerprint bands can be determined by analysis of individuals from large full-sib families.

Several applications of DNA fingerprints for genetic analyses and breed-ing programs in poultry may be feasible. These include: (1) determination of individual identity, family relationships, and levels of relatedness among genetic stocks and selected lines; (2) analyses of evolutionary forces generating genetic differences among selected lines of common origin; (3) identification of specific VNTR loci associated with quantitative trait loci (QTL); (4) im-provement of gene introgression programs by reducing the number of back-cross generations; and (5) prediction of relative heterosis.

Identification

For individual identification, DNA fingerprints obtained by the use of human polycore probes (Jeffreys et al., 1985b, 1986) have allowed detection of multiple polymorphic fragments in DNA from a wide range of species, in-cluding mammals and birds (e.g., Burke and Bruford 1987; Jeffreys et al., 1987a, 1987b; Wetton et al., 1987). It appears that Jeffreys' "core" sequence has been conserved during evolution and therefore cross-hybridizes to min-isatellites in many animal and plant species. Multilocus probes, such as Jef-freys' 33.6 and 33.15 (Jeffreys et al., 1985a, 1985b), bacteriophage M13 (Vassart et al., 1987), and microsatellite probes such as (CAC)$_5$ (Schafer et al., 1988), poly(GAC/TA) (Ali et al., 1986), and poly(GT) (Walmsley et al., 1989; Haberfeld and Hillel 1991; Haberfeld et al., 1991), detect high levels of polymorphism at many loci. The resulting DNA patterns are individual-specific, therefore they are considered as DNA fingerprints. Analysis of DNA fingerprints obtained from 11 unrelated meat-type chickens (Hillel et al., 1989b) using Jeffreys' probe 33.6 (Jeffreys et al., 1985b) revealed an average of 16 large (>7-kb) minisatellite fragments, which is twice that observed in human DNA fingerprints. An average of 6 DNA fragments of medium size (4 – 7 kb) was detected. Approximately one in four bands are shared between pairs of chickens taken randomly from an outbred population. The level of band sharing (see legend of Figure 1 for mathematical definition) for small DNA fragments was higher than for large ones, as reported for human DNA fingerprints (Jeffreys et al., 1985b). An average of 29 bands was detected by probe 33.6 in DNA fingerprints of chickens, a number similar to that reported for humans, other animals, and plants (Jeffreys et al., 1986, 1987b; Jeffreys and Morton, 1987; Burke and Bruford, 1987; Dallas, 1988). The level of band sharing was similar to that for humans, and lower than that for most other vertebrates (Jeffreys and Morton, 1987), indicating that minisatellite loci in chickens are highly polymorphic.

Use of several unrelated probes can increase the number of polymorphic loci that can be detected and, consequently, can increase the saturation level of the avian genome in molecular genetic markers. Ninety-eight percent of loci detected in the chicken genome by Jeffreys' probe 33.6 were not linked to loci detected by probe R18.1, developed from a bovine genomic library (Haberfeld and Hillel, 1991; Haberfeld et al., 1991). Because the number of loci detected by each probe was similar (Haberfeld et al., 1991), information obtained was essentially doubled from use of the two probes.

Population analysis of DNA fingerprint patterns is extremely difficult when individuals are evaluated separately because of the volume of laboratory work required, difficulties in comparing DNA fingerprint lanes from different gels, and complexity of drawing a common DNA fingerprint pattern representing the entire analyzed population. Producing a separate DNA fingerprint for each sampled individual places a constraint on the number of individuals that can be used, and comparisons of individual DNA fingerprint lanes from two or more different populations require either an extensive matrix of pairwise comparisons (Gilbert et al., 1990) or calculation of an index (Kuhnlein et al., 1989) to evaluate population differences.

DNA fingerprint representation of a line or stock by mixing DNA from different individuals has been used to overcome these difficulties. In fingerprints of DNA mixes, all fragments other than those from extremely polymorphic loci are present in each mix. Also, the relative intensity of the same band in different samples or populations may be used to show differences in band frequencies. Using DNA mixes, Dunnington et al. (1990) reported that, in White Rock chickens selected for 31 generations for high or low body weight at 8 weeks of age, approximately 50% of the DNA fingerprint bands were line-specific. Although genetic variability within the selected lines was appreciable as reflected by 54% band sharing between individuals within

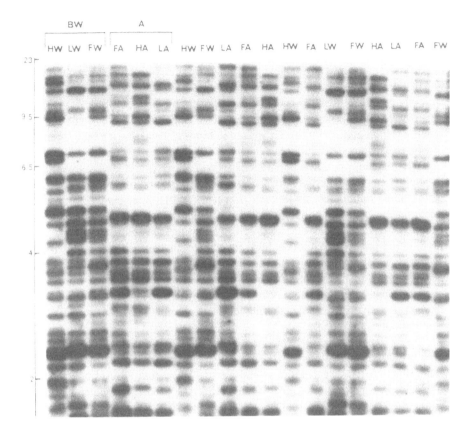

FIGURE 1. The DNA fingerprints produced by Jeffreys' probe 33.6 with DNA mixes from chickens of the following populations: HW = White Rocks selected for 31 generations for high body weight; LW, White Rocks selected for 31 generations for low body weight; FW, F_1 cross between HW and LW lines; HA, White Leghorns selected for 16 generations for high antibody response to sheep erythrocytes; LA, White Leghorns selected for 16 generations for low antibody response to sheep erythrocytes; and FA, F_1 cross between HA and LA lines. The first six lanes represent each of the populations; lanes are then repeated in a different order to facilitate pairwise comparisons. Each lane represents equal amounts of DNA from 10 or more individuals within a line. The level of similarity between two DNA fingerprint patterns was estimated as band sharing, which is the proportion of the common bands out of the total number of bands scored in both patterns (Jeffreys and Morton, 1987; Wetton et al., 1987; Hillel et al., 1989a). The band sharing (BS) between DNA fingerprint patterns for lanes a and b was calculated as BS = 2 (Nab)/(Na + Nb), where Nab is the number of bands shared by lanes a and b, Na is the total number of bands for lane a and Nb is the total number of bands for lane B. (From Dunnington et al., 1991, *Poultry Sci.*, 70, 463–467, with permission.)

lines, the pattern of line specificity was observable even when samples as small as five chickens per line were used in the DNA mixes. It is worth mentioning that, in addition to the sixfold difference in the selected trait, these lines differ in many other traits (Dunnington et al., 1986), therefore these differences in band frequency reflect differences in the entire genome.

In lines of White Leghorn chickens selected for high (HA) or low (LA) antibody response to sheep erythrocytes (Siegel and Gross, 1980; Martin et al., 1990), approximately 25% of the DNA fingerprint bands were line-specific (Hillel et al., 1991b). Expressed in terms of band sharing (see legend of Figure 1), line HA had higher levels (0.69) than line LA (0.58).

When all four selected populations (HW, LW, HA, and LA), as well as F_1 populations from a cross between each pair of selected lines (FW and FA) were studied simultaneously (Dunnington et al., 1991), three levels of genetic relationships between populations were revealed (Figure 1). The greatest genetic distance (i.e., lowest level of band sharing) was between the White Plymouth Rock and White Leghorn breeds (mean band sharing = 0.15). Moderate levels of band sharing (mean band sharing = 0.51) were found between pairs of divergently selected populations (i.e., two lines from the same original base population, but separated by selection into closed populations). Relatively high band-sharing levels were found between parental and F_1 populations, averaging .65.

These experiments confirmed the presence of line-specific DNA finger-print bands in populations of chickens divergently selected for quantitative traits over 16 to 31 generations, and demonstrated that DNA fingerprints obtained from mixing DNA provide an efficient approach for screening and comparing populations of chickens. Mixes of DNA have also been used efficiently for linkage analysis between DNA fingerprint bands and QTLs (Hillel et al., 1989a; Plotsky et al., 1990) and for measuring effects of mutation, genetic drift, and linkage disequilibrium on the production of line-specific DNA fingerprint bands in selected lines (Hillel et al., 1991a).

Although DNA mixes facilitate genetic analyses at the population level and reduce the number of Southern blots required, preparation of DNA from each individual in the group remains a limiting factor. When DNA fingerprints were prepared from mixes of equal amounts of DNA from chickens, turkeys (Hillel et al., 1990a), or Muskovy ducks, (J. Hillel, S. Blum, D. Martin, and J. Powel, unpublished results) of the same sex and age or from DNA from mixes of equal amounts of blood from the same individuals, identical DNA fingerprint patterns were obtained. Therefore, the characterization of a group of individuals from these species can be based on blood mixes, eliminating the need to extract DNA from individual samples, and further reducing the cost of making genotype comparisons of birds by DNA fingerprint analyses.

Evolution

Line-specific DNA fingerprint bands represent genetic differences between lines that were developed over generations of divergent selection with absence of gene flow between them. Line-specific DNA fingerprint bands may result directly from selection, only if in the base population there was linkage between these DNA fingerprint bands (alleles of the VNTR loci) and alleles affecting the selected trait(s) (linkage disequilibrium state). The impediment to gene flow between the selected lines results in each line manifesting differences resulting from cumulative mutations and genetic drift.

To evaluate the relative contribution of selection, mutation, and genetic drift to genetic differences between divergently selected lines, an ideal experimental structure should include population DNA fingerprints from replicated lines for each (opposite) direction of selection, an unselected control population, and the original base population common to all lines. Additionally, DNA fingerprints should be prepared from individuals representing the selected and control lines in each generation of the selection experiment. The size of populations and number of parents selected from the base population to initiate each line must be large enough to minimize founders' effects and inbreeding.

The effect of each of the three evolutionary forces could be calculated from the frequency distribution of line-specific DNA fingerprint bands as follows.

1. **Selection** — When a band is observed in the two replicated lines of only one direction of selection, in the original population, and in the control line, it can be interpreted as the result of linkage disequilibrium between this band and alleles of a QTL controlling the selection trait or a genetically correlated trait in the base population.
2. **Mutations** — Bands absent from the base population, and present in only one of the five lines (four selected and one control), are considered likely to have been generated by mutation in the particular line.
3. **Genetic drift** — Absence of a band in one selected line and presence in the other four lines including the original base population is interpreted as loss due to genetic drift resulting from random sampling in previous generations.

Other, combined situations are also possible: For example, the presence of a band in the original population, the control line and only one replicate of a selected line, could have resulted from selection and subsequent removal of the band from one of the two replicated selected lines due to genetic drift.

This approach was used (Hillel et al., 1991a) to evaluate the relative contribution of mutation, genetic drift, and selection to the presence of line-specific DNA fingerprint bands in the Japanese quail populations developed

by Darden and Marks (1988). These populations included two pairs of replicated lines selected divergently for 4-week body weight for 20 generations,
and a control line randomly bred for 20 generations from the same base
population. DNA of the original base population was not available and was
available only from generation 20. The presence or absence of each DNA
fingerprint band in each of the five lines was interpreted as the result of one
of the three evolutionary forces, or of a combination of these forces. Using
two multilocus probes, 52 polymorphic bands were detected. Based on the
criteria described above, 15 bands were attributed to mutation, 6 to genetic
drift, 4 to selection, 9 to a combination of selection and genetic drift, and 11
to two genetic drift events. Seven other bands were attributed to co-migration
or unexplained factors. Mutation, genetic drift, and selection accounted for
16%, 59%, and 25%, respectively, of the line-specific bands found in Darden
and Marks' five quail lines.

Marker-Assisted Selection

There is considerable interest in linkages between genetic markers and loci
controlling quantitative traits (QTLs). The advantages of using DNA fingerprints in marker-assisted selection are the large number of loci detected by
various probes and their high level of polymorphism. Development and availability of polymorphic DNA markers will enhance mapping and selection
programs, but linkage between QTLs and DNA fingerprint bands can be
exploited independent of a linkage map. Several methods are presented here
that explore this possibility.

As such a linkage search should be made on a large sibship, it is necessary
to pool DNA fingerprint information collected from many gels. The analysis
should be based on association between the presence or absence of each
parental specific DNA fingerprint band (VNTR allele) and the degree to which
any particular quantitative trait is manifested. Bands common to both parents
are omitted from this analysis. The laboratory part of such an experiment and
the statistical analysis of each parental band in each progeny in the analyzed
sibship is labor-intensive and costly. Furthermore, the establishment of a large
full-sib family is not always feasible, and the statistical tests are not sufficiently sensitive to distinguish between false and real association of a band
and a trait due to the large number of bands tested simultaneously. Instead,
it was suggested (Hillel et al., 1989a; Plotsky et al., 1990) that the analysis
should be based on DNA mixes of the two groups having average extreme
performance and of groups along the trait distribution of the analyzed sibship.
When there is no linkage between the QTL and any of the DNA fingerprint
bands, the pattern obtained from mixes of individuals for the different groups
should be identical. However, if there is an association between a DFP band
and the QTL, there should be significant differences in band intensity among
the groups at different locations in the phenotypic distribution. As a

preliminary check for this type of analysis, the mix from the group at each extreme of the distribution can be compared. Any difference in band presence or band intensity reflecting variation in band frequency between the two extreme groups would be attributable to a possible linkage between this band and the relevant QTL.

As a verification of the tests described above and for the purpose of estimation of band associations, a multiple regression analysis can be conducted on a random sample from the analyzed sibship, in which the dependent variable is the value of the trait of interest and the independent variables are the presence or absence of the DNA fingerprint bands. A significant difference in the value of the quantitative trait between the group that has the DNA fingerprint band and the group that does not have it corroborates the association of the QTL with the DNA fingerprint band. The next logical step in this series of analyses is production of the subsequent generation, which can be tested for breeding value based on the presence or absence of the DNA fingerprint band.

This approach was tested on abdominal fat deposition in meat-type chickens (Hillel et al., 1989a; Plotsky et al., 1990). An F_1 sire from a cross between a low-fat (LF) and a high-fat (HF) line (Cahaner, 1988) was mated to LF females. Sixty-seven progeny were divided into seven similarly sized groups ranked from the lowest to the highest abdominal fat percentage. Mixes of DNA from all individuals within a group were used to generate representative DNA fingerprint patterns of each group. Comparison of the two extreme groups showed a clear difference between the DNA fingerprint patterns. Among the sire-specific bands, one band (missing in all dams) was highly intense in the lowest fat percentage group and relatively faint in the highest fat percentage group. The relationship between band frequency and average performance of the above-mentioned seven groups was tested by correlating frequency of the DNA fingerprint band with mean fat percentage in these groups. The frequency of individuals possessing the band in each group was negatively correlated ($r = -0.93, P < 0.05$) with group means of percentage fat. The band frequency in the fattest group was 0.11, whereas that of the leanest group was 1.0. These findings supported the conclusion that the DNA fingerprint band and a locus or loci affecting fat percentage were linked in this paternal family.

Multiple regression analysis was performed in which fat percentage was used as the dependent variable and presence/absence of the DNA fingerprint band was used as the independent variable. Sex, dam, and hatch were included as covariates in this analysis. Association of the DNA fingerprint band with fat percentage was significant ($P = 0.002$). The least-squares estimate for the band effect on fat percentage was -0.39%. The adjusted mean fat percentage of the entire paternal sibship was 1.02% (1.02 g of abdominal fat per 100 g of body weight). Further verification of the association between fat percentages and the DNA fingerprint band was obtained by analysis of progeny from 12 male half-sibs of the 67 individuals used to establish the relationship between the DNA fingerprint band and fat percentage. Seven of these males possessed the DNA fingerprint band and five did not. Fat percentage of each of 358 offspring from random mating of these 12 males was the dependent variable in a covariance analysis, while the independent variable was the presence or absence of the band in the sire of each offspring. Fat percentage was negatively associated ($P = <0.002$) with the DNA fingerprint band of the sires. This example demonstrates an association between a DNA marker and an economically important QTL. The large number of polymorphic loci detected by various DNA fingerprint probes combined with the methodology reported above make detection of genetic linkage between DNA fingerprint bands and QTLs feasible without the existence of a fully saturated linkage map.

The Genomic Selection Approach

The conventional approach for introgression of a single gene trait from an undesirable donor line possessing the desired trait to a commercial recipient line requires a cross between the two lines and numerous subsequent backcross generations. The main purpose of the genomic selection methodology is reduction of the number of required backcross generations. Using conventional backcross breeding schemes, the expected proportions of the recipient genome are increased by 50% in each successive generation. However, these figures are averages, and the actual proportion of the genome in individual progeny after one backcross ranges between 0.5 and 1.0.

Because the breeder is searching for an individual that contains the trait introduced from the donor line but has a genome with maximum similarity to that of the commercial recipient, individuals with maximum similarity to the DNA fingerprint pattern of the recipient grandparent or minimum similarity to the donor grandparent should be selected (Hillel et al., 1990b). This approach assumes that each band represents a segment of the genome, hence the entire DNA fingerprint represents a large proportion of the genome. Based on linkage analyses of DNA fingerprint bands for full-sib families, there are strong indications that this assumption is quite likely to hold in chickens (Hillel et al., 1989b; Haberfeld et al., 1991).

Response to genomic selection using DNA fingerprint information depends on the extent of variability of the recipient genome proportion of the generation for which selection is conducted. The genomic selection is more efficient when this variance is large. Based on theoretical analysis (Hillel et al., 1990b), the variability is inversely proportional to the segregating backcross generation and the mean number of chromosomal segments (N), where a chromosome segment is defined as a chromosome, or part of a chromosome bounded by two crossing-over sites, or a telomere and a crossing-over site. Consequently, genomic selection is more efficient for smaller values of N

and in earlier backcross generations. The large N in poultry requires application of several multilocus probes (at least two) to increase ability to recognize the levels of genomic similarity between backcross progeny and the recipient parents. Genomic selection is more efficient when matings are planned according to the DNA fingerprint patterns of mates, to ensure a large number of heterozygous DNA fingerprint loci in the backcross generations. Therefore, parents should be chosen with the intent of maximizing the total number of unshared DNA fingerprint bands between them, in addition to phenotypic selection for the desired trait. Genomic selection is optimal when the F_1 is produced by a single mating pair, as the selection criterion will be based on distinguishing between genomes of two individuals only. As the number of progeny from a single dam is limited in chickens, several dams from the recipient line are needed to produce a large half-sib BC(1) generation. The DNA fingerprint pattern of these dams should be of maximum similarity, so that several half-sib families will approach the structure of one large full-sib family. Theoretical calculations (Hillel et al., 1990b) have shown that applying genomic selection in only the BC(1) and BC(2) generations is an efficient method for recovering most of the recurrent line genome. More specifically, applying moderate selection intensities (say, 10%) in both generations results in recovery of 99.9% of the recipient genome. Preliminary data from the BC(1) generation in an experiment designed to introduce the naked neck gene from a layer strain to a meat-type commercial sire line support the genomic selection approach in principle (A. Cahaner, A. Yankevich, and J. Hillel, unpublished results).

Heterosis

Intensive efforts are currently underway worldwide to identify DNA fingerprint bands linked to loci affecting economically important quantitative traits in plants and animals. These DNA fragments are intended for use as locus-specific probes in marker-assisted selection programs to utilize the additive genetic effects of traits with moderate to high heritabilities. In poultry, however, there are numerous performance traits that are influenced by nonadditive genetic effects, and commercial egg and meat poultry breeding programs take advantage of heterosis. It is well established that maximum heterosis is expected for traits where there is large nonadditive genetic variation and where crosses involve genetically distant lines. Consequently, it may be possible to improve performance if individuals could be selected on the basis of genetic distance between them, using DNA fingerprints. Recent reports show that the level of similarity between DNA fingerprints (band sharing) of two individuals reflects the genetic similarity between them (Kuhnlein et al., 1990; Reeve et al., 1990), and that band sharing between DNA fingerprints of individuals from two populations can be used as a reliable estimate for genetic distance between them (Kuhnlein et al., 1989). Dunnington et al. (1991) have shown that band sharing between DNA fingerprint patterns of DNA mixes reflects faithfully the genetic distance between lines. From the information and assumptions presented above, one can postulate that maximum heterosis may be expected when the band sharing is minimal between DNA fingerprints of the parental lines (prepared from pooled samples per line) or of the individual parents. Unlike the previous sections in this chapter, empirical evidence to validate this postulate is not available.

Concluding Remarks

During the past three decades, poultry geneticists and poultry breeders have explored the use of genetic markers for evaluating the genotype of an individual and its breeding value (Gelderman, 1975; Soller and Beckmann, 1986). Unfortunately, success has been limited. The discovery of DNA fingerprints by Jeffreys in 1985 contributed significantly to increased availability of highly polymorphic molecular genetic markers. As a result, a wide range of possible applications has been described. There is little doubt that, at present, marker-assisted selection is the most attractive and most promising application (Edwards et al., 1987; Paterson et al., 1988; Lande and Thompson, 1990; Hillel et al., 1990b; Plotsky et al., 1990). Nevertheless, better understanding of the causal factors of genetic differences between selected lines from common origin may provide a basis for using marker-assisted selection at the population level rather than on segregating families. Additionally, use of multilocus DNA markers to evaluate the relative significance of factors that determine speciation could contribute to our understanding of some controversial evolutionary theories. Poultry breeders are expected to benefit from the use of polymorphic DNA markers in the identification of their genetic stocks and possibly in their selection programs. Moreover, potential exists for improvement of gene introgression programs by decreasing the required number of backcross generations and for possible control and prediction of hybrid vigor.

References

Ali, S., Muller, C. R., and Epplen, L. T. (1986). DNA fingerprinting by oligonucleotide probes specific for simple repeats. *Human Genet.*, 74, 239–243.

Burke, T., and Bruford, M. W. (1987). DNA fingerprinting in birds. *Nature*, 327, 149–152.

Cahaner, A. (1988). In: *Leanness in Domestic Birds*, edited by B. Leclercq, and C. C. Whitehead. Butterworths, London, pp. 71–86.

Dallas, J. F. (1988). Detection of DNA "fingerprints" of cultivated rice by hybridization with a human minisatellite DNA probe. *Proc. Natl. Acad. Sci. USA,* 85, 6831–6835.

Darden, J. R., and Marks, H. L. (1988). Divergent selection for growth in Japanese quail under split and complete nutritional environments. 1. Genetic and correlated responses to selection. *Poultry Sci.,* 67, 519–529.

Dunnington, E. A., Gal, O., Plotsky, Y., Haberfeld, A., Kirk, T., Goldberg, A., Lavi, U., Cahaner, A., Siegel, P. B., and Hillel, J. (1990). DNA fingerprints of chickens selected for high and low body weight for 31 generations. *Anim. Genet.,* 21, 247–257.

Dunnington, E. A., Gal, O., Siegel, P. B., Haberfeld, A., Cahaner, A., Lavi, U., Plotsky, Y., and Hillel, J. (1991). Deoxyribonucleic acid fingerprint comparisons between selected populations of chickens. *Poultry Sci.,* 70, 463–467.

Dunnington, E. A., Siegel, P. B., Cherry, J. A., Jones, D. E., and Zelenka, D. J. (1986). Physiological traits in adult female chickens after selection and relaxation of selection for 8-week body weight. *J. Anim. Breed. Genet.,* 103, 51–58.

Edwards, M. D., Stuber, C. W., and Wendel, J. F. (1987). Molecular-marker-facilitated investigations of quantitative-trait loci in maize. I. Numbers, genomic distribution and types of gene action. *Genetics,* 116, 113–125.

Epplen, J. T. (1988). On simple repeated GAT/CA sequences in animal genomes: A critical reappraisal. *J. Hered.,* 79, 409–417.

Gelderman, H. G. (1975). Investigations on inheritance of quantitative characters in animals by gene markers. I. Methods. *Theor. Appl. Genet.,* 46, 319–330.

Gilbert, D. A., Lehman, N., O'Brien, S. J., and Wayne, R. K. (1990). Genetic fingerprinting reflects population differentiation in the California Channel Island fox. *Nature,* 344, 764–767

Haberfeld, A., and Hillel, J. (1991). Development of DNA fingerprint probes: An approach and its application. *Anim. Biotechnol.,* 2, 61–73.

Haberfeld, A., Cahaner, A., Yoffe, O., Plotsky, Y., and Hillel, J. (1991). DNA fingerprints of farm animals generated by microsatellite and minisatellite DNA probes. *Anim. Genet.,* 22, 299–305.

Hillel, J., Avner, R., Baxter-Jones, C., Dunnington, E. A., Cahaner, A., and Siegel, P. B. (1990a). DNA fingerprints from blood mixes in chickens and turkeys. *Anim. Biotechnol.,* 1, 201–204.

Hillel, J., Gal, O., Schaap, T., Haberfeld, A., Plotsky, Y., Marks, H., Siegel, P. B., Dunnington, E. A., and Cahaner, A. (1991a). Genetic factors accountable for line-specific DNA fingerprint bands in quail. In: *DNA Fingerprinting: Approaches and Applications,* edited by T. Burke, G. Dolf, A. Jeffreys, and R. Wolff, Birkhauser Verlag AG, Basel, Switzerland, pp. 263–273.

Hillel, J., Gal, O., Siegel, P. B., and Dunnington, E. A. (1991b). Line-specific DNA fingerprint bands in lines of chickens selected for high or low antibody response to sheep erythrocytes. *Arch. fur Geflugelk.,* 55, 189–191.

Hillel, J., Plotsky, Y., Gal, O., Haberfeld, A., Lavi, U., Dunnington, E. A., Siegel, P. B., Jeffreys, A. J., and Cahaner, A. (1989a). DNA fingerprinting in chickens. 31st British Poultry Breeders Roundtable Conference, 20–22 September 1989, Reading, 1–8.

Hillel, J., Plotsky, Y., Haberfeld, A., Lavi, U. Cahaner, A., and Jeffreys, A. J. (1989b). DNA fingerprints of poultry. *Anim. Genet.,* 20, 145–155.

Hillel, J., Schaap, T., Haberfeld, A., Jeffreys, A. J., Plotsky, Y., Cahaner, A., and Lavi, U. (1990b). DNA fingerprints applied to gene introgression in breeding programs. *Genetics,* 124, 783–789.

Jeffreys, A. J., and Morton, D. B. (1987). DNA fingerprints of dogs and cats. *Anim. Genet.,* 18, 1–15.

Jeffreys, A. J., Hillel, J., Hartley, N., Bulfield, G., Morton, D., Wilson, V., Wong, Z., and Harris, S. (1987a). Hypervariable DNA and genetic fingerprints. *Anim. Genet.,* 18, 141–142.

Jeffreys, A. J., Wilson, V., Kelly, R., Taylor, B. A., and Bulfield, G. (1987b). Mouse DNA "fingerprints": Analysis of chromosome localization and germ-line stability of hypervariable loci in recombinant inbred strains. *Nucleic Acids Res.*, 15, 2823–2836.

Jeffreys, A. J., Wilson, V., and Thein, S. L. (1985a). Hypervariable "minisatellite" regions in human DNA. *Nature*, 314, 67–73.

Jeffreys, A. J., Wilson, V., and Thein, S. L. (1985b). Individual-specific "fingerprints" of human DNA. *Nature*, 316, 76–79.

Jeffreys, A. J., Wilson, V., Thein, S. L., Weatherall, D. J., and Ponder, B. A. J. (1986). DNA "fingerprints" and segregation analysis of multiple markers in human pedigrees. *Am. J. Human Genet.*, 39, 11–24.

Kuhnlein, U., Dawe, Y., Zadworny, D., and Gavora, J. S. (1989). DNA fingerprinting: A tool for determining genetic distances between strains of poultry. *Theor. Appl. Genet.*, 77, 669–672.

Kuhnlein, U., Zadworny, D., Dawe, Y., Fairfull, R. W., and Gavora, J. S. (1990). Assessment of inbreeding by DNA fingerprinting: Development of a calibration curve using defined strains of chickens. *Genetics*, 125, 161–165.

Lande, R., and Thompson, R. (1990). Efficiency of Marker-Assisted Selection in the improvement of quantitative traits. *Genetics*, 124, 743–756.

Martin, A. G., Dunnington, E. A., Gross, W. B., Briles, W. E., Briles, R. W., and Siegal, P. B. (1990). Production traits and alloantigen systems in lines of chickens selected for high or low antibody response to sheep erythrocytes. *Poultry Sci.*, 69, 871–878.

Nakamura, Y., Leppert, M., O'Connell, P., Wolff, R., Holm, T., Culver, M., Martin, C., Fujimoto, E., Hoff, M., Kumlin, E., and White, R. (1987). Variable number of tandem repeat (VNTR) markers for human gene mapping. *Science*, 235, 1616–1622.

Paterson, A. H., Lander, E. S., Hewitt, J. D., Peterson, S., Lincoln, S. E., and Tanksley, S. D. (1988). Resolution of quantitative traits into Mendelian factors by using a complete linkage map of restriction fragment length polymorphisms. *Nature*, 335, 721–726.

Plotsky, Y., Cahaner, A., Haberfeld, A., Lavi, U., and Hillel, J. (1990). Analysis of genetic association between DNA fingerprint bands and quantitative traits by DNA mixes. *Proc. IV World Congress on Genetics Applied to Livestock Production*, 13, 133–136.

Reeve, H. K., Westneat, D. F., Noon, W. A., Sherman, P. W., and Aquadro, C. G. (1990). DNA "fingerprinting" reveals high levels of inbreeding in colonies of the eusocial naked mole-rat. *Proc. Natl. Acad. Sci. USA*, 87, 2496–2500.

Schafer, R., Zischler, H., and Epplen, J. T. (1988). (CAC)$_5$, a very informative probe for DNA fingerprinting. *Nucleic Acids Res.*, 16, 5196.

Siegel, P. B., and Gross, W. B. (1980). Production and persistence of antibodies in chickens selected for sheep erythrocytes. *Poultry Sci.*, 59, 1–5.

Soller, M., and Beckmann, J. S. (1986). Restriction fragment length polymorphisms in poultry breeding. *Poultry Sci.*, 65, 1474–1483.

Vassart, G., Georges, M., Monsieur, E. A., Brocas, H., Lequarre, A. S., and Christophe, D. (1987). A sequence in M13 phage detects hypervariable minisatellites in human and animal DNA. *Science*, 235, 683–684.

Walmsley, R. M., Wikinson, B. M., and Kong, T. H. (1989). Genetic fingerprinting for yeasts. *Bio/Technology*, 7, 1168–1170.

Wetton, J. H., Carter, R. E., Parkin, D. T., and Walters, D. (1987). Demographic study of a wild house sparrow population by DNA fingerprinting. *Nature*, 327, 147–149.

Chapter 17

Sex-Specific DNA Sequences in Galliformes and Their Application to the Study of Sex Differentiation

Shigeki Mizuno, Yasushi Saitoh, Osamu Nomura, Ryota Kunita, Kohei Ohtomo, Katsuhiko Nishimori, Hiroyuki Ono, and Hisato Saitoh

Summary. A substantial fraction of DNA in W chromosomes of chickens, turkeys, and pheasants consists of repetitive families that are unique to this female-specific sex chromosome. Repeating units of these families share common sequence organization, and they all behave as strongly bent DNA in solution. These units are useful probes in sexing of early embryos by DNA/DNA hybridization. Application of this technique to the study of gene expression concerning sex steroid hormone synthesis during sex differentiation of the chicken is described. In addition, a cDNA clone derived from a gene on the Z chromosome and expressed in the chicken ovary is described. Finally, future research involving these sex chromosome-specific DNA sequences is discussed.

Introduction

Sex of most birds is female heterogametic; i.e., their sex chromosomes are ZZ for the male and ZW for the female. Thus, the genetic sex of a bird is determined by an ovum, the sex chromosome content of which is determined after the first meiotic division in the female meiosis. The W chromosome of the chicken is highly heterochromatic, as revealed by the Giemsa-C banding method (Stefos and Arrighi, 1974; Mayr and Auer, 1988), and it forms a conspicuous heterochromatic body in a somatic nucleus (Mizuno, 1991; see Figure 3). These observations have suggested that the DNA in the W chromosome should consist largely of repetitive sequences.

Detection of W Chromosome-Specific Repetitive DNA Sequences in Chickens

In our early attempts to isolate W chromosome-specific repetitive sequences, two results suggested that such sequences should actually be present. First, ^{3}H-labeled total repetitive sequences from genomic DNA of a female chicken were hybridized exhaustively with mercurated, sheared DNA from a male, and unhybridized, labeled DNA fragments were recovered as an unbound fraction from an SH-Sepharose column. The DNA in this fraction was subjected to hydroxyapatite column chromatography and a single-stranded DNA

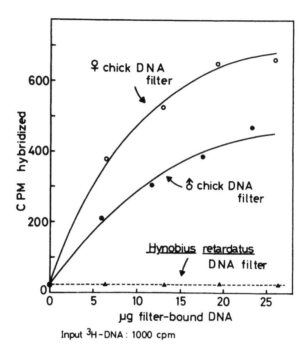

FIGURE 1. Hybridization of ³H-HSU-DNA to the filter-bound DNA from a female or male chicken or from the salamander, *Hynobius retardatus*, as a control.

fraction was obtained. When the labeled DNA in the final fraction (³H-HSU-DNA) was hybridized to filter-bound DNA from the male or female, a significantly higher level of hybridization was attained with the DNA from the female (Tone et al., 1982) (Figure 1). Second, when ³H-labeled total repetitive DNA sequences from a female chicken were hybridized *in situ* to a squash preparation from the female embryo, they hybridized intensely to a relatively small chromosome, interpreted as the W chromosome, and to one end of a relatively large chromosome, interpreted as the Z chromosome (Tone et al., 1982) (Figure 2).

Using various restriction enzymes, we then compared digestion patterns of genomic DNA preparations from the male and female chicken, and found that *Xho*I produced two major female-specific bands of about 0.7 and 1.1 kb (Tone et al., 1982). DNA fragments in these two bands were cloned and characterized further, which revealed that they are major repeating units of the W chromosome-specific repetitive sequences, designated as the *Xho*I family, and that the sequences of these two repeating units are very similar (Tone et al., 1984). *In situ* hybridization with the ³H-labeled, cloned *Xho*I 0.7-kb unit demonstrated that the *Xho*I family sequences are confined to a major heterochromatic body in the nucleus and to a relatively small

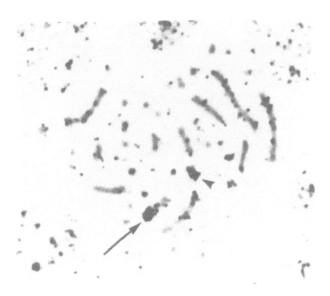

FIGURE 2. *In situ* hybridization of ³H-labeled repetitive DNA [a double-stranded fraction remaining after reassociation of total, sheared, denatured genomic DNA to Cot 2 and removal of the zero-time binding fraction that contains intrastrand duplex structures, using a hydroxyapatite column (Graham et al., 1974)] from the female chicken to a squash preparation of a 72-h female embryo. Hybridization to putative W and Z chromosomes is indicated by an arrowhead and an arrow, respectively.

chromosome in the metaphase set from a female chicken embryo (Figure 3). The same probe did not show any significant hybridization to nuclei and chromosome sets from the male embryo.

*Xho*I Family-Related Repetitive Families in W Chromosomes of Galliformes

Using the cloned *Xho*I 0.7-kb unit as a probe and applying low-stringency hybridization conditions, we found three other *Xho*I family-related repetitive families, namely, the *Eco*RI family in chickens, the *Pst*I family in turkeys, and the *Taq*I family in pheasants. Some properties of these families are listed in Table 1. Although overall sequence similarities of the repeating units of the *Eco*RI, *Pst*I, and *Taq*I families to the 0.7-kb unit of the *Xho*I family are only 68%, 63%, and 57%, respectively, they possess common sequence organization; i.e., they all consist of tandem repeats of basic units containing 21 bp, on average (Figure 4). In most of these basic units, $(A)_{3-5}$ and $(T)_{3-5}$ stretches are present. As these two stretches are separated by six or seven nucleotides, the A and the T stretches appear alternately at nearly every pitch of each DNA helix. As expected from this regularity (Koo et al., 1986;

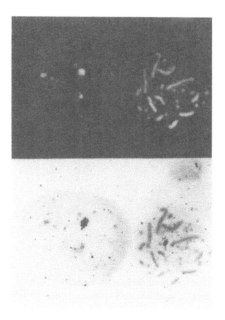

FIGURE 3. *In situ* hybridization of ³H-labeled *Xho*I 0.7-kb unit (from clone pUGD0603) to a nucleus and a set of metaphase chromosomes from a 72-h female chicken embryo: DAPI (4′, 6-diamidino-2-phenylindole)-staining (upper panel) and autoradiogram (lower panel).

Ulanovsky and Trifonov, 1987), all of these repeating units behave as strongly bent molecules in solution (Kodama et al., 1987; Saitoh et al., 1989, 1991). The cloned repeating units of the four families were shown to be repeated 700 to 14,000 times in the female genome, but they are undetectable as such units in the male genomes. The female-genome specificities of *Xho*I, *Eco*RI, and *Pst*I family sequences are so high (Figure 5) that they can be utilized as DNA probes to identify the female sex in dot blot and slot blot hybridization systems. On the other hand, substantial numbers of related sequences are present in the male genome of pheasant for the *Taq*I 0.5-kb unit (Table 1), thus Southern blot hybridization had to be employed to identify the female sex of pheasant using this probe (Figure 5). It was noted that copy numbers of the *Eco*RI 1.2-kb unit are particularly variable among the female chicken population. There seem to be two classes of repetition frequency, a low-frequency class consisting of approximately 700 copies and a high-frequency class consisting of approximately 4000 copies per diploid genome (Saitoh et al., 1991).

Table 1
W Chromosome-Specific Repetitive DNA Families in Galliformes

Species	Repetitive family	Cloned repeating unit (kb)	Average size of internal repeats (bp)	DNA bending	Sequence similarity (%)	Copies per diploid genome			
						Female	(Closely related sequences)	Male[a]	(Closely related sequences)[b]
Gallus. g. domesticus (chicken)	XhoI	0.7	21	Bent	100	14,000	(30,000 as 0.7 kb)	UD	(900 as 0.7 kb)
		1.1	21	Bent		6,000		UD	
	EcoRI	1.2	21	Bent	68	700–4,000	(4,000–9,000)	UD	(900)
Meleagris gallopavo (turkey)	PstI	0.4	21	Bent	63	10,000	(32,000)	UD	(600)
Phasianus versicolor (pheasant)	TaqI	0.5	21	Bent	57	4,000	(20,000)	UD	(8,000)

[a] UD = undetectable.

[b] Detected by DNA dot-blot hybridization. Although restriction fragments like those from the female were not produced from the male, total amount of related sequences was expressed as copy numbers of the female-specific repeating unit.

XhoI family

No.	Sequence	Nucleotides
1	GAAATACCACTTTTCTCTC	19
2	GAAAATCATGCATTTTCATCC	21
3	AAAAATACCACCTGTCTCCC	20
4	AAAAATTCTGCACTTCCTTCCC	22
5	GAAAATACCACTTTTCCCTGG	21
6	GAAATAACACATTTCTACCC	20
7	CAAATATAACACGCTTCACTCA	22
8	CAAAGCACGCATTTTCACCCC	21
9	GAAAGTACCACCTTTCAGCC	20
10	GAAAATTACGCTTTTTCCTCCA	22
11	GAAAATACCACTTCTCAAACA	21
12	GAAATATCACGTTTCGCCAA	20
13	GAAAATAGCACCATTCACCC	20
14	CAAAATCACGCGCGTTTTCTCTCCA	23
15	GAACTACCACTTTTCTCAC	19
16	GGAAATCACACATTTTCTTCCC	22
17	GAAAGTACCACCTTGCACAC	20
18	GAAAATCACGCATTTTCTGCGC	22
19	GAAACAACCCCATTTCACCC	20
20	CAAAATCAGTCTTTTTCTTCCG	22
21	GAAAATACAACTTTTCTAAC	20
22	GAAATCCATGCGTTATCACTCT	22
23	GAAAATCACGTTTTTGCCC	20
24	GAAAATCACGCATTTTCCCTTC	22
25	GTAAATTCCCCATGTCGCCA	20
26	GAAAATAATTCATTTCCTTACC	22
27	GTAAATGCCCCTTTTCACC	19
28	CAAAAATCACGCATTTCCCCCCG	23
29	GAAAAACGCACCTCTGTCC	19
30	AAACATCACGCATTTTCTACCC	22
31	GAAAATACCACTTTTGGCTGG	21
32	GAAATAACACATTTCTCCCC	20
33	CAAATATACCACCTTTCACCC	21
34	CAAAATCACGCATTTTCTCTC	21

Consensus	GAAAATACCACNTTTTCTCCC	21
	A	

EcoRI family

No.	Sequence	Nucleotides
1	AATTCTCTTC	10
2	GAAAAAAACGCTTTTGCCT	19
3	GAAATCACGCGTTTTCTACCC	22
4	AAAAATACCAATATTTTGCCCG	22
5	GAAATCACGCATTTACACCTC	21
6	AAAAACGTCACTTTTCGCCC	20
7	AAAAAGCAAGCATTTTCTCTTA	22
8	GAAAATTTTACCCCC	15
9	AAAACACGCATTTTTACCAT	20
10	AAAAATACCATTTTTCGCACGA	22
11	GAATCATGTATGTTCAGTT	19
12	CAAAACTACTGCATTTCACCCC	22
13	AAAATCACGCTTTTTCCCTCCA	22
14	GAAATGCAACTTTTTAAATA	20
15	GAAATCACGCGTTTCCTCCAC	21
16	GAAAATACCGCCTTTCACCCC	21
17	AAAATCACGCATTTTCTCTCC	21
18	AAAAATACCACCTTTCTCTC	20
19	GAAAAGCACGCATTTCCACACG	22
20	GAAAATACCACTTTTCTCCC	20
21	GAAAATTCCCACATTTTCTTCTC	23
22	GAAAATACCACTTCTCACCG	20
23	GAAAATAACGCCTTTCTACCGC	22
24	AAATACCACCACGTTTCACCCC	21
25	AAAATCACGCATTTTCACCCC	21
26	GAAAATACCGCCTTTCACCCC	21
27	AAAATCACGCATTTTCTTCCC	21
28	GAAAATACCCACCTTTCAAAG	20
29	GAAAATCACTTCATTTTCTCCCC	23
30	GAAAATACCACCTTTCCCCC	20
31	GAAAATCACGCATTTTCTGCGC	22
32	GAAGCAACCCCATTTCTCCCAA	22
33	GAACCACGCATTTTTTTCAC	20
34	GAAAATTGTACGTTCAAAG	19
35	GAAAATCGCACATTTTCTCCCT	22
36	GAACATACCACCTTTCACCC	20
37	AAAAATTCCGCGTTTTCCGCGT	22
38	GAAGGAACCCCGCTTCACAC	20
39	GAATATCACGCATTGTCTACCG	22
40	GAAAATACCACTTTTCTAAC	20
41	GAAAATCACGCACTGTCCGGCC	22
42	GAAAACACAAGTTTTCGCCC	20
43	GAAAATCACGCATTTTCCCCCA	22
44	GAAAAAAAGCACCTCTGTCCA	21
45	CAAGTCACGCATCTTGTACCC	21
46	GAAAATACCACTTTTCTCCT	20
47	GAAAATCCCGCATTTTCACCCCA	23
48	CAAATGCCACTTTTCGCGC	19
49	GAAAAGCATTTCCTTCGG	18
50	GAAAATCACCCTCTTCAACC	20
51	GAAAAACACGTGTTTTCTCCTT	22
52	GAAAATACCATCTTTCGCCT	20
53	GAGAATCACGCATGTTCACTTC	22
54	GAAAATACCCCATTTCCCCA	20
55	GAAAATCTCGCACTGTCTCCTC	22
56	GAAAATACCCTGAGCGCAGGTC	22
57	CGCGCATGCGCAGTGCGATCG	21
58	CAAATCCACGCATACGCGGTG	21
59	CTCCCCCTGCGAG	13

Consensus	GAAAATACCNTTTCNCCCC	21
	CA	

*Pst*I family

No.	Sequence	Nucleotides
1	GTTTCCTCCC	10
2	ACAAATACCATTTTTTCAACC	21
3	AGAAATAGGACGTTTTTCTCCC	22
4	AGAAATACCGGATTTTTGCCCC	22
5	CAAAACATGACATTTTCTCCC	21
6	AGAAATACGAGTTTTCTCCC	20
7	AAAATATGATATTTTGCACC	20
8	AGAAATTCCAGTTTTATCACC	21
9	GAAGACTCTACGTTTTCTACC	21
10	AGAAATACCAATTATCTCC	19
11	GCAAAAATTACATTTTCTCC	20
12	AGAAATACCAGATTTCTTCCC	21
13	TTAAATATGACACCTTTTCC	20
14	AAGAAATAGTAGATTTTTCCCC	22
15	AAAAAATATGACATTTTCTCC	20
16	AGGAAATGCCAGTTTTATCGT	21
17	ATAAATATGACATTTTATACC	21
18	GCAAATATCCGCTTTCTCCC	20
19	AAAAAATATGCCATTTTCTGCC	21
20	AGGAACTGCA	10
Consensus	AGAAATATGNCATTTTCTCCC C	21

*Taq*I family

No.	Sequence	Nucleotides
	C	
1	GACAAAATACCACCATTCTCCC	22
2	ACAGAGATGGCATTTCATCCC	21
3	ACAAGTACTACTTCACACTCC	21
4	ACACGATGATACTTTCCATC	20
5	AAGAATAGGGCATTGGACCAC	21
6	AGAAATACCAGCTTTCTGCCT	21
7	AAGAGATGACATTTTCTCCC	20
8	AGAAATACCACTTTTCTCCC	20
9	AGAAATACAGGAACTTTTCTGC	22
10	CAGAAACACCATGCTCATCCTC	22
11	TCCAGATGTTGTTTTCACCCC	21
12	AAACACTAGGACCTTTCCTC	20
13	TCACTACTCCCTGCTTTTTC	20
14	AAAACTAGATGGTCTTCTCTCCC	23
15	AGAAATACTAGCATTCTCTGC	21
16	AAGATGGGACCTTTTCCAA	19
17	CCAAAGATGGTAGAGTCTCCC	21
18	AGAAATAGCACTTTTCTTC	19
19	ATCAGAACTGTCATTTTCTCC	21
20	AAAAAAAATACTACTTTCAACT	21
21	GTTCCAGGTGACATTTGCAGCC	22
22	AAACATAGGAGAGATTCTCC	20
23	ACAAAGTAACCACTTTTCTCTC ACGT	22
Consensus	AGAAATANNNNATTTTCTCCC C	21

FIGURE 4. Nucleotide sequences of cloned repeating units: *Xho*I 0.7-kb (chicken), *Eco*RI 1.2-kb (chicken), *Pst*I 0.4-kb (turkey) and *Taq*I 0.5-kb (pheasant). Each sequence is arranged to show internal repeats of average 21 bp. A and T clusters are indicated with bold letters, and CpG dinucleotides are underlined.

A Z Chromosome-Linked DNA Sequence of Chicken

It is desirable to have specific DNA markers for the Z chromosome in order to study DNA sequences and genes in this relatively large sex chromosome. In the course of experiments described below, we found that one clone from a cDNA library of the left ovary of the 1-day-old chicken seemed most likely to be derived from a gene on the Z chromosome.

A cDNA library from the left ovary was subjected to differential hybridization with cDNA probes from the left ovary and liver of the 1-day-old chicken and from the 6-day whole embryo. Forty-nine clones, which seemed to be expressed relatively abundantly in the ovary, were selected. Those clones were then classified into 17 independent groups by interclone cross-hybridization (Figure 6). Tissue specificity of these 17 groups was further studied by RNA-dot blot hybridization using poly(A)$^+$ RNA preparations from various tissues. As summarized in Figure 6, the majority of clones (14 of 17) were expressed both in the left ovary and the testis of the 1-day-old chicken.

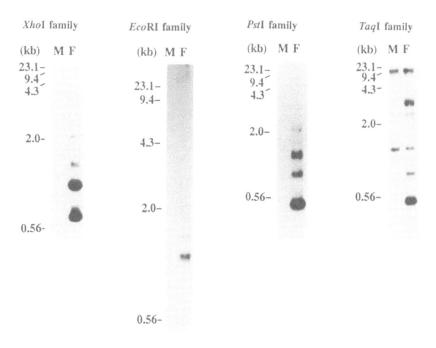

FIGURE 5. Southern blot hybridization of the ^{32}P-labeled repeating units to genomic DNA from male (M) or female (F) chickens (for the *Xho*I and *Eco*RI families), turkeys (for the *Pst*I family), or pheasants (for the *Taq*I family). Genomic DNA was digested with a restriction enzyme that could produce the repeating unit of each family.

Three clones seemed to be expressed only in the ovary. When the cDNA insert from one of these ovary-specific clones, pLON3901, was hybridized to dot blots of total genomic DNA from the male or female chicken, it was revealed that levels of hybridization to the DNA from the male were about twice as high as those to the DNA from the female (Figure 7). Similar results were obtained by Southern blot hybridization to restriction fragments of genomic DNA from the male or female (Figure 8). These results were confirmed using genomic DNA preparations from different breeds of chicken, thus it seemed very likely that the cDNA sequence of pLON3901 was derived from a gene in the Z chromosome. Further characterization of this clone with respect to its Z chromosome linkage and nucleotide sequence is in progress in the authors' laboratory.

Application of the W Chromosome-Specific DNA Probe in the Study of Sex Differentiation

The sex of an individual chicken embryo could be identified by hybridizing the *Xho*I family probe to DNA prepared from the extraembryonic membrane.

Mizuno et al. 265

Group	Clone	Size(b.p.)	OVARY	TESTIS	Ov	Li	Ki	Br	Mu	Bl	Em
1	pLOA0411	2150	+ +	−	100	ND	ND	ND	ND	ND	ND
	pLON2301	240	+	−							
	pLON2901	500	+ +	−							
	pLON3801	470	+ +	−							
	pLON4301	630	+ +	−							
2	pLOA0511	1850	+ +	+ +	100	ND	ND	ND	ND	ND	ND
	pLON0101	320	+ +	+ +							
	pLON0701	320	+ +	+ +							
	pLON3301	1100	+ +	+ +							
	pLON3701	760	+ +	+ +							
	pLON4001	300	+ +	+ +							
3	pLON0301	1450	+ +	+	100	<1	<1	ND	ND	ND	2
	pLON3101	700	+ +	+							
	pLON3601	1450	+ +	+							
	pLON4701	850	+ +	+							
4	pLON0601	320	+ +	+	100	<1	<1	ND	ND	ND	5
	pLON1601	250	+ +	+							
	pLON2501	470	+ +	+							
	pLON4101	800	+ +	+							
5	pLON0501	1050	+ +	+ +	100	5	18	<1	24	ND	7
	pLOB0902	750	+ +	+ +							
	pLON0801	1250	+ +	+ +							
	pLON1101	1450	+ +	+ +							
	pLON1201	800	+ +	+ +							
	pLON1401	470	+ +	+							
	pLON1701	380	+ +	+ +							
	pLON1801	460	+ +	+ +							
	pLON2101	1350	+ +	+ +							
	pLON2401	1350	+ +	+ +							
	pLON2801	1450	+ +	+ +							
	pLON4201	650	+ +	+ +							
	pLON4401	350	+	+							
6	pLOA0602	450	+ +	+ +	100	ND	12	3	9	ND	3
	pLON2001	670	+ +	+ +							
7	pLON0901	350	+	+	100	<1	<1	<1	<1	25	30
8	pLON2601	830	+ +	+ +	100	ND	34	<1	ND	ND	<1
9	pLON3901	2150	+ +	−	100	ND	ND	ND	ND	ND	ND
10	pLON3001	1210	+ +	+ +	100	ND	4	ND	<1	ND	<1
	pLON3201	1700	+ +	+ +							
	pLON4501	270	+	+							
	pLON4801	370	+ +	+							
11	pLON1001	1050	+ +	+	100	ND	ND	ND	ND	ND	ND
	pLON3401	450	+	−							
12	pLON1301	390	+	−	100	ND	ND	ND	ND	ND	ND
13	pLON2701	750	+	+	100	ND	ND	ND	ND	ND	ND
14	pLON0401	530	+ +	+	100	3	22	4	37	ND	20
15	pLON1901	550	+	+	100	ND	ND	ND	ND	ND	ND
16	pLON3501	630	+ +	+	100	ND	ND	ND	ND	ND	ND
17	pLON4601	950	+	+	100	ND	ND	ND	ND	ND	ND

FIGURE 6. List of selected clones from the cDNA library of the left ovary of a 1-day-old chicken. Intensities of colony hybridization with the ^{32}P-labeled cDNA probe prepared from the left ovary or testis are indicated with + or −. Results of RNA dot blot hybridization between poly(A)$^+$RNA from ovary (Ov), liver (Li), kidney (Ki), brain (Br), muscle (Mu), or blood (Bl) from a 1-day-old chicken or from a whole embryo incubated for 6 days (Em) and the ^{32}P-labeled cDNA insert of each clone are expressed as relative levels of hybridization. ND = not detected.

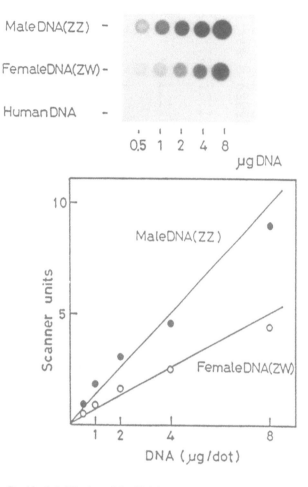

FIGURE 7. Dot blot hybridization of the ^{32}P-labeled cDNA insert of pLON3901 to genomic DNA of male or female chicken, or of human. Results of densitometric scanning are shown in the lower panel.

Examples are shown in Figure 9, in which the digoxigenin-labeled *Xho*I 0.7-kb probe was hybridized to slot blots of DNAs from 72-h embryos. Hybridized DNA was detected following reaction with alkaline phosphatase-conjugated anti-digoxigenin antibody. Positive signals meant that those embryos were females.

We applied this sexing method to the study of gene expression that is involved in sex steroid hormone synthesis in male and female chicken embryos. One cDNA clone, pLOA0511, which belonged to group 2 in Figure 6, was identified as a putative cDNA sequence for chicken P-450c17 (steroid 17α-hydroxylase/17,20 lyase) because of its sequence similarity to the bovine

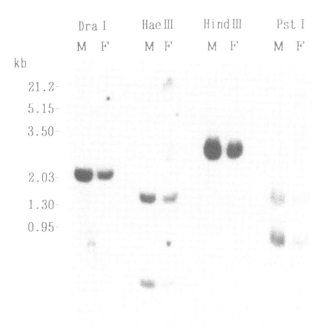

FIGURE 8. Southern blot hybridization of the ^{32}P-labeled cDNA insert of pLON3901 to the restriction fragments of genomic DNA from male (M) or female (F) chicken.

and human P-450c17 cDNA clones (Ono et al., 1988). Recently, we confirmed this identification by ligating the cDNA insert of pLOA0511 into a high-efficiency cDNA expression vector, pcDL-SRα296 (Takebe et al., 1988), and expressing it in COS-7 cells (K. Nishimori and S. Mizuno, unpublished results). When ^{14}C-pregnenolone was added to the culture for 24 to 72 h, it was converted to 17α-hydroxypregnenolone by the 17α-hydroxylase reaction and to dehydroepiandrosterone by the 17,20-lyase reaction. These products were not detected at zero time of incubation or when COS-7 cells were transfected with the vector alone. Similarly, another substrate, ^{14}C-progesterone, was converted to 17α-hydroxyprogesterone and androstenedione.

We were interested in examining the timing of expression of the gene for P-450c17 in male and female chicken embryos during their development, because P-450c17 is a key enzyme in the metabolic pathway from cholesterol to sex steroid hormones (testosterone and estradiol-17β) (Figure 10). In this investigation, we also directed our attention to the expression of the aromatase (P-450arom) gene, because aromatase is essential in the conversion of androstenedione or testosterone to estradiol-17β (Figure 10). Individual embryos

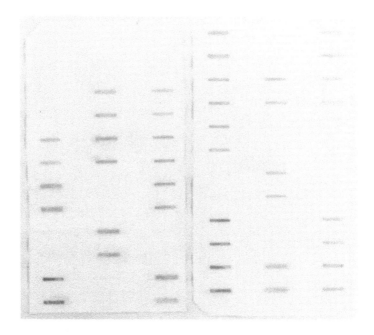

FIGURE 9.   Slot blot hybridization of the digoxigenin (Dig)-labeled *Xho*I 0.7-kb unit to DNA preparations from the extraembryonic membrane of individual 72-h chicken embryos. Female-specific signals were produced following reaction with alkaline phosphatase-conjugated sheep anti-Dig IgG. One-twentieth of the amount of DNA extracted from each extraembryonic membrane was applied to each slot. Those slots where DNAs from males had been placed gave little or no signal.

at day 2 to day 9 of incubation were stored frozen in a buffer containing 4 *M* guanidinium thiocyanate and 1% 2-mercaptoethanol until their sexes were determined using DNA from the extraembryonic membrane. Total RNA was then extracted from individual embryos at day 3 to day 9 or from a group of five male or female embryos at day 2. Total RNA was also extracted from gonads plus adrenals of individual embryos at day 11 and day 13. The first-strand cDNA was synthesized from the RNA by reverse transcriptase using oligo(dT)$_{12-18}$ as a primer, then the specific cDNA sequence was amplified by the polymerase chain reaction (PCR). Primers (20 nucleotides each) for PCR were selected and synthesized from the sequence of P-450c17 cDNA, which had been determined in the authors' laboratory (Ono et al., 1988), and from that of the chicken P-450arom published by McPhaul et al. (1988).

Figure 11 shows examples of the PCR analysis. PCR products of expected sizes, i.e., 465 bp for the P-450c17 cDNA and 412 bp for the P-450arom cDNA, were detected by ethidium bromide staining as shown in Figure 11 or by Southern blot hybridization with ^{32}P-labeled cDNA probes when

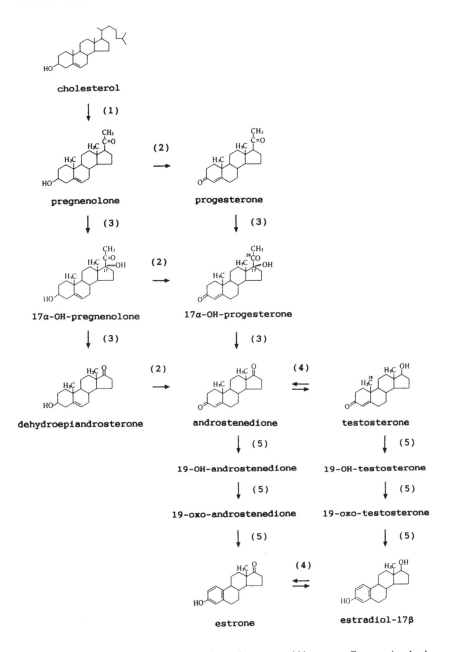

FIGURE 10. Metabolic pathway from cholesterol to sex steroid hormones. Enzymes involved are as follows: (1) cholesterol side-chain cleavage cytochrome P-450 (P-450scc); (2) 3β-hydroxysteroid dehydrogenase/$\Delta^{5\to4}$isomerase; (3) 17α-hydroxylase cytochrome P-450 (P-450c17); (4) 17β-hydroxysteroid dehydrogenase; and (5) aromatase cytochrome P-450 (P-450arom).

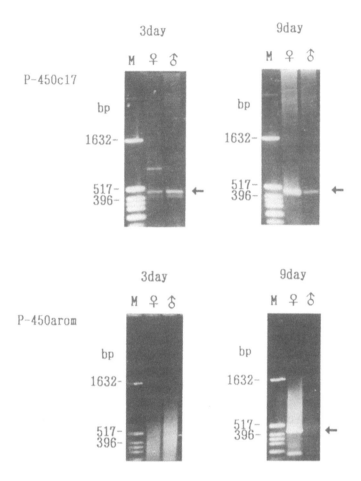

FIGURE 11. P-450c17 and P-450arom cDNA sequences amplified by PCR and detected by ethidium bromide staining. Typical results from total RNA preparations from day-3 and day-9 chicken embryos are shown. The PCR product of expected size for each cDNA is indicated by an arrow. M = molecular size markers.

ethidium bromide-stained bands were ambiguous. We concluded that these bands represented amplified cDNA sequences from the corresponding species of mRNA, because production of these bands was dependent on the reverse transcriptase reaction and the bands were undetectable when the total RNA was first digested with RNase A or genomic DNA was used instead of RNA. As shown in Figure 11, PCR-products of P-450c17 cDNA were detected in the day 3 and day 9 embryos of both sexes. PCR products of P-450arom cDNA were detected in the day-9 female embryo but not in the day-3 female embryo or in the day-3 or day-9 male embryos. We repeated similar experiments using at least 10 RNA samples for each sex at various days of

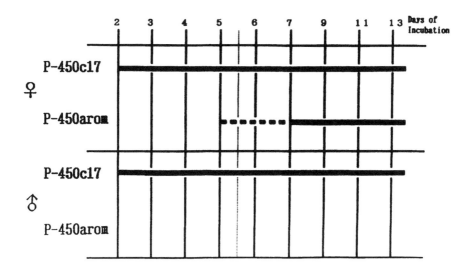

FIGURE 12. Detection of P-450c17 and P-450arom mRNAs in chicken embryos at different days of incubation. Amplified cDNA sequences by PCR were detected in all (thick horizontal bar) or some (thick discontinuous bar) of the embryos examined.

incubation (O. Nomura, K. Nishimori, and S. Mizuno, unpublished results). Our tentative conclusion from these and other experiments is illustrated in Figure 12. Transcription of the P-450c17 gene seems to be started at as early as day 2 of incubation in both sexes, whereas a substantial level of transcription of the P-450arom gene seems to start during day 5 or 6 only in the female embryo. Thus, timing of the start of transcription of genes involved in sex steroid hormone synthesis does not seem to be coordinated, but might be categorized into two classes: one starting much earlier than the beginning of morphological differentiation of the gonads (about 5.5 days of incubation; Romanoff, 1960), and the other starting during or soon after the differentiation of the gonads. Five genes are known to be involved in the synthesis of estrogen from cholesterol (Figure 10), i.e., genes for P-450scc, 3β-hydroxysteroid dehydrogenase/$\Delta^{5\rightarrow4}$ isomerase, P-450c17, 17β-hydroxysteroid dehydrogenase, and P-450arom. Although sex steroid synthesis probably occurs after sex determination has been initiated, we think it important to know exactly how expression of these five genes is programmed in the development of male and female embryos.

Future Research Involving Sex Chromosome-Specific DNA Sequences

Identification of the sex of an individual embryo at an early stage of development using the sex chromosome-specific DNA probes described earlier is the most direct and important application of these probes. They provide opportunities to identify the processes of sexual differentiation such as those described above for steroid hormone synthesis. It would also be feasible to construct male- or female-derived cDNA libraries from embryos at an early stage of development and to identify sex-specific clones by applying suitable subtraction and amplification procedures. By this approach, genes involved in sex determination may be identified.

Identification of sex of embryos should also be useful when a tissue transplantation experiment is to be conducted between opposite sexes. Identification of the sex at the cellular level is a prerequisite to the production of somatic or germline chimeras between opposite sexes and to the identification of cells derived from one particular sex in a chimeric tissue.

Searching for other sex chromosome-linked DNA sequences by random hybridization or chromosome walking should be possible using the sex chromosome-specific repetitive or unique DNA probes. The putative Z-linked sequence described above should be a useful landmark in searching for genes on this hitherto unexplored sex chromosome at the level of the DNA sequence. To this end, information of the chromosomal site of the sex chromosome-linked specific DNA sequences by *in situ* hybridization and construction of a locus-specific DNA library using the technique of chromosome microdissection would be very helpful.

Roles of sex chromosome-specific DNA sequences in the structure and behavior of sex chromosomes and sex chromatin are other interesting subjects. We have been working on nuclear proteins that bind with high affinity to the bent-repetitive DNA sequences of the chicken W chromosome (Harata et al., 1988). Because 70–90% of the DNA in the W chromosome consists of bent-repetitive sequences, we speculate that formation of the W-heterochromatic body might be facilitated by association of such proteins with chromatin fiber containing the repetitive sequences.

An interesting behavior of chicken sex chromosomes has been observed in female meiosis; i.e., a recombination nodule is observed in the terminus of the pairing region of the synaptonemal complex between Z and W chromosomes at the pachytene stage, suggesting the presence of homologous sequence regions in both chromosomes that are responsible for the recombination (Solari et al., 1988). Recently, Hutchison and LeCiel (1991) observed by *in situ* hybridization using the W chromosome-specific *Xho*I and *Eco*RI family probes that the W lampbrush chromosome forms a complex that looked like an end-to-end pairing with the Z lampbrush chromosome in chicken oocytes. However, *Xho*I and *Eco*RI family sequences do not seem to be present in the vicinity of the pairing region. Thus, the properties of DNA sequences that are localized in the pairing region remain to be elucidated.

Finally, an ambitious but unpredictable prospect is that one may be able to introduce properly constructed sequences consisting of gene(s) and W chromosome-specific repetitive sequences into the W chromosome of chicken by homologous recombination in a female embryonic stem cell and raise a chimeric chicken utilizing a currently available technique (Petitte et al., 1990). If germline transmission of the stem cell were attained, it would be possible to raise a transgenic chicken in which female sex-linked expression of the transgene may be expected.

References

Graham, D. E., Neufeld, B. R., Davidson, E. H., and Britten, R. J. (1974). Interspersion of repetitive and non-repetitive DNA sequences in the sea urchin genome. *Cell*, 1, 127–137.

Harata, M., Ouchi, K., Ohata, S., Kikuchi, A., and Mizuno, S. (1988). Purification and characterization of W-protein. A DNA-binding protein showing high affinity for the W chromosome-specific repetitive DNA sequences of chicken. *J. Biol. Chem.*, 263, 13952–13961.

Hutchison, N. J., and LeCiel, C. (1991). Gene mapping in chickens via fluorescent in situ hybridization to mitotic and meiotic chromosomes. *J. Cell. Biochem.*, Suppl. 15E, 205.

Kodama, H., Saitoh, H., Tone, M., Kuhara, S., Sakaki, Y., and Mizuno, S. (1987). Nucleotide sequences and unusual electrophoretic behavior of the W chromosome-specific repeating DNA units of the domestic fowl, *Gallus gallus domesticus*. *Chromosoma*, 96, 18–25.

Koo, H.-S., Wu, H.-M., and Crothers, D. M. (1986). DNA bending at adenine•thymine tracts. *Nature*, 320, 501–506.

McPhaul, M. J., Noble, J. F., Simpson, E. R., Mendelson, C. R., and Wilson, J. D. (1988). The expression of a functional cDNA encoding the chicken cytochrome P-450arom (aromatase) that catalyzes the formation of estrogen from androgen. *J. Biol. Chem.*, 263, 16358–16363.

Mizuno, S. (1991). Evolutionary implication of sex chromosome-specific repetitive DNA sequences in birds belonging to the order, Galliformes. In: *New Aspects of the Genetics of Molecular Evolution*, edited by M. Kimura and N. Takahata. Japan Science Society Press, Tokyo/Springer-Verlag, Berlin, pp. 213–226.

Ono, H., Iwasaki, M., Sakamoto, N., and Mizuno, S. (1988). cDNA cloning and sequence analysis of a chicken gene expressed during the gonadal development and homologous to mammalian cytochrome P-450c17. *Gene*, 66, 77–85.

Petitte, J. N., Clark, M. E., Liu, G., Gibbins, A. M. V., and Etches, R. J. (1990). Production of somatic and germline chimeras in the chicken by transfer of early blastodermal cells. *Development*, 108, 185–189.

Romanoff, A. L. (1960). In: *The Avian Embryo. Structural and Functional Development.* Macmillan, New York, pp. 819–822.

Saitoh, H., Harata, M., and Mizuno, S. (1989). Presence of female-specific bent-repetitive DNA sequences in the genomes of turkey and pheasant and their interactions with W-protein of chicken. *Chromosoma*, 98, 250–258.

Saitoh, Y., Saitoh, H., Ohtomo, K., and Mizuno, S. (1991). Occupancy of the majority of DNA in the chicken W chromosome by bent-repetitive sequences. *Chromosoma,* 101, 32–40.

Solari, A. J., Fechheimer, N. S., and Bitgood, J. J. (1988). Pairing of ZW genosomes and the localized recombination nodule in two Z-autosome translocations in *Gallus domesticus. Cytogenet. Cell Genet.,* 48, 130–136.

Stefos, K., and Arrighi, F. E. (1974). Repetitive DNA of *Gallus domesticus* and its cytological locations. *Exp. Cell Res.,* 83, 9–14.

Takebe, Y., Seiki, M., Fujisawa, J., Hoy, P., Yokota, K., Arai, K., Yoshida, M., and Arai, N. (1988). SRα promoter: An efficient and versatile mammalian cDNA expression system composed of the simian virus 40 early promoter and the R-U5 segment of human T-cell leukemia virus type 1 long terminal repeat. *Mol. Cell. Biol.,* 8, 466–472.

Tone, M., Nakano, N., Takao, E., Narisawa, S., and Mizuno, S. (1982). Demonstration of W chromosome-specific repetitive DNA sequences in the domestic fowl, *Gallus g. domesticus. Chromosoma,* 86, 551–569.

Tone, M., Sakaki, Y., Hashiguchi, T., and Mizuno, S. (1984). Genus specificity and extensive methylation of the W chromosome-specific repetitive DNA sequences from the domestic fowl, *Gallus gallus domesticus. Chromosoma,* 89, 228–237.

Ulanovsky, L. E., and Trifonov, E. N. (1987). Estimation of wedge components in curved DNA. *Nature,* 326, 720–722.

Chapter 18

Aspirations of Breeders in the Poultry Industry as Manipulation of the Avian Genome Continues

Alan Emsley

Summary. Poultry-breeding companies historically have been involved in the development and application of new genetic technology for industrial purposes. Comprehensive challenges demand comprehensive solutions, and the few remaining breeders in the industry will assess the potential for novel methods of manipulating the avian genome to improve their individual competitiveness and the marketability of their products. However, meager funds available from the breeding industry for speculative research ultimately aimed at lower food cost to the consumer will necessitate continued strong public funding.

Introduction

Modern commercial use of layers is barely 100 years old, and the application of genetic research to bring about their improvement is less than 70 years old. Remarkable genetic progress in productivity during that period is testimony to the adoption of sound breeding practices founded on rigorous research in quantitative methods of manipulating the avian genome. While the application of genetics is aimed at improvement of the commercial product, the sector of the industry in which breeding is applied is separated from commercial production by considerable time and space (Figure 1). Most genetic selection occurs at the "pureline" level, where preferential mating of elite individuals is expected to shift gene frequencies in a favorable direction. The "grandparent" and "parent" levels are produced in turn from this elite gene pool by carefully structured matings *between* lines. Their purpose is primarily to multiply the foundation gene pool in order to supply sufficient commercial stock to poultry producers. Each step in the process, pureline through commercial stock production, takes about 1 year. The interval between selection decisions and on-farm evaluation of derived commercial product is almost 5 years.

This relatively young science moved from the laboratory into the commercial poultry world because there was demand from producers for improved performance. Entrepreneurs saw the application of the science to their own experience in poultry husbandry and business management as an opportunity to improve their competitiveness as poultry-breeding organizations. The successful application of the science is, therefore, a credit to poultry farmers and entrepreneurs. Farmers persistently ask for "profitability through more marketable eggs or more broiler and turkey meat in a given production period,

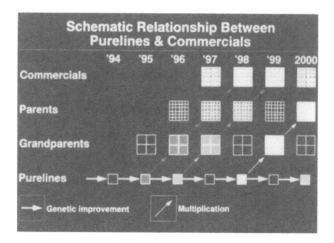

FIGURE 1. Schematic representation of the relationship between selection of pureline stock and incorporation of this genetic progress into production stock. Pureline stocks are multiplied to obtain grandparents, which are crossed to obtain the parent population. A second generation of crossing among parent stocks produces the commercial bird.

lower mortality, and better feed conversion." Industry pioneers continue to risk themselves financially to apply and refine the research that promises to bring about these improvements.

As in the past, the available research is not universally and homogeneously adopted by all who seek their fortunes in the industry. In the early days of the poultry-breeding industry, breeders learned from the experiences of one another. There was open exchange of ideas and even of breeding stock. Sometimes, university researchers worked in the facilities of poultry-breeding companies and published jointly with industry scientists. The poultry-breeding industry was hungry for new understanding. As one geneticist who spent many years in the industry put it, "We felt that there was more to be lost by *not* sharing information than by participating in open and frank discussions."

For many reasons, including genetic progress, the industry experienced a rapid rate of improvement in productivity and, as each company's market widened and overlapped with another's, an increased level of competition between primary breeders resulted. Simultaneously, and, some would say, *consequentially*, there was a reduction in the *number* of breeding companies competing on the world market, from hundreds to dozens and, more recently in the egg industry, to four. In the last quarter-century, the evolution of the industry has made it more difficult for the openness of the past to be seen as anything but a threat to a breeding company's survival.

Today, poultry breeders are no less interested in improving their market acceptability than in decades past. Their survival depends on it. Today, poultry breeders are no less willing than before to incorporate new methods of manipulating the avian genome if they hold out promise of greater competitiveness. But breeders hope that even as research in molecular genetics rapidly advances, those interested in applying that research to the poultry industry will temper the excitement of discovery with some hard reality. Forging a strong and long-lasting union between industry and nonindustry researchers will require a good understanding of one another's priorities. Although each breeder must speak to his own private concerns regarding priorities, there are some common concerns that I shall address below.

The Marketplace

Primary breeders in today's poultry industry are responsible for all of the genetic improvement that will affect future performance, but they are largely unable to profit from these improvements except by capturing market share from one another. A breeding company must make decisions about its performance goals long before its products reach the marketplace. Figure 1 illustrates the relationship between primary breeder and the producer and shows how genetic changes resulting from manipulation of the avian genome would take several years to appear in the commercial product and be assessed by the producer and/or consumer.

Stiff competition between breeders results in low profit margin per parent or grandparent chick sold by the primary breeders. Higher performance per commercial chick results in a decline in the number of multiplier females required for the same amount of commercial product unless the total market expands.

With only a handful of breeders remaining in the three major industry sectors, eggs, broiler meat, and turkey meat, the consequences of each change in methods of manipulating the genome can be much more serious. Ultimately, the marketplace will decide whether the decision was the right one.

But breeding decisions must be made long before the market has had a chance to judge their outcome. The wisdom of those decisions will depend very much on the breeder's clear understanding of the potential for improvement of aggregate genotype from each new technology tool.

Defining the Goal

In attempting to provide a genetically suitable product for the future marketplace, a breeding company must listen to the needs of the people in the industry and to the consumer.

K. N. May of Holly Farm & Foods, Inc., stated that "the broiler industry will surely be much larger in the year 2000, as will the turkey industry, but the egg industry will likely be the size that it is today or only modestly larger. The number of firms involved in each of these businesses will be smaller, but the average size of each will be larger" (May, 1990).

Don Bell, poultry specialist with the University of California Extension Service, at the 1991 International Poultry Trade Show in Atlanta, pointed to the consumer's increasing interest in "nutritious, safe, convenient, versatile, and competitively priced food products." He felt that the egg producer, for example, in response to further development of shell-less egg products, will place more emphasis upon "strains of chickens that produce more egg mass and have better feed efficiencies."

Each breeder also knows that what the producer wants is a comprehensive blend of acceptable performance in many traits. The breeder would like to believe that the relative economic importance of each of those component traits can be reasonably well estimated several years ahead of the appearance of the product on the market. In the layer marketplace with which I am most familiar, questions such as the following are frequently asked to enable the breeder to detect any shift in priorities.

1.　What will be the most economical egg size for producers?
2.　How important will feed conversion be? How will the producer measure it?
3.　Will earlier sexual maturity be desirable? Will the producer be able and willing to control the environment in order to take advantage of any genetic changes in the age at first egg?
4.　To what extent can vaccination be relied upon to provide adequate disease protection? What diseases have the greatest economic impact?
5.　What new developments might there be in automated egg handling that could change the demands on egg shell integrity?
6.　What refinements of ration formulation are anticipated, and what previously undetected genetic differences between commercial products might be exposed by marginal levels of some nutrients? What new developments might there be in automated feed handling and cage design that could affect the availability of adequate nutrients to each hen?
7.　Under what conditions of temperature, housing, bird density, and disease exposure will tomorrow's layer be expected to perform? How will issues of animal welfare influence this? How will the trends toward larger production complexes affect the producer's ability to manage the birds and keep disease organisms at bay?
8.　Relative to the "whole-egg" marketplace, how important is "further processing" of eggs going to be? What properties of the egg will this market demand from the producer and his product?
9.　How will human health and food safety issues influence the production environment and the nature of the product itself?

A similar list of questions could be constructed for the broiler and turkey markets. For many reasons, the poultry industry has been divided into several sectors, where the product of one level was the resource for the next. The advantages of specialization at each level, however, often served the interests of that level without maximizing the net return of the whole system. As profit margins per unit of product decreased, opportunities for business success often required expansion by the commercial producer and investment in his supplier. Through this integration, more widespread in the meat industry than in the layer industry, the commercial producer became keenly interested in how performance of *both* reproductive and production levels needed to be optimized to maximize the net return of the combined operations. Therefore, the list of questions in a market appraisal system for the meat industry is even longer and needs to address traits of both generations.

Geographic Diversity

It is certain that market priorities will never be the same for all regions of the world. With added emphasis on food safety and versatility in the developed world, it is likely that a larger proportion of eggs will be industrially broken for further processing and pasteurization. Under this kind of demand, the most economical egg size will be that which maximizes, for a given amount of feed, the quantity and quality of egg mass, or egg-component mass, to the processor. The creation of a major new market niche, such as the further processing of eggs, could well place demands for improvement in features of the layer that heretofore had not been emphasized or had been considered relatively less important. Quality of egg albumen is a good example of a characteristic that could prove much more important in the fast-growing further-processing market. But in many parts of the world, the number of marketable eggs produced by each hen in her lifetime will continue to be a simple and useful measure of productivity.

Further processing of poultry meat has undergone a revolution in the recent past, and consumer demand propels the industry through an annual growth of 4% per year. But in many marketplaces, the whole bird will continue to be the form of choice.

The need for higher and higher production without consideration of production efficiency will be questioned. Harry C. Mussman, Deputy Assistant Secretary for Science and Education, U. S. Department of Agriculture, said recently that "the emphasis in the developed world is on increasing efficiency, not production. There is a corresponding move toward food safety, quality, and the environment" (Mussman, 1990).

In the *developing world*, where the need for high-quality nutrients in human food production is very great, there are substantial opportunities for market growth in egg and poultry meat consumption but without the automation and environmental controls found elsewhere. Here, the challenge for geneticists is to evaluate the bird's productivity under conditions that are less predictable. Breeding companies must also keep in mind that developing nations may well resist importation of "value-added" products in favor of increases in their own domestic production. Because of more favorable labor costs in these countries and the high expense of the processing technology, they may also postpone moves toward domestic "further processing."

Thus, for both the layer and the poultry meat industries, there is more than one definition of what the market wants. Each breeding company must consider the strengths and weaknesses of its own pool of genetic resources and devise a strategy appropriate for meeting these divergent market needs.

Diseases and Vaccination

Infectious diseases occur in the chicken because a pathogen is presented to the bird that is unable to repel it or reverse its adverse effects. As production facilities become larger and more complex, often with many different ages of birds in close proximity, the chances for disease exposure will increase in the coming years. With good management and sanitation programs, this exposure can be minimized but will become more and more difficult to eliminate. Thus, there will be increasing need for the chicken to be able to defend itself against pathogens, but by what means — vaccines or genetic manipulation, or some combination of the two? The provision of a means of genetically improving disease resistance that will outlast the evolution of pathogenicity in the virus or bacterium itself and with no compromise of the performance levels with other traits would be of significant benefit to the industry. Can the tools of molecular biology be used to develop birds with these attributes?

Automation

One visit to the International Poultry show in Atlanta or a glance through any one of a dozen trade journals will confirm the ever-increasing promotion of automation in the commercial marketplace for layers and broilers.

Even in some developing countries, where labor cost has traditionally been very low, the potential for cost reduction through automation is being examined. Each innovation is aimed at improving egg and bird handling, feed provision, light and temperature control, ventilation, and other environmental factors that can influence the complete expression of a flock's genetic potential. Industry geneticists are keenly aware that a flock is made up of individuals with slightly different productivity schedules and, therefore, with differing needs at any given time in the production cycle. The risk of automation is that a flock will be treated as a homogeneous group regardless of how true that really is.

By designing the commercial bird as a cross, the industry geneticist has improved the chances for performance uniformity and buffered the bird against some of the stresses of the production environment. By breeding for strong shells and strong skeletal structure, he has attempted to protect the product until it reaches the consumer or some food processing site where its components can be suitably retrieved. Nevertheless, the demands continue for breeders to provide a bird that is insensitive to management stresses. Can the need for improvement in the uniformity of performance in the commercial bird within a single environment and over different environments be balanced with the need to maintain genetic variation necessary for further improvement at the pureline level? Alternatively, can future technology provide economically feasible ways to achieve greater uniformity in the commercial bird and develop alternative ways to obtain the necessary genetic variation at the pureline level? Regardless of how genetic change is made or the techniques used to alter the genome, the performance of the bird must meet the expectations of the scrutinizing eye of the producer.

In summary, then, the marketplace is the primary source of information about future product needs, and it is the judge, jury, and executioner when the product is put on trial. Shifts in the relative importance of different traits can and do occur, and they force the breeder to promptly reassess his company's strategy to meet the demand if he hopes to remain competitive.

The Means at Hand

The challenges for the future are (1) to retain long-term potential for genetic change; (2) to maximize the rate of change; and (3) to find ways to overcome biological constraints. Under our present selection methodology, however, biological reality dictates that only a fraction of the desired genetic change is possible each year. Several questions are repeatedly asked by each company's geneticists as they help devise a competitive strategy:

1. Under field conditions, how much of the variation that we observe in each commercial performance trait is attributable to heritable factors and how much to nonheritable (''environmental'') factors?
2. How effectively is the selection program improving the properties of the derived *commercial* population when it is evaluated under various field environments?
3. What are the most suitable data to gather on the population under selection in order to permit selection decisions to improve most *efficiently* the field performance of commercials? Involved here are the kinds, sources, and timing of performance measures.
4. How antagonistic are the genetic relationships between traits of interest?
5. What is the net economic impact of changes in any one of these commercial traits?
6. How much time does it take to evaluate the individual breeder candidates within the populations under selection?
7. What decision-making tools are necessary to analyze efficiently the information gathered on the populations under selection?

In summary, the breeder is continually seeking measures that are more heritable, free of negative correlations to other traits, and obtained at minimum cost. The cost can often be expressed in time and, for many traits, selection systems that can be implemented early in life and in both sexes would be highly desirable.

Funding

Funds for research are limited and must be spent wisely. Every opportunity to spend money is not equally capable of bringing about improvement in the competitiveness of a company's products. Priorities will be set so that the breeding company's investments have the highest opportunity for maximizing the competitive position of its products. This requires a clear statement of objectives by all researchers involved, keen attention to good experimental design, and careful analyses of alternatives because every project funded is at the expense of some other possible research.

How Many Different Products?

Providing distinctly different commercial products is not without challenge to the genetic program that seeks to make best use of a finite amount of facilities for data gathering and breeding value assessment. If the facilities are devoted solely to the improvement of enough lines to produce a *single* commercial bird, the rate of genetic improvement will, on the whole, be greater than if the facilities have to house more lines for additional products. So, unless the breeding company is willing to invest in more facilities, the *number* of products being developed can affect the success of any *one* of the products.

In many market niches, the definition of what is desirable does not change dramatically from one year to another. So the industry geneticist more often is faced with fine-tuning each product and refining the relative emphasis on change in the constituent traits. In this way, rather than by orchestrating a major shift in direction of the breeding program, he hopes to bring his product more in line with anticipated demand. For that reason, the most progressive breeding companies do not plan spectacular changes, but instead, adopt strategies for gradual and balanced genetic improvement.

In summary, while sophisticated methods and technology will be needed for genetic improvement in the future, it must be recognized that the efficiency of the program itself is an important factor in the competitiveness of a breeding company and its products. That efficiency is dependent on the answers to questions listed above and requires that the geneticist be on the lookout for earlier predictors of total productivity and faster ways of estimating accurate breeding values.

Table 1
Economics of the Egg Industry in the United States Assuming
an Average Per-Capita Consumption of 235 Eggs

Company	Market	Gross revenue
Commercial selling eggs	200 million people	$3.5 billion
Parent selling commercials	200 million chicks	$90 million
Grandparent selling parents	2 million chicks	$10.5 million

New Technology

Five years from now, the selection decisions that will affect performance in the year 2000 will have been made. By that time, it is unlikely that there will have been any major change in the techniques by which genetic improvements are brought about in layer breeding programs. Quantitative methods using selection of individuals with the highest calculated breeding value will continue to be used. Faster computers equipped with greater amounts of relatively inexpensive memory will permit better use of the data and of the facilities in which they are gathered.

On the other hand, new technology will more and more be a factor in poultry-breeding programs. The promise implied is *faster genetic improvement* of important features of good performance, but the timetable is not so clear. Sometimes the expense seems staggering. It remains to be seen whether the promise can be fulfilled in a cost-effective way. When one examines the breakdown of income for different sectors of the poultry industry (Tables 1 and 2), it is striking how little is available to the primary breeder to do the necessary research and development, marketing, production, and technical support of his products.

Of course, the breeding companies that see most clearly their own needs for new technology and that discern most quickly which tools are most useful will gain at least a temporary competitive edge. But it is likely that each breeding company has different needs and will adopt different strategies for using the technology, thereby moderating the advantages of others.

Despite development of new techniques aimed at probing the genetic process, we are still in the dark about the place of these techniques in poultry selection programs. Plant breeders are making dramatic progress in this area, but it is likely that we are many years away from the application of similar technology to the commercial poultry product. Nevertheless, poultry-breeding companies will continue to monitor developments in this area closely, and, like their predecessors, will adopt new technology when it appears to be promising.

At the very least, biotechnology will teach poultry geneticists much more about cellular biology in chickens and may lead to a renewed definition of the underlying statistical model and its assumptions. As Richard Forsythe put

Table 2
Economics of the Broiler Industry in the United States
Assuming an Average Per-Capita Consumption
of 31 kg (70 lb)

Company	Market	Gross revenue
Commercial selling meat	200 million people	$11.2 billion
Parent selling commercials	3 billion chicks	$1.0 billion
Grandparent selling parents	50 million chicks	$100 million

it, "the heart and soul of poultry science research will be the use of biotechnology to understand and regulate growth, to ensure product safety, to enhance gene expression and to expand the understanding of neuroendocrine and behavioral processes in reproduction" (Forsythe, 1990).

Poultry producers and the consuming public have benefitted from keen competition between breeding companies. The real challenge for the primary breeders today, in readiness for the next millennium, is to continue to optimize genetic, financial, and physical resources in order to ensure that the company's products can *remain* competitive. Those researching the new technology will be partners in meeting this challenge.

Acknowledgments

The author wishes to acknowledge the generosity of Don Bell in the preparation of visual aids for the oral presentation of this chapter. Portions of this chapter were also presented at the I Simposio Tecnico do Producao de Ovos-APA, March 13, 1991, Sao Paulo, Brazil.

References

Forsythe, Richard H. (1990). Poultry science in the year 2000: Academic inputs. *Poultry Sci.*, 69, 2107–2109.

May, K. N. (1990). Poultry science in the year 2000: Industry outlook. *Poultry Sci.*, 69, 2103–2106.

Mussman, Harry C. (1990). Poultry science in the year 2000: The federal government viewpoint. *Poultry Sci.*, 69, 2114–2117.

Chapter 19

Future Interactions Between Molecular Biologists and the Poultry Industry, Illustrated by a Comparison of the Genome Mapping and Gene Targeting Paradigms

Ann M. Verrinder Gibbins

Summary. Molecular biologists and members of the poultry industry have much to offer each other, but each group must learn to understand the language, expectations, and constraints of the other. Strategies for mapping the chicken genome and for the production of transgenic birds that have undergone site-directed alterations of the genome are reviewed in detail as they exemplify two widely different types of research that could have important implications for the industry. Conceptually, the poultry industry is familiar with strategies involving mapping and selection, albeit by classical methods, but the development of transgenic birds, especially those produced by gene targeting, represents a radically new departure. The genome mapping technology is likely to afford incidental technological benefits for the industry throughout its development, but the time frame for the fruition of gene targeting strategies is uncertain and necessitates the introduction of several novel technological developments. Both programs rely heavily on experience gained in other species, and both will provide benefits that will be unlikely or impossible to achieve by conventional practices in the industry. More effective communication between molecular biologists and members of the poultry industry must be developed, and the two groups must unite to improve the quality of reports in the popular press on our joint work and to prepare the public for the innovations that the industry should soon experience as a result of research in avian molecular biology.

Introduction

As has been well illustrated in other chapters in this book, we have learned during the past 10 years that molecular biology can provide us with information that is of much more than academic interest. The burgeoning biotechnology industries rely on an interaction between scientists who are adept at manipulating the components of the cell and entrepreneurs who can exploit these manipulations to give a novel product or process. Traditional animal production industries have never been slow to recognize the value of technological advances and have embraced expanding knowledge, particularly in the areas of health maintenance and nutrition. But the recent increase in our knowledge of molecular biology has occurred at such a rapid rate that even the aficionados must struggle to stay abreast of the latest developments. As in other rapidly expanding fields such as computer technology, the discipline must constantly coin new terms as novel concepts are introduced, and so develops a jargon of its own that is confusing for nonspecialists. Even if nonspecialists have an understanding of the vocabulary, where can they find an objective appraisal of the latest state of the art or of its potential? Other than a few notable exceptions, such as articles in *Nature* or *Science*, most publications of original data appear some time after the research has been completed. Important recent developments are missed in a rapidly developing field, even if electronically based literature search systems are employed. And then there is the problem of accuracy and objectivity in the publication of interpretive commentaries on molecular biology and its application. Although many journalists provide sterling service in explaining scientific advances to the nonspecialist, there are too many instances of inadequate reports. The journalist may be ill-prepared for the assignment, sometimes because of poor communication with an uncooperative scientist, or may distort the issue to obtain notoriety or to follow a particular policy of the publication. Because of these problems, a nonspecialist, such as a traditional member of the poultry industry, frequently has great difficulty in arriving at an informed and rational conclusion about the potential of specific aspects of molecular biology.

On the other hand, many molecular biologists have been trained in an environment that is remote from the pressures of the profit incentive and certainly remote from poultry production. Only recently has there been wide-spread encouragement of interaction between basic biologists and members of industry. This communication has arisen for several reasons — clearly, molecular biologists can provide industry with novel procedures, and many molecular biologists find that their involvement in an applied program is extremely rewarding. Some molecular biologists have found, however, that the idea that they nurtured with such enthusiasm in the confines of their laboratory has little relevance when exposed to the realities of the poultry industry. Since many molecular biology projects are long-term, it is beneficial for the scientists to interact with industry at an early stage of a project so that applicable systems can be used from the outset.

Ambitious projects can now be contemplated by avian biologists because of the development of many sophisticated techniques, but these programs are extremely expensive in terms of labor costs, animal maintenance, data processing, and the purchase of reagents and consumable supplies. In these days of fiscal restraint, it is very difficult for molecular biologists to fund their research programs by using only grants from traditional sources, such as national agencies that support basic research. There has been increased pressure by governments for cooperation between industry and academia, so that matching funds are now available from government sources in several countries if productive partnerships can be established between industry and academia in which industry is prepared to commit some financial support. In order to initiate such programs, there must be effective dialogue between the partners. It is not enough to expect members of the poultry industry to understand molecular biology and its jargon — molecular biologists must understand the poultry industry. Superficially, this seems a trivial issue, but, as many of us on both sides of the feather curtain know, penetrating this curtain requires perseverance! Not only do we have to understand each others' capabilities and objectives, but we must also wrestle with issues of confidentiality. The poultry industry is highly competitive, with stocks and strategies that are not divulged willingly. Increasingly, the business of molecular biology is highly competitive too, again with stocks and strategies that are not divulged willingly — perhaps the feather curtain would not seem so impenetrable at times if lack of communication were due only to an uncertain vocabulary. The only way of dealing with these seemingly unavoidable problems is for each side to recognize the constraints of the other and to respect the underlying cause, yet strive for a good working relationship.

A major area of difficulty in establishing a good working relationship between molecular biologists and the poultry industry is that members of the industry must often come to terms with the apparently erratic and variable nature of much research in molecular biology. Molecular biologists can scan a scientific paper or attend a seminar in the morning and, as a result of something that they read or heard, can make a major change in the strategy of their research project by the afternoon. Because of the ready availability of many of the reagents involved, the new experiments may even be in progress that day. This must appear to be frivolous to an industry that relies on the careful development of a long-term breeding plan, but molecular biologists could be foolish to deny themselves the advantages of the latest development. An important component in any collaborative agreement between the biologists and industry, therefore, must be an element of flexibility allowing rapid change in an agreed strategy. A further difficulty relates to research that requires a relatively long period of development, for which it is usually not possible to predict a precise outcome, let alone a completion date. Such high-risk programs are hard for the industry to justify, and yet there are aspects of poultry production that could benefit from innovative technology and that may never be developed if scientists have to rely only on conventional sources of federal funding available for basic research. Finally, the poultry industry should not look askance at avian molecular biologists who wish to draw on information or models derived from other species, ranging from nematodes to humans. The great lesson that has been learned over the past 25 years is the commonality of fundamental molecular biology and the fact that avian molecular biologists can benefit enormously from the experience gained with other species as well as provide our own contributions to the pool of basic knowledge.

Two important research endeavors that should have dramatic consequences for the poultry industry in the future, and that incorporate elements of everything that has been referred to in this volume, are the plans to map the chicken genome and the strategies for the production of transgenic birds with a view to gene targeting, which is the precise, preplanned modification of a specific region of the genome. These two areas of research will be reviewed in detail because they provide excellent examples of the problematic issues that have been discussed. The genome mapping program will rely heavily on experience gained in other species, and will change frequently in concept and methodology throughout its existence. During the life of the program the likelihood of technological spinoffs at all stages of the program is high. In addition, many different laboratories will be involved to varying degrees in different aspects of the program because of its multicomponent structure and because much data can be derived using equipment and methods that are readily available. Conceptually, the poultry industry is familiar with strategies involving mapping and selection, albeit by classical methods. Gene targeting strategies, however, are unprecedented and require several novel, specific technological developments, with no certain date for completion. Although each of these developments en route to successful gene targeting in the bird will be of value to embryologists and developmental biologists, only at the end of the program is the poultry industry likely to benefit. But the modifications that might be made to the bird at that stage could be unique and dramatic, and impossible to achieve by any other means in many cases. Although the experience of many avian biologists will be brought to bear on the problem of which regions of the genome should by modified by gene targeting, the development of the technology to allow these changes to be made will be in the hands of relatively few scientists.

Genome mapping and gene targeting strategies will be reviewed to illustrate the lessons that might be learned from these paradigms about future cooperation between molecular biologists and the poultry industry. The two areas of research were not chosen because they were deemed to be especially meritorious among the wealth of excellent programs reported in this volume, but because they represent very different aspects of avian molecular biology and different types of collaborative strategies.

Mapping the Chicken Genome

Traditionally, the relative positions of loci in a genome have been derived by observing the results of segregation of pairs of loci at meiosis. In order to achieve this goal, the loci being mapped must be polymorphic, so that individuals that are heterozygous for the loci in question can be found in a population or can be constructed by crosses. The chosen individuals must be heterozygous for both loci being studied, so that the results of segregation of the two pairs of alleles relative to each other can be observed. The ideal genetic marker for mapping purposes, therefore, is highly polymorphic, and should be easy to detect in all tissues and at all stages of development of an organism. Until recently, however, genome mapping depended on the availability of polymorphic forms of phenotypic markers. These markers have limited value because they may only be expressed at restricted times during development and to varying degrees of penetrance, and different types of tests may be required for each marker that is studied.

The introduction of restriction fragment length polymorphisms (RFLPs) as genome markers (Botstein et al., 1980) represented a radically new approach. When DNA of different members of a population is digested by a restriction endonuclease, fragments of different length may result for a particular locus depending on alterations in the base-pair sequence of the restriction endonuclease recognition site, or on additions to or deletions from the DNA of the region studied. These fragments can be detected by using a labeled probe consisting of DNA or RNA that is complementary in sequence (i.e., homologous) to a portion of the fragments and, therefore, can bind to the fragments by base pairing. Such DNA markers offer the advantage of being available for analysis at any stage in the development of an organism from embryo to adult, and are available in either sex even for sex-limited traits. The RFLP markers are co-dominant and distributed widely throughout the genome, and each is detected by only a minor variation of a standard, uniform technique. RFLPs have contributed to the development of genome maps of increasing complexity in the human (White et al., 1985; Donis-Keller et al., 1987). This knowledge has already allowed the identification, isolation, and elucidation of function of a number of genes whose malfunction is responsible for certain human diseases. Using a variety of combinations of restriction endonucleases and probes by trial and error, specific alleles of one or more RFLP markers may be identified that always co-segregate with the uncharacterized disease-causing locus in the family under study. Ideally, all individuals in the family who have the disease will have a certain RFLP allele, and those without the disease will have other alleles. In other words, the chosen RFLP marker is "tightly linked" to the disease-associated locus, which means that the two loci are so close to each other that recombination occurs rarely between them and so they must lie near to each other on the chromosome. If a series of cloned, overlapping fragments of DNA is available, the disease-associated gene can be isolated by searching the DNA for likely sequences in the vicinity of the RFLP locus, gradually moving away from the RFLP locus progressively by using techniques known as chromosome walking and jumping. This search is usually laborious, since much DNA may have to be screened, but an illustration of the efficacy of the approach is provided by the isolation of the cystic fibrosis gene (Rommens et al., 1989). RFLP linkage maps have been assembled for certain plants, such as *Arabidopsis thaliana* (Nam et al., 1989), rice (McCouch et al., 1988), potato (Gebhardt et al., 1989), *Brassica rapa* (Song et al., 1990), *B. oleracea* (Slocum et al., 1990), and *B. napus* (Landry et al., 1991), based on the meiotic segregation of both RFLP and morphological markers. Strategies to exploit the full power of such linkage maps have been devised (Lander and Botstein, 1986). The value of RFLP linkage maps to both animal (Beckman and Soller, 1987) and plant (Tanksley et al., 1989) breeding has been discussed, including strategies for mapping Mendelian factors underlying quantitative traits by co-selection of quantitative traits, and RFLPs (Lander and Botstein, 1989). Lander and Botstein (1989) demonstrated that parental strains should be chosen to maximize the chance that the progeny of crosses show segregation for quantitative trait loci (QTLs) having relatively large phenotypic effects, thus reducing the number of progeny that must be analyzed. In addition, they promoted the concept of interval mapping whereby information is integrated from genetic markers spaced throughout the genome using a lod score analysis. Lod scores are the ratios of log likelihoods, which indicate how much more probable it is that the data arose assuming the presence of a linked QTL relative to assuming no linked QTL. The use of a lod score method of linkage analysis to map a dominant major locus for a quantitative trait should be practical in a wide class of situations (Boehnke, 1990).

Certain quantitative traits in the tomato have been "resolved into Mendelian factors" (i.e., ascribed to single gene loci) by using an RFLP linkage map (Paterson et al., 1988; Tanksley and Hewitt, 1988). Such approaches allow the rapid introgression of agriculturally important traits into domestic stocks from exotic relatives. In the initial experiments with the tomato (Paterson et al., 1988), QTLs affecting mass per fruit, soluble solids concentration, and fruit pH were mapped to 20-cM regions of the genome in backcross generation 1 (BC1) of a cross between a wild Peruvian species, *Lycopersicon chmielewskii* (CL), and the domestic tomato, *L. esculentum* (E). In subsequent experiments, a technique named substitution mapping was introduced that allowed the QTLs to be mapped to intervals of as little as 3 cM (Patterson et al., 1990). BC1 individuals were selected that carried different CL segments associated with elevated soluble solids concentration but not with reduced mass per fruit. Self-pollinated progeny (BC1F2) were grown from each of these BC1 plants, and six BC1F2 individuals were selected that carried the desired CL segments but that carried a minimum of other CL segments. These individuals were backcrossed to E, and one plant was selected from each BC2F1 population that retained the desired CL segment but was free of other CL segments. The resulting six BC2 individuals, which had different genotypes from each other and had been derived from five different BC1 plants, were used in the substitution mapping. The CL segments associated with elevated soluble solids concentration that are retained by these six plants were characterized in detail, using all available polymorphic markers, thus allowing delineation of regions where the different CL fragments overlapped or were unique. The phenotypic effects of each CL segment were determined in segregating self-pollinated progeny from the six BC2F1 plants. Phenotypic effects shared by the different CL segments were attributed to loci in regions shared by the segments, while phenotypic effects unique to a segment were attributed to a region unique to that segment. As is rightly claimed (Paterson et al., 1990), substitution mapping will be useful in narrowing the gap between linkage mapping and physical mapping of QTLs. The size of the region to which a QTL can be mapped will be determined mainly by the density of previously mapped genetic markers in the region surrounding the QTL. Important findings of this research were that genes with desirable effects (increased soluble solids) mapped to different regions than genes with undesirable effects (reduced fruit mass), indicating that close linkage rather than pleiotropy appears to be responsible for the previously observed association between these traits, and indicating the potential for separating these traits in selection programs. Although preliminary, this result provides hope for the separation of many quantitative traits that have been assumed to be inextricably linked in a wide range of species. In chickens, for example, classical selection for increased egg number results in the reduction of individual egg mass, and classical selection for increased body size results in a decline in the rate of egg production.

Another notable result of the mapping experiments in the tomato was that the QTLs that were studied continued to show the desired phenotype when segregated away from much of the remainder of the donor genome, giving little indication of the necessity for epistatic interactions between the QTLs and their native genome to produce the desired trait. This result is encouraging for schemes that are designed to cause the introgression of a trait into a domestic species. In further experiments, the genetic basis of quantitative variation in the same traits was studied in a cross between the domestic tomato and another wild species, *L. cheesmanii* (Paterson et al., 1991). The relationships between the inheritance patterns of RFLPs and phenotype were determined for F_2 progeny, in which all three possible gene dosages at a locus would be represented, and for F_3 families (derived from self-pollinated F_2) grown in different environments from each other and from the F_2 progeny. In all, 29 putative QTLs of varying modes of inheritance were detected, some in all three environments, some in two, and some only in one, indicating that studies performed using a single environment could underestimate the number of QTLs associated with a trait. The results also suggested that, for traits of low heritability, the phenotype of F_3 progeny could be predicted more accurately in the F_2 parent by using RFLP markers rather than the phenotype as an indicator. RFLP markers or the phenotype were equally effective as predictors for a trait with intermediate heritability.

In addition to the examples cited above, linkage maps have been used to identify RFLP markers that are linked to various genes that confer resistance to diseases in tomato such as gray leaf spot (Behare et al., 1991) and root knot nematode infestation (Klein-Lankhorst et al., 1991). Reports are beginning to emerge of linkage of RFLP markers to traits of importance in livestock animals, such as the association of structural variation around the prolactin gene with various aspects of milk and cheese production in Holstein dairy cattle (Cowan et al., 1990), but the establishment of these relationships between DNA markers and traits in animals is still fortuitous and is not based on a systematic survey of an entire genome using a predetermined set of mapped markers.

Despite these successes, RFLPs have limitations as potential, ubiquitous genome markers, primarily because of their low degree of polymorphism and the laborious nature of their identification. A more polymorphic source of variability is provided by hypervariable "minisatellite" DNA sequences, which have been exploited for DNA fingerprinting (Jeffreys, 1987). The polymorphism in length of the DNA fragments derived from each minisatellite locus depends on the highly variable number of tandem repeats of certain core sequences. Using a probe that hybridizes to a range of slightly different core sequences under low-stringency hybridization conditions, members of a family of hypervariable loci can be probed simultaneously, giving the complex DNA fingerprint pattern that is unique to the individual. Although admirable for identification purposes, the fingerprint pattern is usually too complex to resolve for linkage studies (for exceptions, see Chapter 16), and so probes for individual hypervariable minisatellite loci, known as variable number of tandem repeat (VNTR) loci, have been identified for use under stringent hybridization conditions (Nakamura et al., 1987). These single, hypervariable loci can then be treated as conventional genetic markers, and will be valuable components of a complete linkage map.

Unfortunately, many VNTR markers tend to be clustered near the ends of chromosomes (Royle et al., 1988), in association with other tandem and dispersed repeat elements (Armour et al., 1989). For this reason, other types of hypervariable sequences, which are widely distributed throughout the genome, have been investigated as potential markers. For instance, "simple sequences" of monotonously repeated short nucleotide motifs have been detected that are usually less than 100 base pairs long, comprise up to 5% of a typical eukaryotic genome, and are normally found within unique DNA stretches (Tautz and Renz, 1984). Tautz has shown (Tautz, 1989) that the simple sequences are highly polymorphic in length and that simple sequence length polymorphisms (SSLPs) should act as general hypervariable DNA markers for genome mapping. Simple sequences are detected in libraries of cloned DNA by hybridization to synthesized probes that consist of simple repeated sequences. Selected clones are sequenced and unique sequences are identified that lie immediately adjacent to the repeated motif, on either side. These sequences can then be used as primers in the polymerase chain reaction (PCR; reviewed by Erlich et al., 1991), and the polymorphism is revealed by the generation of amplified fragments of different length from different alleles. The advent of PCR technology marked a turning point in the strategies used for detecting polymorphic markers, and heralded the arrival of a plethora of confusing acronyms that are used to identify different types of markers! Conventional RFLP mapping depends on a laborious Southern blotting technique in which carefully purified DNA is digested to completion by restriction endonucleases to produce fragments that are separated by electrophoresis through agarose gels and transferred to membranes. Specific fragments are revealed by hybridization of the membrane-bound DNA to labeled, homologous probes. In contrast, PCR technology is not only rapid and amenable to automation, but also requires very little DNA. So little DNA is needed that DNA sequences have even been analyzed in single sperm and diploid cells (Li et al., 1988), allowing the determination of the genetic distance between two loci by coamplification of linked DNA segments from individual sperm (Cui et al., 1989). An added advantage of PCR technology is that primers can be synthesized anew from reported sequences, whereas probes must be passed from group to group. A confusion has now arisen in the literature regarding the term "RFLP" — as well as referring to the polymorphism revealed by the conventional Southern blotting strategy that is described above, restriction fragment length polymorphism (interpreted literally) can also result when PCR-amplified products are cleaved by restriction endonucleases. In addition, the term "probe" is now used loosely, sometimes referring to the labeled sequences used in the analysis of Southern blots and sometimes referring to the use of primers in PCR. The introduction of PCR technology encouraged the development of new genome mapping strategies, one of which entails marking DNA sites throughout the genome by defining sequences that could be used as specific primers to amplify these sites — the sequence tagged sites (STSs; Olson et al., 1989). This approach has been modified by concentration on those regions of the genome that are expressed as messenger RNA, with the development of the concept of expressed sequence tags (ESTs; Adams et al., 1991).

Currently, polymorphisms that are revealed by PCR and that are of value in genome mapping can be divided into three general categories (Cox and Lehrach, 1991). The first category includes all of the methods that involve amplification of various classes of monotonously repeated short sequences, resulting in polymorphism similar to the SSLP described previously, using primers of known sequence that are homologous to unique regions of DNA that border the repetitive elements. The "simple sequences" of Tautz (1989) have been termed short tandem repeats (STRs) or microsatellites by others (Litt and Luty, 1989; Weber and May, 1989). Tandemly reiterated trimeric or tetrameric highly polymorphic sequences have been shown to occur frequently in the human genome — the combined frequency of trimeric and tetrameric STRs could be as high as one locus per 20 kilobase pairs (kbp) of DNA (Edwards et al., 1991). In addition, the small size of the amplified units allows the simultaneous detection of many fluorescently labeled fragments in automated, multiplex PCR assays (Edwards et al., 1991). Numerous dinucleotide repeat sequences were identified in the mouse genome by scanning databases of known DNA sequences, and many were shown to be polymorphic by PCR amplification (Love et al., 1990). The genome map that resulted from this study was used to identify two genes that influence the onset of autoimmune type 1 diabetes in mouse (Todd et al., 1991). Based on the combined use of conventional RFLP markers for crude mapping and PCR-amplified dimeric STRs for more detailed analysis, linkage of human facioscapulohumoral dystrophy to a dimeric STR has been achieved (Wijmenga et al., 1990).

The second category of PCR-amplified markers is detected by using primers that are homologous to certain repetitive elements that are found ubiquitously in mammalian DNA. These are members of the class of short interspersed repeat DNAs (SINES) (Nelson et al., 1989) or long interspersed repeat DNAs (LINES) (Ledbetter et al., 1990), and the marker system is given the general name of interspersed repetitive sequence PCR (IRS-PCR). One common class of human SINES includes the so-called *Alu* sequences; these SINE-generated polymorphic variants have also been called "alumorphs" (Sinnett et al., 1990). The polymorphism that is detected by using SINE or LINE sequences as primers depends on the amplification of unique DNA sequences that are bordered, by chance, by identical primer sequences, in reverse orientation to each other.

The third class of PCR-amplified polymorphic elements is similar to that generated using IRS-PCR except that the primers are of arbitrary sequence. These marker systems have been termed random amplified polymorphic DNA (RAPD; Williams et al., 1990, 1991) or arbitrarily primed PCR (AP-PCR; Welsh and McClelland, 1990; Welsh et al., 1991). As for IRS-PCR, amplification of marker DNA depends on the chance that identical primer sequences can hybridize to genomic DNA in pairs with reverse orientation to each other and close enough to each other to allow the PCR to take place. Randomly synthesized oligonucleotide 10mers, which are sequences of DNA consisting of 10 bases, are tested as suitable primers, one by one. The advantage of the RAPD mapping technique is that it is indeed rapid, because no preliminary knowledge of the genome is required in terms of cloned or sequenced DNA, and the technique lends itself to automation. In addition, a universal set of primers can be used for genomic analysis in a wide variety of species, distinguishing it from IRS-PCR. As for IRS-PCR markers, most RAPD markers are dominant: the polymorphism is reflected in specific DNA segments of a particular length being amplified from one individual but not from another. Unfortunately, it may not be possible to distinguish whether the DNA segments are amplified from heterozygous or homozygous loci, thus less mapping information may be derived from individual RAPD markers than from co-dominant markers such as RFLPs. However, strategies are being designed to overcome this shortcoming (Williams et al., 1991), and dominant markers are still useful for constructing linkage maps in F_2, backcross, and recombinant inbred populations. The overriding power of the technique is the speed with which a usable genomic map can be constructed, and markers can apparently be distinguished throughout the genome. RAPD markers have already been identified that are linked to traits of agricultural importance, such as resistance to *Pseudomonas* spp. in tomato (Martin et al., 1991).

The ultimate objective of mapping is to combine three different types of map — the classical genetic map, the DNA marker map (using RFLP, VNTR, SSLP, or RAPD markers, for instance, or identified gene loci), and a physical map that consists of overlapping fragments of cloned DNA. Many novel techniques are emerging for assigning specific DNA markers to locations on chromosomes, including differential PCR amplification in somatic cell hybrids (Iggo et al., 1989), the cloning of defined regions of the genome by microdissection of chromosomes and enzymatic amplification (Ludecke et al., 1989), oligonucleotide-primed *in situ* DNA synthesis (PRINS) in fixed metaphase chromosomes (Gosden et al., 1991), and the use of yeast artificial chromosomes (YACs) (Coulson et al., 1991), to name but a few. Established techniques for assignment of DNA markers to chromosomes, such as *in situ* hybridization, are beginning to be used to good effect for the chicken (Hutchison and Le Ciel, 1991; Lakshmanan et al., 1991; Ponce de Leon et al., 1991; Tereba et al., 1991; and Chapter 4). By building a detailed genomic map, access can be gained to any region of the genome that has been identified by linkage studies. Extensive international efforts are under way to build a 5- and eventually 1-cM map of the human genome — the most recent status of the map was depicted in the form of a chart by M. Chipperfield, B. Maidak, and P. Pearson in *Science*, 254, 247–262 (1991). The strategies and technology developed for this enterprise will benefit the burgeoning schemes to map the genomes of important crop and livestock species. In addition, heterologous probes of use in more than one species will undoubtedly be identified, and regions of synteny homology, whereby specific loci appear to occur in similar linked groups in different species, can be expected between animals (Nadeau, 1989), so that mapping studies in one species will benefit those in another.

The chicken genome map is rudimentary at present (see Chapter 5), but excellent plans are underway to coordinate the construction of high-resolution chicken genome maps in America (contact, L. B. Crittenden, Michigan State University) and in Europe. The American plans include the construction of a backcross reference mapping population using highly divergent inbred lines to develop the cross in order to ensure that the F_1 is heterozygous at as many loci as possible, following a strategy that has proved successful for the mouse. A bank of DNA from the parents, F_1s, and backcross progeny will be available to collaborators for the assignment of markers, and data will be compiled centrally.

Gene Targeting

It is likely that transgenic birds will be introduced into the poultry industry during the next decade. Among these first transgenic birds, most probably, will be those designed to have improved resistance to specific diseases (see Chapters 9 and 15). Much of the technology used to produce transgenic organisms has resulted in the random insertion of exogenous DNA sequences into the genome, with the fortuitous production of the desired phenotype. Of great value would be the opportunity to engineer the genome very precisely, in a predetermined manner, even to the alteration of a single base pair. In this way, the structural sequences of a gene could be altered to give a product that is more stable, more biologically active, or more diverse in its properties. The regulatory regions of a gene could be altered to provide different levels of production of the product, or production at different stages of development or in different tissues than normal. Finally, exogenous DNA sequences could be introduced into specific sites in the genome so that their expression would be assured, either by inserting them adjacent to appropriate promoters or by placing a gene that bears its own promoter in a region of the genome that is known to be transcriptionally active. Such precise engineering of the mouse genome has been achieved through the development of chimeric intermediates; biologists are working toward this long-term objective with the bird, also.

The development of animals of modified genotype through chimeric intermediates depends on the incorporation of modified donor cells into a recipient embryo at an early stage of development, with the expectation that some of those cells will contribute to the development of gametes. Thus, transferred donor cell descendants may form ova or spermatozoa in the mature chimera, and their genetic material may be passed on to the next generation, in which case the intermediate is said to be a germline chimera. If the genetic material of the donor cells is modified by recombinant DNA techniques during the transfer of cells from the donor to the recipient embryo, the genetic modification may be inherited by offspring of the chimera. Chimeras have been formed by the surgical fusion of portions of different avian (Rawles, 1945; and Chapter 7) or amphibian embryos (Volpe, 1963), or by aggregation of two mouse blastocysts (Tarkowski, 1961). In addition, mouse (Gardner, 1968) and chicken (Petitte et al., 1990; and Chapter 6) chimeras have been formed by injection of dispersed cells from one embryo into another embryo. In particular, the latter methodology lends itself to the development of organisms of modified genotype because the donor cells can be genetically altered during transfer.

A major modification of the technology for the production of chimeric mice resulted from the development of immortalized lines of pluripotent mouse embryonic stem (ES) cells (Evans and Kaufman, 1981; Martin, 1981). Cultured ES cells are probably homologous to cells of the embryonic epiblast and are equivalent to cells in an embryo prior to segregation of somatic and germ cell lines (Evans, 1981). The ES cells are injected into the blastocyst, which is a hollow, spherical structure containing a compact disk of approximately 15 cells, the inner cell mass, that is attached to the inner surface of the cavity wall. The inner cell mass develops into the embryo, and the ES cells are deposited against this structure, into which they become integrated. Cultured ES cells were shown to be capable of contributing to the formation of germline chimeras (Bradley et al., 1984), even after the stable incorporation of exogenous DNA following transfection mediated by calcium phosphate (Gossler et al., 1986) or retroviral vectors (Robertson et al., 1986). By introducing DNA constructs that include an exogenous gene, or specially modified sequences of an endogenous gene, flanked by DNA sequences that are homologous to the targeted site, the genome of transfected cells of many organisms, including the mouse (Smithies et al., 1985; Thomas et al., 1986), can be modified by homologous recombination. Unfortunately, most recombination in mammalian mitotic cells is nonhomologous (Folger et al., 1984), i.e., the exogenous DNA will integrate in the genome of the host cell at sites other than those targeted, but strategies have been developed for the selection of the few cells that have undergone homologous recombination; these manipulations have been used to select specifically modified mouse ES cells (Thomas and Capecchi, 1987; Capecchi, 1989). A variety of germline chimeric mice have been produced following the injection into mouse blastocysts of selected ES cells that have undergone specific gene targeting events mediated by homologous recombination (Joyner et al., 1989; Schwartzberg et al., 1989; Thompson et al., 1989; Zijlstra et al., 1989; Zimmer and Gruss, 1989; Stanton et al., 1990; Thomas and Capecchi, 1990; Chisaka and Capecchi, 1991; Soriano et al., 1991).

The factors affecting the efficiency of gene targeting in mouse ES cells are still being defined, but could include the length of the homologous and nonhomologous sequences carried in the targeting vector, the transcriptional activity of the targeted gene, the number of copies of the targeted gene in the ES cell genome, and the precise location of the targeted region within the gene or within the genome (Capecchi, 1989). Gene targeting has been used to insert DNA sequences as long as 12 kbp into the mouse genome (Mansour et al., 1990), and also to delete sequences of at least 15 kbp (Mombaerts et al., 1991). This variety of modifications illustrates the great potential of gene targeting in the specific engineering of a genome. Available evidence suggests that the targeted region may not need to be undergoing transcription in the treated cells (Smithies et al., 1985; Johnson, et al., 1989), and single copy genes may be targeted as efficiently as multicopy genes (Zheng and Wilson, 1990). In general, linear constructs appear to be more effective in mediating homologous recombination, and cleavage of a circular construct within the region of homology to the genome may improve the frequency of directed gene modification (Smithies et al., 1985). It is too soon to be able to delineate precise guidelines for the development of targeting constructs and strategies because the reported studies of gene targeting in the mouse were performed using a range of variables by a number of different laboratories, resulting in widely varying efficiencies of homologous recombination.

Because of the preponderance of random insertion events compared to correct modifications mediated by homologous recombination, a major concern is the development of methods for detecting appropriately modified cells.

Correctly targeted mouse ES cells have been selected by a variety of strategies, including those that depend on whether or not the gene encoding hypoxanthine phosphoribosyl transferase (HPRT) is expressed. The karyotypically stable mouse ES cell lines that are in general use for the production of chimeras are male, and so bear one copy of the *hprt* gene, which is X-linked and expressed in ES cells grown in culture. Cells may be selected for HPRT activity by growth in medium containing hypoxanthine, aminopterin, and thymidine (HAT medium), which kills HPRT$^-$ cells, or may be selected for lack of HPRT by growth in medium containing 6-thioguanine, which kills HPRT$^+$ cells. Correct gene targeting events are rare, and so it is extremely unlikely that both copies of an autosomally located gene would be modified simultaneously. However, a single targeting event in an X-linked gene that encodes an enzyme could be revealed by the presence or absence of activity of that enzyme in the cell. The ease with which HPRT activity can be selected for or against, and the presence of only one copy of the *hprt* gene per cell, explain why site-directed modifications of the *hprt* gene have played a prominent role in the development of gene targeting strategies in mouse ES cells (for example, Thompson et al., 1989). Selection for and against HPRT activity is the basis of the double-replacement strategy (suggested by Evans, 1989), which is for use in HPRT$^-$ ES cells for targeting genes that do not give a readily selectable product. A gene that is to be modified is first "marked" by the insertion into it of the *hprt* gene, using homologous recombination. HPRT$^+$ cells are selected, then a second targeting event is effected whereby the *hprt* gene is replaced concomitant with the desired modification being introduced into the target gene. Appropriately targeted cells will have lost HPRT activity and can be readily selected. The double-replacement strategy has also been named "hit and run" (Hasty et al., 1991) or "in-out" (Valancius and Smithies, 1991). Another important feature of selection for or against HPRT activity is that the selection conditions do not appear to exert a deleterious effect on the surviving cells; suboptimal culture conditions could probably impair the ability of the surviving cells to incorporate into a host embryo and form a germline chimera (Evans, 1989). This factor should be considered when strategies are designed involving selection based on the activity of enzymes other than HPRT. Another direct method of screening for correct targeting events in genes that code for cell-surface determinants could be to sort the treated cells physically by using fluorescence-activated cell sorting (FACS). Although FACS is effective with ES cells, the cell determinant of interest must be expressed in ES cells for this strategy to work (Evans, 1989).

A general strategy for the selection of transfected cells has been to include the gene encoding resistance to neomycin, *neo*, in the construct. Transfected ES cells will be resistant to the neomycin analog, G418. Selection for transfection can be combined with selection for correct gene targeting by embedding a promoterless *neo* gene in the homologous sequences and designing the construct such that the *neo* gene will be expressed from the promoter of the targeted gene, either as an independent protein (Doetschman et al., 1988) or as a fusion protein (Dorin et al., 1989; Sedivy and Sharp, 1989). Rather than select for a correctly targeted ES cell clone, a method has been devised for selection *against* nonhomologous integrants (Mansour et al., 1988). This method, named positive and negative selection (PNS), requires the construction of targeting vectors in which the modifying sequences and a marker gene that is to be selected *for*, such as *neo*, are embedded in regions of homology to the genome, with a marker that is to be selected *against*, such as the herpes thymidine kinase gene, or the diphtheria toxin A-fragment gene (Yagi et al., 1990), placed distal to the homologous sequences. Correct recombination of these sequences with the genome should result in the inclusion of *neo*, and growth of the cells in the presence of G418, but the exclusion of the thymidine kinase gene and the resultant ability of the cells to survive in the presence of gancyclovir, or the exclusion of the diphtheria toxin A-fragment (which is toxic itself, without the application of drugs). With the increasing sophistication of PCR methodology, more use is being made of direct screening strategies that rely on detecting specific modifications to genomic DNA sequences that occur on correct gene targeting (Joyner et al., 1989; Zimmer and Gruss, 1989).

From the foregoing account, it is clear that all of the elements are in place to allow the production of precisely designed mice by using site-directed modifications of ES cells in culture and the injection of selected cells into blastocysts to give germline chimeras. Despite the fact that numerous lines of mutant mice have been produced by this procedure, it is fair to say that much more work needs to be done before the methodology becomes routine, particularly in the identification of more ES lines that contribute to the germline, and in the refinement of strategies for the design of effective gene targeting constructs and for the selection of correctly modified cells. However, these exciting developments in the production of specifically modified mice provide a stimulus for similar developments with the bird. The processes that must be developed in order to allow the production of birds with specifically modified genotype via chimeric intermediates are as follows:

1. Isolation and long-term culture of pluripotent avian ES cells
2. Development of constructs capable of undergoing homologous recombination with the genome
3. Selection of ES cells that have undergone correct gene targeting events
4. Introduction of these selected cells into a recipient genome to generate germline chimeras
5. Incorporation of appropriately modified birds into commercial poultry breeding strategies

As noted previously, germline chicken chimeras have been produced by the transfer of dispersed blastodermal cells into recipient embryos, and transfected donor cells have been incorporated into chimeric embryos (Petitte et al., 1990; Brazolot et al., 1991; and Chapter 6). Strategies have been developed for the incorporation of appropriately modified birds into commercial poultry breeding schemes (J. Hillel, A. M. Verrinder Gibbins, and R. J. Etches, submitted for publication), thereby addressing a common criticism of molecular biologists by animal breeders that the incorporation of individual, genetically modified birds will be impractical because of the rapid rate of normal progress in poultry breeding. Notwithstanding this promising start, progress must be achieved in developing immortalized, pluripotent ES cells and in designing gene targeting strategies for avian cells. These experiments are beginning in several laboratories throughout the world, including our own, and surely deserve the qualifier, "the more the merrier," for the more ES lines that can be produced, the more likely it will be that karyotypically stable lines that can contribute to the germline in chimeras will be identified. Gene targeting strategies should follow those used in the production of modified ES cells in the mouse, except that it is not known whether both male (ZZ) and female (ZW) avian ES cell lines will be produced, and it may not be possible to use W-linked markers in an analogous manner to the X-linked *hprt* locus in mouse. This may not be a serious impediment, because other, more general strategies are being developed for the mouse, as noted above. Despite the current uncertainties, these challenges offer the possibility of exciting rewards for avian biologists and the poultry industry.

Considerations for the Future

What lessons can be learned from the development of genome mapping strategies over the past decade? First, 10 years ago no one could have predicted the remarkable technological developments that have taken place during this period or the marked change in our understanding of the structure of DNA and the molecular basis of the determination of traits. Who would have predicted that DNA is so polymorphic within a population? We forget that such a view was heretical 10 years ago. The introduction of a succession of different types of DNA markers quickly over only a few years, and of different strategies for their use, emphasizes the importance of remaining flexible in approach, and not becoming enmeshed in a ponderous framework of research that is tied to a rapidly outdated technique. Perhaps we shall find, as may be the case for the human genome map (see the commentary by Roberts, 1990), that few molecular biologists are willing to devote their time to assembling an overall detailed map of the chicken or turkey genome — more probably, different research groups will concentrate on regions of the genome that contain the genes that are of particular interest to them. Indeed, because of the speed with which a battery of markers may be identified, it may be more appropriate to develop new markers rapidly for each particular breeding stock being studied, with the assurance that the chosen markers will be polymorphic and informative in that particular population (S. Tingey, personal communication). This does not negate the invaluable efforts of a few dedicated researchers who are attempting to coordinate the previously fragmentary attempts at mapping the chicken genome with phenotypic and molecular markers. These programs must continue in order to provide a basic framework for a generic chicken genome map, but the detailed mapping of local regions of interest in a particular research project is more likely to be achieved by rapid saturation with novel polymorphic DNA markers in a specific stock. The value of the latter approach could ultimately be the relative freedom that the individual group would experience from patent restrictions. As with many other endeavors in biotechnology, genome mapping is beginning to feel the constraints of patent protection, and this spectre will surely appear in chicken genome mapping before long. The patenting of individual markers for specific traits is acceptable and provides some compensation for the considerable financial outlay that may have been required to identify the marker, but the blanket patenting of an entity as embracing as a genome map hampers progress in the scientific community. The conflict between patent protection and freedom of communication in the scientific community could be a major impediment in the development of good working relationships between the poultry industry and academics. A fundamental aspect of academic life is the training of students, and it is unfortunate if graduate students, in particular, should be constrained in communication of their research because of the possibility of it being patented. The problem of patent protection in the rapidly changing world of molecular biology is a conundrum that will not be solved easily; not the least consideration is that our rapidly developing knowledge may allow many patent strictures to be sidestepped relatively easily.

The pioneering work of the groups of Hillel (see Chapter 16) and Kuhnlein (Kuhnlein et al., 1991) in developing strategies for identifying molecular markers associated with quantitative traits of the chicken, and in using molecular markers to investigate other aspects of poultry production, indicates that investment by industry in long-term programs such as chicken genome mapping should produce benefits even at a relatively early stage. The review of the development of mapping strategies emphasizes the invaluable insight that research in other species can give to programs in poultry biology and production, therefore the industry should continue to be sympathetic to supporting evidence for research proposals that is derived from apparently unlikely sources.

And what can we learn from the strategies for the development of birds of precisely modified genotype? Again, the experiential framework for the program is based on work developed with another species, namely, the mouse. Here, the conceptual gap between what has been achieved in another species and what might be possible in the bird is wide — the mouse blastocyst and the chicken blastoderm in the freshly laid egg are of radically different sizes (60 as opposed to about 60,000 cells, respectively) and are maintained in very different environments (the mammalian reproductive tract and the shelled egg), requiring ingenious modifications in order to adapt the strategy developed for one species to the biological constraints of the other. It is our experience that even our own colleagues may have difficulty in understanding the concept of the production of modified birds through chimeric intermediates, let alone understanding the added manipulations that mediate gene targeting by homologous recombination! Of all the problems that will be encountered in this work, this problem of a lack of understanding should be the easiest to overcome and merely requires very careful explanation, but molecular biologists are usually asked to present ideas to industry in as simple a form as possible, preferably on one sheet of paper. If we are serious in our intent to foster better communication between molecular biologists and the poultry industry, both groups should be prepared to do more preparatory work. After all, many years of research and significant benefits to the industry may ride on an initial exploration of a research proposal. Therefore both groups should be prepared to invest effort in the early encounter.

Besides being conceptually unfamiliar for the industry, the production of commercially useful birds of modified genotype, by any means, is of uncertain time frame. All that we can say is that we shall see commercially useful transgenic birds "before long." To be more precise would be unrealistic, but the industry encourages molecular biologists to commit themselves more definitely in order to arrange their priorities and to decide whether or not they should have "a foot in the door." If the maximum benefit is to be reaped from innovative technology of uncertain outcome, whether it be site-directed modification of genes or any of the other exciting developments outlined in this volume, molecular biologists must be clear and frank in the appraisal of the potential of their research, and the poultry industry must be willing to support some radically new concepts.

The concerns of the public about the deliberate engineering of animals to give modified genotypes must be recognized. Even though the development of transgenic poultry has been slower than the production of experimental transgenic pigs, sheep, and fish, the poultry industry should soon be deciding how best to prepare the public for the commercial use of transgenic birds. The more predictable the outcome of the technology that is used in the formation of these animals, the more likely it is that they will be acceptable. For example, the public reacted with abhorrence to the reports of malformed pigs that had been produced by microinjection of exogenous genes into one-celled zygotes. This technology results in the random insertion of the exogenous DNA in the genome, which may cause unexpected phenotypes, but the experiments were also performed at a time when the control of expression of exogenous genes in transgenic animals was less understood than at present. The more that can be done to ensure that transgenic livestock and poultry will be healthy and comfortable by carefully designing the genetic modifications, the better the outcome will be for the animals and for the industry.

Gene targeting strategies using chimeric intermediates exemplify an approach in which the genotype of the animal is carefully designed, and have the added advantage that the major genetic manipulations and selection procedures that are required to produce the founder animal are conducted on cells in culture and not on animals in the barn. As stated earlier, no genetic engineering strategies in birds can be performed sensibly without a thorough understanding of the molecular genetics and physiology underlying the trait of interest.

Who knows what remarkable developments in technology and in our understanding of biological processes will emerge in the next 10 years? For the molecular biologist, such speculation brings nothing but excitement, but the poultry industry views the future with mixed feelings. Without doubt, the industry would welcome new tools to improve selection strategies and to develop novel birds with commercially valuable characteristics. But, as explained by Alan Emsley in Chapter 18, individual breeding companies operate within a complex hierarchical system, with the expectation and necessity of steady annual progress. New technology may eventually augment or supplant some traditional methods used to achieve this steady progress, but, currently, the industry is understandably reluctant to contemplate any new procedure that will disrupt the pattern of steady progress, even if the long-term gains achieved by adopting the innovation might be considerable.

A major problem that molecular biologists and the poultry industry must tackle together is that of communication between the two groups and, just as important, appropriate communication of both groups with the public. We should foster the concept of short-term fellowships in each other's institutions, with the obligation for profound communication on both sides, and perhaps without any expectation of a specific collaborative program arising from the encounter, although this would be a welcome bonus. Each group would have to be careful whom they chose as their representatives, so that the potential of the latest developments in molecular biology and the objectives of the breeding companies would be conveyed by knowledgeable practitioners, actively engaged in the program under discussion. Better understanding of the role of molecular biology in poultry breeding would then allow the industry and scientists to begin to advise the public, prospectively, about the innovations that will surely be experienced as a result of molecular biology during the next decade, and both groups must lobby vigorously to improve the quality of popular reports of our collaborative work so that the public is well informed.*

---

* In an attempt to foster regular communications between avian biologists in research laboratories and industry who are interested in the development of strategies for the manipulation of the avian genome, an informal, nonprofit newsletter, *MAGazine*, has been initiated by myself and Rob Etches (first issue, October 1991). Currently, we have subscribers from 13 different countries. For more information, please contact us.

Acknowledgments

I appreciate extensive discussions with Guodong Liu on gene targeting. As a molecular biologist, I must add that I have thoroughly enjoyed my interaction with members of the poultry industry over the past five years, and thank them for their forbearance while I have struggled to understand the complexities of the industry. The Ontario Egg Producers' Marketing Board has been generous and remarkably farsighted in supporting our research, for which we are very grateful.

References

Adams, M. D., Kelley, J. M., Gocayne, J. D., Dubnick, M., Polymeropoulos, M. H., Xiao, H., Merril, C. R., Wu, A., Olde, B., Moreno, R. F., Kerlavage, A. R., McCombie, W. R., and Venter, J. C. (1991). Complementary DNA sequencing: Expressed sequence tags and human genome project. *Science,* 252, 1651–1656.

Armour, J. A. L., Wong, Z., Wilson, V., Royle, N. J., and Jeffreys, A. J. (1989). Sequences flanking the repeat arrays of human minisatellites: Association with tandem and dispersed repeat elements. *Nucleic Acids Res.,* 17, 4925–4935.

Beckman, J. S., and Soller, M. (1987). Molecular markers in the genetic improvement of farm animals. *Bio/technology,* 5, 573–576.

Behare, J., Laterrot, H., Sarfatti, M., and Zamir, D. (1991). Restriction fragment length polymorphism mapping of the Stemphylium resistance gene in tomato. *Mol. Plant-Microbe Interactions,* 4, 489–492.

Boehnke, M. (1990). Sample-size guidelines for linkage analysis of a dominant locus for a quantitative trait by the method of lod scores. *Am. J. Human Genet.,* 47, 218–227.

Botstein, D., White, R. L., Skolnick, M., and Davis, R. W. (1980). Construction of a genetic linkage map in man using restriction fragment length polymorphisms. *Am. J. Human Genet.,* 32, 314–331.

Bradley, A., Evans, M., Kaufman, M. H., and Robertson, E. (1984). Formation of germline chimaeras from embryo-derived teratocarcinoma cell lines. *Nature,* 309, 255–256.

Brazolot, C. L., Petitte, J. N., Etches, R. J., and Verrinder Gibbins, A. M. (1991). Efficient transfection of chicken cells by lipofection, and introduction of transfected blastodermal cells into the embryo. *Mol. Reprod. Devel.,* 30, 304–312.

Capecchi, M. (1989). Altering the genome by homologous recombination. *Science,* 244, 1288–1292.

Chisaka, O., and Capecchi, M. R. (1991). Regionally restricted developmental defects resulting from targeted disruption of the mouse homeobox gene *hox1.5. Nature,* 350, 473–479.

Coulson, A., Kozono, Y., Lutterbach, B., Shownkeen, R., Sulston, J., and Waterston, R. (1991). YACS and the *C. elegans* genome. *BioEssays,* 13, 413–417.

Cowan, C. M., Dentine, M. R., Ax, R. L., and Schuler, L. A. (1990). Structural variation around prolactin gene linked to quantitative traits in an elite Holstein sire family. *Theor. Appl. Genet.,* 79, 577–582.

Cox, R. D., and Lehrach, H. (1991). Genome mapping: PCR based meiotic and somatic cell hybrid analysis. *BioEssays,* 13, 193–198.

Cui, X., Li, H., Goradia, T. M., Lange, K., Kazazian, H. H., Galas, D., and Arnheim, N. (1989). Single-sperm typing: Determination of genetic distance between the ᴳgamma-globin and parathyroid hormone loci by using the polymerase chain reaction and allele-specific oligomers. *Proc. Natl. Acad. Sci. USA,* 86, 9389–9393.

Doetschman, T., Maeda, N., and Smithies, O. (1988). Targeted mutation of the HPRT gene in mouse embryonic stem cells. *Proc. Natl. Acad. Sci. USA,* 85, 8583–8587.

Donis-Keller, H., Green, P., Helms, C., Cartinhour, S., Weiffenbach, B., Stephens, K., Keith, T. P., Bowden, D. W., Smith, D. R., Lander, E. S., Botstein, D., Akots, G., Rediker, K. S., Gravius, T., Brown, V. A., Rising, M. B., Parker, C., Powers, J. A., Watt, D. E., Kauffman, E. R., Bricker, A., Phipps, P., Muller-Kahle, H., Fulton, T. R., Ng, S., Schumm, J. W., Braman, J. C., Knowlton, R. G., Barker, D. F., Crooks, S. M., Lincoln, S. E., Daly, M. J., and Abrahamson, J. (1987). A genetic linkage map of the human genome. *Cell,* 51, 319–337.

Dorin, J. R., Inglis, J. D., and Porteus, D. J. (1989). Selection for precise chromosomal targeting of a dominant marker for homologous recombination. *Science,* 243, 1357–1360.

Edwards, A., Civitello, A., Hammond, H. A., and Caskey, C. T. (1991). DNA typing and genetic mapping with trimeric and tetrameric tandem repeats. *Am. J. Human Genet.,* 49, 746–756.

Erlich, H. A., Gelfand, D., and Sninsky, J. J. (1991). Recent advances in the polymerase chain reaction. *Science,* 252, 1643–1651.

Evans, M. J. (1981). Origin of mouse embryonal carcinoma cells and the possibility of their direct isolation into tissue culture. *J. Reprod. Fert.,* 62, 625–631.

Evans, M. J. (1989). Potential for genetic manipulation of mammals. *Mol. Biol. Med.,* 6, 557–565.

Evans, M. J., and Kaufman, M. H. (1981). Establishment in culture of pluripotential cells from mouse embryos. *Nature,* 292, 154–156.

Folger, K., Thomas, K., and Capecchi, M. R. (1984). Analysis of homologous recombination in cultured mammalian cells. *Cold Spring Harbor Symp. Quant. Biol.,* 49, 123–197.

Gardner, R. L. (1968). Mouse chimeras obtained by the injection of cells into the blastocyst. *Nature,* 220, 596–597.

Gebhardt, C., Ritter, E., Debener, T., Schachtschabel, U., Walkemeier, B., Uhrig, H., and Salamini, F. (1989). RFLP analysis and linkage mapping in *Solanum tuberosum. Theor. Appl. Genet.,* 78, 65–75.

Gosden, J., Hanratty, D., Starling, J., Fantes, J., Mitchell, A., and Porteous, D. (1991). Oligonucleotide-primed in situ DNA synthesis (PRINS): A method for chromosome mapping, banding, and investigation of sequence organization. *Cytogenet. Cell Genet.,* 57, 100–104.

Gossler, A., Doetschman, T., Korn, R., Serfling, E., and Kemler, R. (1986). Transgenesis by means of blastocyst-derived embryonic stem-cell lines. *Proc. Natl. Acad. Sci. USA,* 83, 9065–9069.

Hasty, P., Ramirez-Solis, R., Krumlauf, R., and Bradley, A. (1991). Introduction of a subtle mutation into the *Hox-2.6* locus in embryonic stem cells. *Nature,* 350, 243–246.

Hutchison, N. J., and Le Ciel, C. (1991). Gene mapping in chickens via fluorescent in situ hybridization to mitotic and meiotic chromosomes. *J. Cell. Biochem.,* Suppl. 15E, 205.

Iggo, R., Gough, A., Xu, W., Lane, D. P., and Spurr, N. K. (1989). Chromosome mapping of the human gene encoding the 68-kDa nuclear antigen (p68) by using the polymerase chain reaction. *Proc. Natl. Acad. Sci. USA,* 86, 6211–6214.

Jeffreys, A. J. (1987). Highly variable minisatellites and DNA fingerprints. *Biochem. Soc. Trans.,* 15, 309–317.

Johnson, R. S., Sheng, M., Greenberg, M. E., Kolodner, R. D., Papaioannou, V. R., and Spiegelman, B. M. (1989). Targeting of nonexpressed genes in embryonic stem cells via homologous recombination. *Science,* 245, 1234–1236.

Joyner, A. L., Skarnes, W. C., and Rossant, J. (1989). Production of a mutation in mouse *En-2* gene by homologous recombination in embryonic stem cells. *Nature*, 338, 153–156.

Klein-Lankhorst, R., Rietveld, P., Machiels, B., Verkerek, R., Weide, R., Gebhardt, C., Koornneef, M., and Zabel, P. (1991). RFLP markers linked to the root knot nematode resistance gene *Mi* in tomato. *Theor. Appl. Genet.*, 81, 661–667.

Kuhnlein, U., Zadworny, D., Gavora, J. S., and Fairfull, R. W. (1991). Identification of markers associated with quantitative trait loci in chickens by DNA fingerprinting. In: *DNA Fingerprinting: Approaches and Applications*, edited by T. Burke, G. Dolf, A. J. Jeffreys, and R. Wolffe. Birkhauser Verlag, Basel, pp. 274–282.

Lander, E. S., and Botstein, D. (1986). Strategies for studying heterogeneous genetic traits in humans by using a linkage map of restriction fragment length polymorphisms. *Proc. Natl. Acad. Sci. USA*, 83, 7353–7357.

Lander, E. S., and Botstein, D. (1989). Mapping Mendelian factors underlying quantitative traits using RFLP linkage maps. *Genetics*, 121, 185–199.

Landry, B. S., Hubert, N., Etoh, T., Harada, J. J., and Lincoln, S. E. (1991). A genetic map for *Brassica napus* based on restriction fragment length polymorphisms detected with expressed DNA sequences. *Genome*, 34, 543–552.

Lakshmanan, N., Ponce de Leon, F. A., Smyth, J. R., and Smith, E. (1991). Physical mapping of the *ev7* locus in chicken by chromosomal in situ hybridization. *Poultry Sci.*, 70, Suppl. 1, 69.

Ledbetter, S. A., Nelson, D. L., Warren, S. T., and Ledbetter, D. H. (1990). Rapid isolation of DNA probes within specific chromosome regions by interspersed repetitive sequence polymerase chain reaction. *Genomics*, 6, 475–481.

Li, H., Gyllensten, U. B., Cui, X., Saiki, R. K., Erlich, H. A., and Arnheim, N. (1988). Amplification and analysis of DNA sequences in single sperm and diploid cells. *Nature*, 335, 414–417.

Litt, M., and Luty, J. A. (1989). A hypervariable microsatellite revealed by *in vitro* amplification of a dinucleotide repeat within the cardiac muscle actin gene. *Am. J. Human Genet.*, 44, 397–401.

Love, J. M., Knight, A. M., McAleer, M. A., and Todd, J. A. (1990). Towards construction of a high resolution map of the mouse genome using PCR-analysed microsatellites. *Nucleic Acids Res.*, 18, 4123–4130.

Ludecke, H.-J. Senger, G., Claussen, U., and Horsthemke, B. (1989). Cloning defined regions of the human genome by microdissection of banded chromosomes and enzymatic amplification. *Nature*, 338, 348–350.

Mansour, S. L., Thomas, K. R., and Capecchi, M. R. (1988). Disruption of the proto-oncogene *int-2* in mouse embryo-derived stem cells: A general strategy for targeting mutations to non-selectable genes. *Nature*, 336, 248–252.

Mansour, S. L., Thomas, K. R., Deng, C., and Capecchi, M. R. (1990). Introduction of *lacZ* reporter gene into the mouse int-2 locus by homologous recombination. *Proc. Natl. Acad. Sci. USA*, 87, 7688–7692.

Martin, G. (1981). Isolation of a pluripotent cell line from early mouse embryos cultured in medium conditioned by teratocarcinoma stem cells. *Proc. Natl. Acad. Sci. USA*, 78, 7634–7638.

Martin, G. B., Williams, J. G. K., and Tanksley, S. D. (1991). Rapid identification of markers linked to a *Pseudomonas* resistance gene in tomato using random primers and near-isogenic lines. *Proc. Natl. Acad. Sci. USA*, 88, 2336–2340.

McCouch, S. R., Kochert, G., Yu, Z. H., Wang, Z. Y., Khush, G. S., Coffman, W. R., and Tanksley, S. D. (1988). Molecular mapping of rice chromosomes. *Theor. Appl. Genet.*, 76, 815–829.

Mombaerts, P., Clarke, A. R., Hooper, M. L., and Tonegawa, S. (1991). Creation of a large genomic deletion at the T-cell antigen receptor B-subunit locus in mouse embryonic stem cells by gene targeting. *Proc. Natl. Acad. Sci. USA*, 88, 3084–3087.

Nadeau, J. H. (1989). Maps of linkage and synteny homologies between mouse and man. *TIG,* 5, 82–86.

Nakamura, Y., Leppert, M., O'Connell, P., Wolff, R., Holm, T., Culver, M., Martin, C., Fujimoto, E., Hoff, M., Kumlin, E., and White, R. (1987). Variable number of tandem repeat (VNTR) markers for human gene mapping. *Science,* 235, 1616–1622.

Nam, H.-G., Giraudat, J., den Boer, B., Moonan, F., Loos, W. D. B., Hauge, B. M., and Goodman, H. M. (1989). Restriction fragment length polymorphism linkage map of *Arabidopsis thaliana. The Plant Cell,* 1, 699–705.

Nelson, D. L., Ledbetter, S. A., Corbo, L., Victoria, M. F., Ramirez-Solis, R., Webster, T. D., Ledbetter, D. H., and Caskey, C. T. (1989). *Alu* polymerase chain reaction: A method for rapid isolation of human-specific sequences from complex DNA sources. *Proc. Natl. Acad. Sci. USA,* 86, 6686–6690.

Olson, M., Hood, L., Cantor, C., and Botstein, D. (1989). A common language for physical mapping of the human genome. *Science,* 245, 1434–1435.

Paterson, A. H., Damon, S., Hewitt, J. D., Zamir, D., Rabinowitch, H. D., Lincoln, S. E., Lander, E. S., and Tanksley, S. D. (1991). Mendelian factors underlying quantitative traits in tomato: Comparison across species, generations, and environments. *Genetics,* 127, 181–197.

Paterson, A. H., Deverna, J. W., Lanini, B., and Tanksley, S. D. (1990). Fine mapping of quantitative trait loci using selected overlapping recombinant chromosomes, in an interspecies cross of tomato. *Genetics,* 124, 735–742.

Paterson, A. H., Lander, E. S., Hewitt, J. D., Peterson, S., Lincoln, S. E., and Tanksley, S. D. (1988). Resolution of quantitative traits into Mendelian factors, using a complete linkage map of restriction fragment length polymorphisms. *Nature,* 335, 721–726.

Petitte, J. N., Clark, M. E., Liu, G., Verrinder Gibbins, A. M., and Etches, R. J. (1990). Production of somatic and germline chimeras in the chicken by transfer of early blastodermal cells. *Development,* 108, 185–189.

Ponce de Leon, F. A., Li, Y., and Smith, E. (1991). Reassignment of the *evl* locus by high resolution chromosomal *in situ* hybridization. *Poultry Sci.,* 70, Suppl. 1, 95.

Rawles, M. E. (1945). Behavior of melanoblasts derived from the coelomic lining in interbreed grafts of wing skin. *Physiol. Zool.,* 18, 1–16.

Roberts, L. (1990). Whatever happened to the genetic map? *Science,* 247, 281–282.

Robertson, E., Bradley, A., Kuehn, M., and Evans, M. J. (1986). Germ-line transmission of genes introduced into cultured pluripotential cells by retroviral vector. *Nature,* 323, 445–448.

Rommens, J. H., Iannuzzi, M. C., Kerem, B.-S., Drumm, M. L., Melmer, G., Dean, M., Rozmahel, R., Cole, J. L., Kennedy, D., Hidaka, N., Zsiga, M., Buchwald, M., Riordan, J. R., Tsui, L. C., and Collins, F. S. (1989). Identification of the cystic fibrosis gene: Chromosome walking and jumping. *Science,* 245, 1059–1065.

Royle, N. J., Clarkson, R. E., Wong, Z., and Jeffreys, A. J. (1988). Clustering of hypervariable minisatellites in the proterminal regions of human autosomes. *Genomics,* 3, 352–360.

Schwartzberg, P. L., Goff, S. P., and Robertson, E. J. (1989). Germ-line transmission of a *c-abl* mutation produced by targeted gene disruption in ES cells. *Science,* 246, 799–803.

Sedivy, J. M., and Sharp, P. A. (1989). Positive genetic selection for gene disruption in mammalian cells by homologous recombination. *Proc. Natl. Acad. Sci. USA,* 86, 227–231.

Sinnett, D., Deragon, J.-M., Simard, L. R., and Labuda, D. (1990). Alumorphs — Human DNA polymorphisms detected by polymerase chain reaction using *Alu*-specific primers. *Genomics,* 7, 331–334.

Slocum, M. K., Figdore, S. S., Kennard, W. C., Suzuki, J. Y., and Osborn, T. C. (1990). Linkage arrangement of restriction fragment length polymorphism loci in *Brassica oleracea. Theor. Appl. Genet.,* 80, 57–64.

Smithies, O., Gregg, R. G., Boggs, S. S., Koralewski, M. A., and Kucherlapati, R. S. (1985). Insertion of DNA sequences into the human chromosomal beta-globin locus by homologous recombination. *Nature*, 317, 230–234.

Song, K. M., Suzuki, J. Y., Slocum, M. K., Williams, P. H., and Osborn, T. C. (1990). A linkage map of *Brassica rapa* (syn. *campestris*) based on restriction fragment length polymorphism loci. *Theor. Appl. Genet.*

Soriano, P., Montgomery, C., Geske, R., and Bradley, A. (1991). Targeted disruption of the *c-src* proto-oncogene leads to osteopetrosis in mice. *Cell*, 64, 693–702.

Stanton, B. R., Reid, S. W., and Parada, L. F. (1990). Germ line transmission of an inactive *N-myc* allele generated by homologous recombination in mouse embryonic stem cells. *Mol. Cell. Biol.*, 10, 6755–6758.

Tanksley, S. D., and Hewitt, J. D. (1988). Use of molecular markers in breeding for soluble solids in tomato — A re-examination. *Theor. Appl. Genet.*, 75, 811–823.

Tanksley, S. D., Young, N. D., Paterson, A. H., and Bonierbale, M. W. (1989). RFLP mapping in plant breeding: New tools for an old science. *Bio/technology*, 7, 257–264.

Tarkowski, A. K. (1961). Mouse chimaeras developed from fused eggs. *Nature*, 190, 857–860.

Tautz, D. (1989). Hypervariability of simple sequences as a general source for polymorphic DNA markers. *Nucleic Acids Res.*, 17, 6463–6471.

Tautz, D., and Renz, M. (1984). Simple sequences are ubiquitous repetitive components of eukaryotic genomes. *Nucleic Acids Res.*, 12, 4127–4138.

Tereba, A., McPhaul, M. J., and Wilson, J. D. (1991). The gene for aromatase (P450$_{arom}$) in the chicken is located on the long arm of chromosome 1. *J. Heredity*, 82, 80–81.

Thomas, K. R., and Capecchi, M. R. (1987). Site directed mutagenesis by gene targeting in mouse embryo-derived stem cells. *Cell*, 51, 503–512.

Thomas, K. R., and Capecchi, M. R. (1990). Targeted disruption of the murine *int-1* proto-oncogene resulting in severe abnormalities in midbrain and cerebellar development. *Nature*, 346, 847–850.

Thomas, K. R., Folger, K. R., and Capecchi, M. R. (1986). High frequency targeting of genes to specific sites in the mammalian genome. *Cell*, 44, 419–428.

Thompson, S., Clarke, A. R., Pow, A. M., Hooper, M. L., and Melton, D. W. (1989). Germ line transmission and expression of a corrected HPRT gene produced by gene targeting in embryonic stem cells. *Cell*, 56, 313–321.

Todd, J. A., Aitman, T. J., Cornall, R. J., Ghosh, S., Hall, J. R. S., Hearne, C. M., Knight, A. M., Love, J. M., McAleer, M. A., Prins, J.-B., Rodrigues, N., Lathrop, M., Pressey, A., Delarato, N. H., Peterson, L. B., and Wicker, L. S. (1991). Genetic analysis of autoimmune type 1 diabetes mellitus in mice. *Nature*, 351, 542–547.

Valancius, V., and Smithies, O. (1991). Testing an "in-out" targeting procedure for making subtle genomic modifications in mouse embryonic stem cells. *Mol. Cell. Biol.*, 11, 1402–1408.

Volpe, E. P. (1963). Interplay of mutant and wild-type pigment cells in chimeric leopard frogs. *Develop. Biol.*, 8, 205–221.

Weber, J. L., and May, P. E. (1989). Abundant class of human DNA polymorphisms which can be typed using the polymerase chain reaction. *Am. J. Human Genet.*, 44, 388–396.

Welsh, J., and McClelland, M. (1990). Fingerprinting genomes using PCR with arbitrary primers. *Nucleic Acids Res.*, 18, 7213–7218.

Welsh, J., Petersen, C., and McClelland, M. (1991). Polymorphisms generated by arbitrarily primed PCR in the mouse: application to strain identification and genetic mapping. *Nucleic Acids Res.*, 19, 303–306.

White, R., Leppert, M., Bishop, D. T., Barker, D., Berkowitz, J., Brown, C., Callahan, P., Holm, T., and Jerominski, L. (1985). Construction of linkage maps with DNA markers for human chromosomes. *Nature*, 313, 101–105.

Wijmenga, C., Frants, R. R., Brouwer, O. F., Moerer, P., Weber, J. L., and Padberg, G. W. (1990). Location of facioscapulohumeral muscular dystrophy gene on chromosome 4. *Lancet,* 336, 651–653.

Williams, J. G. K., Kubelik, A. R., Livak, K. J., Rafalski, J. A., and Tingey, S. V. (1990). DNA polymorphisms amplified by arbitrary primers are useful as genetic markers. *Nucleic Acids Res.,* 18, 6531–6535.

Williams, J. G. K., Rafalski, J. A., and Tingey, S. V. (1991). Genetic analysis using RAPD markers. In: *Methods in Enzymology,* in press.

Yagi, T., Ikawa, Y., Yoshida, K., Shigetani, Y., Takeda, N., Mabuchi, I., Yamamoto, T., and Aizawa, S. (1990). Homologous recombination at *c-fyn* locus of mouse embryonic stem cells with use of diphtheria toxin A-fragment gene in negative selection. *Proc. Natl. Acad. Sci. USA,* 87, 9918–9922.

Zheng, H., and Wilson, J. H. (1990). Gene targeting in normal and amplified cell lines. *Nature,* 344, 170–173.

Zijlstra, M., Li, E., Sajjadi, F., Subramani, S., and Jaenisch, R. (1989). Germ-line transmission of a disrupted beta$_2$-microglobulin gene produced by homologous recombination in embryonic stem cells. *Nature,* 342, 435–438.

Zimmer, A., and Gruss, P. (1989). Production of chimaeric mice containing embryonic stem (ES) cells carrying a homeobox *Hox 1.1* allele mutated by homologous recombination. *Nature,* 338, 150–153.

INDEX

T - #0365 - 071024 - C2 - 234/156/16 - PB - 9780367402587 - Gloss Lamination